Next-Generation Fertilizers for Sustainable Agricultural Practices and Food Security

Next-Generation Fertilizers for Sustainable Agricultural Practices and Food Security

Edited by

Shah Fahad
Department of Agronomy, Abdul Wali Khan University Mardan, Mardan, Khyber Pakhtunkhwa, Pakistan

Shah Saud
College of Life Science, Linyi University, Linyi, Shandong, China

Taufiq Nawaz
Department of Biology/Microbiology, South Dakota State University, Brookings, South Dakota, USA

Sajid Fiaz
Institute of Molecular Biology and Biotechnology, The University of Lahore, Lahore, Pakistan

Iqbal Munir
Institute of Biotechnology & Genetic Engineering, The University of Agriculture, Peshawar, Khyber Pakhtunkhwa, Pakistan

WILEY

This edition first published 2026

Registered Offices
John Wiley & Sons, Inc., 111 River Street, Hoboken, NJ 07030, USA
John Wiley & Sons Ltd, New Era House, 8 Oldlands Way, Bognor Regis, West Sussex, PO22 9NQ, UK

For details of our global editorial offices, customer services, and more information about Wiley products visit us at www.wiley.com.

The manufacturer's authorized representative according to the EU General Product Safety Regulation is Wiley-VCH GmbH, Boschstr. 12, 69469 Weinheim, Germany, e-mail: Product_Safety@wiley.com.

Library of Congress Cataloging-in-Publication Data Applied for:

Hardback ISBN: 9781394383344

Cover Design: Wiley
Cover Image: © sarayut_sy/Shutterstock

Set in 9.5/12.5pt STIXTwoText by Straive, Pondicherry, India

Contents

List of Contributors

Shabbar Abbas
Department of Biosciences
COMSATS University Islamabad
Islamabad, Pakistan

Sundas Abbas
Regenerative Climate Smart Agriculture
SAWiE (Sustainable Agriculture Water and Intelligence Ecosystem)
Faisalabad, Pakistan

Banzeer Ahsan Abbasi
Department of Botany
Rawalpindi Women University
Rawalpindi, Pakistan

Iftikhar Ali Ahmad
College of Resources and Environment/Microelement Research Center/Key Laboratory of Arable Land Conservation (Middle and Lower Reaches of Yangtze River), Ministry of Agriculture and Rural Affairs
Huazhong Agricultural University
Wuhan, China

Arshad Ali
Institute of Chemistry
University of Sargodha
Sargodha, Pakistan

Muhammad Anas
Department of Soil Science & Plant Nutrition, Selçuk University, Konya, Türkiye

Department of Plant Sciences, Faculty of Biological Sciences
Quaid-i-Azam University
Islamabad, Pakistan

UNESCO-UNISA Africa Chair in Nanoscience and Nanotechnology (U2ACN2)
College of Graduate Studies
University of South Africa
Muckleneuk Ridge, Pretoria, South Africa

Muhammad Anwar-ul-Haq
Institute of Soil and Environmental Sciences
University of Agriculture Faisalabad
Faisalabad, Pakistan

Nasir Assad
Institute of Chemistry
University of Sargodha
Sargodha, Pakistan

Yasir Assad
Department of Zoology
Hazara University Mansehra
Mansehra, Pakistan

Syeda Zainab Batool
Institute of Chemistry
University of Sargodha
Sargodha Pakistan

Khadija Bibi
Department of Botany, Faculty of Sciences
Ghazi University
Dera Ghazi Khan, Pakistan

Guangda Ding
College of Resources and Environment/
Microelement Research Center/Key
Laboratory of Arable Land Conservation
(Middle and Lower Reaches of Yangtze
River), Ministry of Agriculture and
Rural Affairs
Huazhong Agricultural University
Wuhan, China

Shah Fahad
Department of Agronomy
Abdul Wali Khan University Mardan
Mardan, Khyber Pakhtunkhwa, Pakistan

Rao Muhammad Faisal Iqbal
Department of Chemistry
The University of Lahore
Lahore, Pakistan

Aliza Falak
Department of Plant Science
Faculty of Biological Sciences
Quaid-i-Azam University
Islamabad, Pakistan

Anam Fatima
Department of Plant Sciences
Faculty of Biological Sciences
Quaid-i-Azam University
Islamabad, Pakistan

Sharia Fatima
Institute of Chemistry
University of Sargodha
Sargodha, Pakistan

Iqra Ghafoor
Institute of Soil and Environmental Science
University of Agriculture Faisalabad
Faisalabad, Pakistan

Hansa Gul
Department of Zoology
University of Mianwali
Mianwali, Pakistan

Sharjeel Haider
College of Resources and Environment/
Microelement Research Center/Key
Laboratory of Arable Land Conservation
(Middle and Lower Reaches of Yangtze
River), Ministry of Agriculture and
Rural Affairs
Huazhong Agricultural University
Wuhan, China

Institute of Soil and Environmental
Sciences
University of Agriculture Faisalabad
Faisalabad, Pakistan

Erdogan E Hakki
Department of Soil Science & Plant Nutrition
Selçuk University
Konya, Türkiye

Salma Hameed
Department of Environmental Sciences
University of Jhang
Jhang, Pakistan

Ali Hasnain
Department of Agronomy, Faculty of
Agricultural Sciences
University of the Punjab
Lahore, Pakistan

Sameel Hassan
Department of Agriculture & Forest
Science (DAFNE)
University of Tuscia
Viterbo, Italy

Wajih ul Hassan
Department of Agronomy, Faculty of Agricultural Sciences
University of the Punjab
Lahore, Pakistan

Bushra Huma
Hubei Key Laboratory of Soil Environment and Pollution Remediation, College of Resources and Environment
Huazhong Agricultural University
Wuhan, China

Rahib Hussain
Sustainable Development Study Centre
Government College University Lahore
Lahore, Pakistan

Javed Iqbal
Department of Botany
Bacha Khan University
Charsadda, Pakistan

Hina Jabeen
Knowledge Center
SAWiE (Sustainable Agriculture Water and Intelligence Ecosystem)
Faisalabad, Pakistan

Alevcan Kaplan
Department of Pharmaceutical Botany
Dicle University
Diyarbakır, Turkey

Amjid Khan
UNESCO-UNISA Africa Chair in Nanoscience and Nanotechnology (U2ACN2)
College of Graduate Studies
University of South Africa
Muckleneuk Ridge, Pretoria, South Africa

Haris Ahmed Khan
Department of Biotechnology
University of Mianwali
Mianwali, Pakistan

Muhammad Nauman Khan
Department of Botany
University of Chakwal
Punjab, Pakistan

Muhammad Rizwan Khan
Department of Plant Sciences
Faculty of Biological Sciences
Quaid-i-Azam University
Islamabad, Pakistan

UNESCO-UNISA Africa Chair in Nanoscience and Nanotechnology (U2ACN2)
College of Graduate Studies
University of South Africa
Muckleneuk Ridge, Pretoria, South Africa

Rashid Abbas Khan
Department of Biology, Faculty of Sciences
Prince Sonkla University
Songkhla, Thailand

Waseem Ahmed Khattak
Department of Plant Sciences
Faculty of Biological Sciences
Quaid-i-Azam University
Islamabad, Pakistan

UNESCO-UNISA Africa Chair in Nanoscience and Nanotechnology (U2ACN2)
College of Graduate Studies
University of South Africa
Muckleneuk Ridge, Pretoria, South Africa

Rashid Mahmood
Department of Soil Science
Faculty of Agricultural Sciences
University of the Punjab,
Lahore, Pakistan

Saif Ullah Masood
Department of Agronomy, Faculty of Agricultural Sciences
University of the Punjab
Lahore, Pakistan

Muhammad Mehran
College of Resources and Environment/
Microelement Research Center/Key
Laboratory of Arable Land Conservation
(Middle and Lower Reaches of Yangtze
River), Ministry of Agriculture and
Rural Affairs
Huazhong Agricultural University
Wuhan, China

Nosheen Mirza
Department of Soil and Environmental
Sciences
Ghazi University, Dera Ghazi Khan
Pakistan

Tehmeena Mukhtar
Department of Environmental Sciences
University of Jhang, Jhang, Pakistan

Venuste Munyaneza
College of Resources and Environment/
Microelement Research Center/Key
Laboratory of Arable Land Conservation
(Middle and Lower Reaches of Yangtze
River), Ministry of Agriculture and
Rural Affairs
Huazhong Agricultural University
Wuhan, China

Faisal Nadeem
Department of Soil Science
Faculty of Agricultural Sciences
University of the Punjab
Lahore, Pakistan

Soil and Water Testing Laboratory for
Research, Gujranwala, Pakistan

Muhammad Naeem-ul-Hassan
Institute of Chemistry
University of Sargodha, Sargodha
Pakistan

Zahida Nasreen
Department of Zoology
University of Mianwali, Mianwali
Pakistan

Sofia Nazar
Department of Zoology
University of Sargodha
Sargodha, Pakistan

Tauqeer Ahmed Qadri
Department of Biosciences
COMSATS University Islamabad
Islamabad, Pakistan

Ziaur Rehman
Department of Chemistry
Quaid-i-Azam University
Islamabad, Pakistan

Sobia Riaz
Department of Environmental Sciences
University of Jhang, Jhang, Pakistan

Tayyaba Samreen
Institute of Soil and Environmental Sciences
University of Agriculture
Faisalabad, Pakistan

Muhammad Saqlain
Department of Environmental Sciences
University of Jhang, Jhang, Pakistan

Rubab Sarfaraz
Institute of Agriculture and
Life Sciences
Gyeongsang National University
Jinju, South Korea

Sidra Sohail
State Key Lab of Soil Erosion and
Dryland Farming on the Loess
Plateau

Institute of Soil and Water Conservation
Northwest A & F University
Xianyang China

Sudenaz Soylu
Department of Field Crops
Faculty of Agriculture
Selcuk University
Konya, Türkiye

Tooba
Institute of Chemistry
University of Sargodha
Sargodha, Pakistan

Sher Ullah
Institute of Chemistry
University of Sargodha
Sargodha, Pakistan

Abdul Wahab
Department of Biosciences
COMSATS University Islamabad
Islamabad, Pakistan

Faiza Waseem
Department of Soil Science, Faculty of Agricultural Sciences
University of the Punjab
Lahore, Pakistan

Asma Zafar
College of Natural Resources and Environment
Northwest A&F University
Yangling, China

Iram Zahra
Institute of Chemistry
University of Sargodha
Sargodha, Pakistan

Dua e Zainab
Key Laboratory of Green and Low Carbon Agriculture on Dryland in Northwest China, Ministry of Agriculture and Rural Affairs
College of Natural Resource and Environment, Northwest A & F University
Xianyang, China

Foreword

Agriculture stands at a critical crossroads in human history. With the global population projected to exceed nine billion in the coming decades, the demand for food, fiber, and fuel continues to rise, while natural resources such as land, water, and soil fertility face mounting pressures. Fertilizers have long been essential in supporting global food production, but conventional practices often carry heavy environmental costs. The emergence of next-generation fertilizers represents a pivotal opportunity to reshape the agricultural paradigm, enabling humanity to achieve food security while safeguarding the planet.

Next-Generation Fertilizers for Sustainable Agricultural Practices and Food Security provides an illuminating exploration of how science, technology, and innovation are converging to address these pressing challenges. The book offers readers a comprehensive journey—beginning with the foundations of advanced fertilizer production and characterization, and progressing through groundbreaking applications such as nanotechnology, bioactive formulations, and precision nutrient management. It does not stop at science alone; it thoughtfully addresses soil health, crop resilience, economic impacts, and even the regulatory and policy frameworks that will govern the adoption of these technologies.

What makes this work particularly valuable is its balance between research depth and practical relevance. The chapters present not only the underlying scientific principles but also real-world implications for farmers, agronomists, researchers, and policymakers. This holistic approach ensures that the insights shared within these pages resonate across disciplines and sectors, making the book a vital reference for all those invested in the future of sustainable agriculture.

In an era where climate change and resource scarcity demand urgent innovation, this book serves as both a guide and an inspiration. It underscores the importance of collaboration across science, technology, and practice in forging agricultural systems that are resilient, efficient, and sustainable. Readers will come away not only with knowledge but also with a renewed sense of responsibility and optimism for what the future of agriculture can achieve.

It is with great anticipation that I commend this volume to students, researchers, professionals, and decision-makers alike. The vision articulated here is clear: through next-generation fertilizers, we can shape an agricultural future that nourishes both humanity and the earth itself.

Shah Fahad
Assistant Professor
Abdul Wali Khan University, Pakistan

Preface

Agriculture has always stood at the heart of human progress, sustaining civilizations and shaping societies. Yet today, the sector faces unprecedented challenges—feeding a growing global population, mitigating the effects of climate change, and safeguarding natural resources for future generations. Within this context, fertilizers remain indispensable, but traditional approaches often come with significant ecological costs, including soil degradation, nutrient runoff, and greenhouse gas emissions.

This book, *Next-Generation Fertilizers for Sustainable Agricultural Practices and Food Security*, emerges from the pressing need to rethink how we nourish crops and manage soils in a more sustainable and responsible way. Next-generation fertilizers represent a transformative step forward, combining scientific innovation with practical solutions that optimize crop nutrition while minimizing environmental harm. From controlled-release and bio-based formulations to precision nutrient delivery systems, these advancements are reshaping the agricultural landscape with tools that are both efficient and environmentally sound.

Our aim in this volume is to bridge the gap between research and practice. We explore not only the science and technology behind these fertilizers—nanotechnology, biotechnology, and data-driven agriculture—but also their real-world applications and successes. By integrating next-generation fertilizers with digital farming tools such as soil health monitoring and precision agriculture, farmers and scientists can work together to build systems that are productive, resilient, and sustainable.

The chapters of this book are carefully structured to provide a holistic view of this evolving field. **Chapter 1** introduces innovative approaches and characterization methods in next-generation fertilizer production. **Chapter 2** examines their revolutionary impact on modern farming practices. **Chapter 3** explores bioactive-based fertilizers and their role in harnessing biostimulants, nutrients, and signaling molecules. **Chapter 4** highlights the promise of nanofertilizers, while **Chapter 5** looks into technologies that enhance nutrient use efficiency for sustainable crop production.

Chapter 6 takes a global perspective, addressing the role of these fertilizers in feeding the world, and **Chapter 7** discusses how advanced technologies can bolster crop resilience to both biotic and abiotic stresses. **Chapter 8** emphasizes soil health and balanced fertilizer use, complemented by **Chapter 9**, which considers synergistic bioactivities and their effects on crops and soil systems. **Chapter 10** shifts focus to the economic dimension, exploring how next-generation fertilizers can transform global agriculture.

In **Chapter 11**, the emphasis moves to enhancing the nutritional quality and functionality of crops. **Chapter 12** reviews the regulatory frameworks and policy implications that guide the safe and effective use of these innovations. Finally, **Chapter 13** provides insights into integrated strategies such as green manures and reduction practices, underlining how fertilizer optimization can align with sustainable agricultural goals.

This book is intended for a diverse audience: farmers seeking practical guidance, agronomists and crop scientists pursuing innovation, policymakers designing sustainable strategies, and students or researchers exploring the cutting edge of agricultural science. Each chapter highlights both the opportunities and challenges of adopting these technologies, offering insights into how next-generation fertilizers can enhance nutrient use efficiency, strengthen crop tolerance to stresses, and improve the nutritional quality of food.

Ultimately, this book is a call to action—to harness science, technology, and collaboration in reshaping the future of agriculture. It is our hope that the knowledge shared here will inspire continued innovation and guide a global transition toward farming systems that meet human needs while protecting the planet.

The Editors

1

Innovative Approaches and Cutting-Edge Methods in Next-Generation Fertilizer Production and Characterization

Erdogan E Hakki[1], Waseem Ahmed Khattak[2], Muhammad Anas[1], Javed Iqbal[3], Banzeer Ahsan Abbasi[4], Ziaur Rehman[5], and Shah Fahad[6]

[1] *Department of Soil Science & Plant Nutrition, Selçuk University, Konya, Türkiye*
[2] *Department of Plant Sciences, Faculty of Biological Sciences, Quaid-i-Azam University, Islamabad, Pakistan*
[3] *Department of Botany, Bacha Khan University, Charsadda, Pakistan*
[4] *Department of Botany, Rawalpindi Women University, Rawalpindi, Pakistan*
[5] *Department of Chemistry, Quaid-i-Azam University, Islamabad, Pakistan*
[6] *Department of Agronomy, Abdul Wali Khan University Mardan, Mardan, Khyber-Pakhtunkhwa, Pakistan*

1.1 Introduction

1.1.1 Background and Significance

Fertilizer is the foundation of agriculture, producing the nutrients that plants need to feed the world and boost farmers' pocketbooks (Ameen & Raza, 2017). The development of synthetic fertilizers in the green revolution altered agricultural systems and increased food production in numerous developing countries (Ameen & Raza, 2017). The reliance on conventional fertilizers has raised issues of inefficacy and environmental concerns. Traditional nitrogen, phosphorus, and potassium (NPK) fertilizers have been reported to exhibit low nutrient use efficiency (NUE), and a significant portion of these fertilizers is lost to the environment due to volatilization, leaching, and runoff. It is assumed that 30–50% of the applied nitrogen is used by the plants, and the rest is released to the environment, leading to water pollution and greenhouse gases (GHGs). These inefficiencies have led to the search for new-generation fertilizers, which could offer more controlled and targeted release of nutrients. Advances in material science, nanotechnology, and biotechnology have provided new opportunities in the design and delivery of fertilizers, helping to fit better with crop needs and cycles. The development of precision farming has added serious momentum to the demand for fertilizers that are tailored specifically to the instantaneous conditions of soil and crop. These advances represent a major shift in how fertilizers are viewed—not as simple productivity boosters, but as a fundamental part of a more sustainable agricultural system (Pretty, 2018).

1.1.2 Need for Advanced Fertilizer Technologies

With the inefficiency and environmental degradation of conventional fertilizers and with the amount of energy used for their production, there is now a focus on alternative solutions. Fertilizers do not meet site-specific crop nutrient needs, and this can result in excessive or insufficient application and the consequent waste of resources and the damage to soil health (Singh, 2018). The global population will reach 9.7 billion by 2050, and food production needs to grow by more than 60% to meet demand, meaning the way in which we provide nutrients to the world's crops needs to change. The emerging fertilizer systems, such as controlled-release fertilizer (CRF), nano-fertilizer, and biofertilizer, belong to the next-generation fertilizers and have opened a new era for the intelligent and efficient fertilization system. CRFs employ physical or chemical barriers (polymer coatings or nanoencapsulation) controlling the timing and quantity of nutrient release such that it coincides with the crop demand (Li & Li, 2024). Nano-fertilizers employ nanoscale carriers to improve nutrient uptake at the cellular level, which results in reduced losses and greater efficacy even when used at a low dose. Biofertilizers use microorganisms like nitrogen-fixing bacteria or fungus-solubilizing phosphate to enhance the nutrient status of the soil naturally. The ideas are designed to not just drive up crop yields, but to promote soil fertility over the long term, reduce water usage, and bolster the wallet against the ravages of the climate. Especially in export-oriented agriculture, there is a growing need for organic and residue-free production, and in this sense, next-generation fertilizers become the strategic tools in the views on balancing productivity and environmental protection (Kundius, 2021).

1.1.3 Environmental and Agricultural Challenges

Contemporary agricultural systems are confronted by a convergence of problems that require environmentally friendly and sustainable fertilizing technologies (Pretty, 2018). The most prominent of these is the effect on the environment from the use of traditional fertilizer. Overuse of synthetic nitrogen fertilizers is a major driver of nitrous oxide releases, an extremely potent GHG with a climate-warming potential nearly 300 times that of carbon dioxide. Nitrates leached into groundwater from these sources threaten human health through methemoglobinemia, or "blue baby syndrome," particularly in rural areas that use shallow wells. The issue is also a problem in waterways. Fertilizer leaching from farms causes eutrophication in lakes, rivers, and coastal shores, including the harm of algal blooms, oxygen depletion, and fish death. The economic and environmental costs of such pollution are staggering, affecting, as it does, drinking water supplies, fisheries, and biodiversity. Phosphate rock, the source of phosphorus for phosphorus fertilizers, is also nonrenewable and is characterized by a geopolitical bottleneck: nearly all reserves are in the hands of few countries. Sustainability of long-term P supply is, therefore, a great concern. Climate change introduces further complications. Geographic shifts in precipitation, greater fluctuations in temperature, and the prevalence of severe weather events disrupt historical fertilizer application timings and lower NUE. Nutrient dissolution and distribution are hampered by drought, while floods take with them any fertilizers applied to the soil surface. This mandates the creation of formulations, which can resist and accommodate climatic variations. There are implications for soil health: long-term use of high-dose chemical fertilizers can destabilize the structure of the soil, lower microbial diversity, and reduce important micronutrient levels. This loss of soil fertility makes it difficult for the land to sustain a healthy crop stand. Salinization and acidification, usually secondary to overuse of fertilizers, are additional complications, especially in semiarid and arid areas. Economically, the increased price of chemical fertilizers, due to fluctuations in global energy sources and the depletion of raw materials, is a financial strain on farmers, especially poor smallholders in the developing world. Both under-fertilizing and excessive debt are common. This financial strain highlights the need for biodegradable, low-input fertilizer alternatives that sustain or increase productivity. To help tackle these myriad challenges, the next-generation fertilizers are being formulated not only to enhance NUE, but also to make broader contributions to sustainability. For instance, waste and biomass-based fertilizers (liquid and solid, made from agricultural residues and other industrial wastes) provide a circular solution that decreases reliance on virgin resources. Likewise, amendments like fertilizers with soil conditioners, microbial inoculants, or bio-stimulants can be used to rejuvenate degraded soils and enhance pest and disease tolerance. The significance of sustainable fertilizer use is being acknowledged in global policy frameworks. The United Nations' Sustainable Development Goals (SDGs), including SDG 2 (Zero Hunger), SDG 12 (Responsible Consumption and Production), and SDG 13 (Climate Action), emphasize the importance of agricultural innovation that maximizes resource efficiency and enables food security. Fertilizer technologies that serve these goals are increasingly seen as key enablers of sustainable agricultural transitions (Olvermann et al., 2023).

1.2 Classification of Next-Generation Fertilizers

As world agriculture advances under the demanding pressures of food production and the environment, a range of new-generation fertilizers is now taking shape. These fertilizers are an alternative to conventional fertilizers as they can be more effective in supplying nutrients, cause less environmental damage, and are more conducive to sustainable agriculture. According to composition, release mechanisms, and action mode, the new generation of fertilizers seems conveniently grouped into four main types, which include CRFs, nano-fertilizers, bio-based and organic fertilizers, and smart/responsive fertilizers. Both have their unique advantages, depending on the specific crop requirements, the soil, and the environment (Shah & Wu, 2019).

1.2.1 Controlled-Release Fertilizers (CRFs)

CRFs are one of the first advancements in advanced fertilizer technology. They are designed to release nutrients slowly over time in line with crop growth stages and nutrient requirements. CRFs are generally coated or encapsulated with materials, such as sulfur and polymers, or resins, to control the rate of release of nutrients. CRFs ultimately aim to reduce nutrient loss from leaching, volatilization, and denitrification known to be significant limitations associated with traditional fertilizers, particularly in high rainfall or irrigated regions. CRF's efficacy has been widely reported. They improve NUE and crop fertilizer ratio and also minimize environmental hazards from overfertilization. CRFs are particularly highly efficient in intensive production systems, such as in horticulture and greenhouse cultivation, where nutrient management requires high accuracy. But the price is still the problem preventing it from being more widely used, especially in the smallholder farming system. However, continuous research on biodegradable coatings and cheap encapsulation materials is making CRFs easier to use. CRFs are also being combined with new technologies (e.g., bio-inoculants or slow-release micronutrients) to form hybrid release solutions that maximize both macro- and micronutrient delivery (Lodhi et al., 2023).

1.2.2 Nano-fertilizers

Nano-fertilizer is a major emerging area of research innovation in the fertilizer sector due to the innovation of nanotechnology (Anas et al., 2024b). These fertilizers are formulated based on nanoparticulates or nanoscale materials through which better nutrient flow can be observed at the cellular level than in regular formulations. Nanoparticle cation reduces the fertilizer dosage greatly because of its high specific surface area, which intensifies nutrient solubility, reactivity, and absorption by plants. Nano-fertilizers can be categorized according to the composition of the functional material, for example, nano-urea, nano-zinc, and nano-phosphorus, or based on the delivery system, that is, nanoencapsulation, nano-emulsions, and coated nanocarriers. Controlled and targeted nutrient delivery is one of the important advantages of nano-fertilizers (Solanki et al., 2015). Sprays like foliar or root application of nanoparticles can reach plant cells better, minimizing losses and delivering nutrients where they are most needed. Second, nano-fertilizers are engineered to respond to a defined physiological or environmental trigger (e.g., pH, light, or enzymatic activity), making for precision nutrition. For instance, nitrogen encapsulated in the nano is released gradually to synchronize the crop absorption throughout several weeks to months, thus reducing the gush of nitrogen into surrounding water bodies. While they offer potential, there are obstacles related to regulation and safety for nano-fertilizers. Worries about the accumulation of nanoparticles in soil and plants, potential toxicity, and long-term ecological effects have led to the cry for a careful risk assessment and regulation. Yet institutions and companies worldwide, among them in countries like India and China, are ramping up nano-fertilizer production and application trials, suggesting their increasing relevance in contemporary farming. If wisely applied, nano-fertilizers might be pivotal in the cut of fertilizer input and maximal crop yield (Mahesha et al., 2023).

1.2.3 Bio-based and Organic Fertilizers

A "Green alternative," the bio-based and organic fertilizers, presents an environmentally friendly option to synthetic inputs by recycling nutrients derived from natural substances such as animal manure, compost, green manure, crop residues, seaweed extracts, and microbial inoculants. These fertilizers add value to the soil not only by supplying necessary macro- and micronutrients, but also by improving soil structure, water-holding capacity, and microbial community activity (Dhaliwal et al., 2019). Organic and bio-based fertilizers nourish the soil, which in turn benefits long-term soil fertility and biological diversity, unlike synthetic fertilizers that can degrade the quality of the soil over time.

The major one of these is microbial biofertilizers, and they are further classified as nitrogen-fixing bacteria (e.g., Rhizobium and Azotobacter), phosphate-solubilizing microorganisms (e.g., Bacillus and Pseudomonas), and mycorrhizal fungi. These organisms enhance nutrient acquisition through developing symbiotic associations with the plant roots by rendering unattainable nutrients to the crop plants (Anas et al., 2025). Moreover, composted bio-solids and anaerobic digester digestate are increasingly being used as nutrient-dense organic amendments for organic and sustainable farming systems. There are several benefits of bio-based fertilizers, and one of the most important is that they fit with organic and regenerative agriculture. They act as a tool for sequestering carbon in soils, reducing reliance on fossil fuels, and bring about a circular economy model by converting a waste into a valuable agricultural input. Nevertheless, their nutrient release curves are generally slow and irregular and have to be managed carefully. Advancements in composting methods, strain selection, and fermentation technologies are being developed to standardize and improve the quality of these fertilizers. In areas where synthetic fertilizers are too costly or ecologically unacceptable, biofertilizers can offer an economic and environmentally friendly choice (Sahoo et al., 2012).

1.2.4 Smart and Responsive Fertilizers

Smart and responsive fertilizers are the most sophisticated type of next-generation fertilizers, which have been formulated to react to specific environmental or physiological stimuli (Moradi et al., 2024). They include sensors, stimuli-responsive polymers, and nanosensors that can detect soil or plant cues and release nutrients when required. The goal is simply to provide "on-demand nutrition," that is, to adjust feeding in response to changing levels of soil moisture, pH, temperature, and crop development stage. One well-known group, for example, is fertilizers locked inside hydrogels that grow and shrink depending on the moisture in soil so that they only release nutrients when they are required. Other approaches include coatings with a pH-dependent solubility for release in acidic or alkaline conditions or temperature-sensitive materials that adjust the rate of nutrient release as a function of root metabolism (Heuchan et al., 2019). Some are even "smart fertilizers," containing diagnostic components that monitor nutrient concentrations or root-zone conditions, allowing farmers to better time fertilizing. These fertilizers are an important element of modern precision agriculture, especially in data-based and sensor-based farm management. Integration with technologies such as Internet of Things (IoT), GPS, and artificial intelligence (AI) escalates their power to transform nutrient management (Dayana et al., 2025). Smart fertilizers could, for instance, be connected to automatic irrigation applications that automatically adapt the fertilizer dose as a function of weather forecasts, evapotranspiration estimation, or real-time crop sensing. Although smart fertilizers are the most efficient, they remain underdeveloped and are not yet commercialized. High production cost, complicated formulation, as well as low familiarity by farmers are some burdens (Figure 1.1). However, the possibility of such products is rapidly becoming more plausible through continuing efforts in material science, synthetic biology, and nanotechnology. Governments and research organizations are beginning to recognize the strategic role of this new science in climate-smart agriculture, eschewing investment in pilot projects and field demonstrations, and public–private collaboration to facilitate adoption (Dayana et al., 2025).

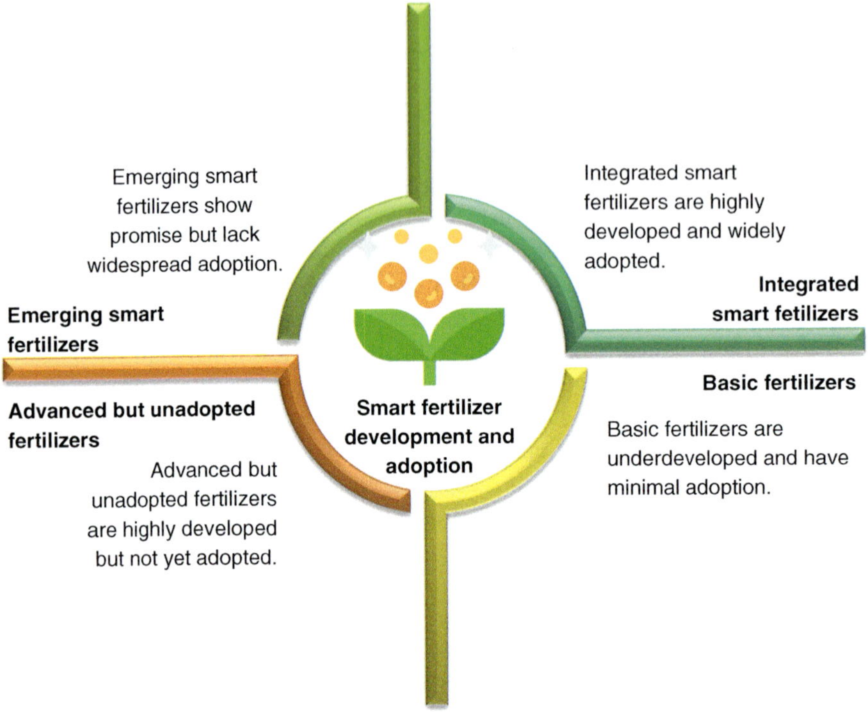

Figure 1.1 Small fertilizer development and adoption.

1.3 Innovative Synthesis and Fabrication Methods

The new-generation fertilizers have come to light as a consequence of the need for higher NUEs as well as from advances in material technologies and engineering. New synthesis and fabrication approaches have the potential to provide fertilizers with the desired characteristics such as controlled release and targeted delivery, environmental responsiveness, and improved compatibility with sustainability in agriculture. This chapter intends to give an overview of some of the most groundbreaking techniques that are ubiquitously used to produce advanced fertilizers, such as nanoencapsulation, green chemistry, sol–gel and hydrothermal methods, advanced digital fabrication such as 3D printing, and the like. All these methodologies are intended to develop smart, efficient, and environmentally friendly fertilizers for the enhancement of both plant demands and environmental constraints.

1.3.1 Nanoencapsulation Techniques

Nanoencapsulation is at the forefront of new-generation fertilizer technology, which facilitates encapsulation and controlled delivery of nutrients into nanocarriers. This process permits the gradual and continuous release of the active nutrients to improve the efficiency of assimilation for the uptake and to curtail nutrient loss as a result of leaching or volatilization. Nanoencapsulation is often performed using materials like biopolymers (e.g., chitosan, alginate, and starch) and synthetic polymers (e.g., polyvinyl alcohol, PVA, and polylactic acid, PLA). The nanocarriers are designed to protect the nutrient payload against environmental degradation and to release nutrients in response to targeted signals in growth environments, such as pH, heat, or enzyme activity in the rhizosphere. For example, nitrogen fertilizers coated by the use of chitosan nanoparticles were shown to have extended release and decreased leaching in acid soils. Likewise, phosphorus coated with a silicon-based nanoparticle may enhance solubility and has been reported to give improved localization in the root zone. Nanoencapsulation also allows the concurrent delivery of macro- and micronutrients, or the integration of microbial inoculants and plant growth regulators (PGRs), thus constituting multipurpose delivery systems. Emulsification, coacervation, spray drying, and electrospinning are some of the methods used to produce these nanostructures. Significantly, nanoencapsulation is well suited to the precision agriculture guidelines, as it allows the effective delivery of fertilizers at microdoses when induced and consequently reducing the environmental load (Sabarivasan & Murali Arthanari, 2025).

1.3.2 Green Chemistry Approaches

Green chemistry tools implicate minimization of the environmental impact of fertilizer synthesis, specifically removal of hazardous solvents, and reduction in energy requirements and feedstock access. By using environment-friendly materials and renewable feedstocks, the synthesis degradation of the fertilizer product, from cradle to grave, is consistent with the requirements of sustainability (Sanghi & Singh, 2012). In this respect, green synthesis approaches have become crucial for the production of bio-derived and nano-equipped fertilizers. For instance, nanoparticles can be synthesized through metabolic pathways of plant extracts/microbial cultures, and in addition to their potential for reducing the use of chemicals, these bio-derivatives render bioactive functionalities currency in ameliorating plant–microbe interactions. Rice husk, banana peel, and sugarcane bagasse agricultural by-products are also employed in the preparation of silica-, potassium-, and carbon-based nanomaterials. These methods not only address the value of the waste but also save on the carbon footprint of fertilizer production. New solvents such as ionic liquids and supercritical fluids are being considered for fertilizer production, as they offer lower toxicity and higher efficacy. Biofertilizers, nutrient-dense biochar, can also be prepared using enzymatic and microbial fermentation processes. Green chemistry is of particular importance for organic farming systems and for providing fertilizers suitable for environmentally sensitive areas. Lastly, green synthesis techniques serve to fill the void between high-performance fertilizers and environmentally friendly stewardship (Kamal et al., 2025).

1.3.3 Sol–Gel and Hydrothermal Processes

Sol–gel and hydrothermal processes are complex routes commonly used to prepare inorganic and hybrid nano-fertilizers (Arora & Khosla, 2024). Such approaches allow for accurate tailoring of particles' morphology, size distribution, surface functionalization, and porosity, all of which are relevant to modulate nutrient release kinetics and bioavailability. The sol–gel process refers to the formation of a system from a colloidal solution (sol), which subsequently becomes a solid gel phase; metal alkoxides or inorganic salts are the usual ones. This method is well suited to produce silica-, titanium-, and iron-based nano-fertilizers with high surface areas and modulated porosities. For instance, silica porous matrices with doping of certain micronutrient elements, such as zinc or copper, through the sol–gel process have been used to obtain mesoporous nanoparticles with high-capacity adsorption (Eivazzadeh-Keihan et al., 2020). These nanomaterials gradually release nutrients at an appropriately controlled rate condition through environmental stimulation, so that the absorption increases and the number of applications decreases. Furthermore, sol–gel synthesis facilitates the co-entrapment of various nutrients and bio-stimulants together in one single matrix exhibiting pronounced synergistic effects on plant growth. In contrast, hydrothermal synthesis is the crystallization of compounds at high temperatures from a hot, high-pressure aqueous solution. This process permits the synthesis of nanoparticles, which are highly crystalline and possess novel structural and surface characteristics. It is especially ideal for the preparation of metal oxides and hydroxides, such as nano-Fe_2O_3, ZnO, and $Mg(OH)_2$, which are very important in plant metabolism (Gobinath, 2018). Hydrothermal conditions can be tuned for surface charges and functional groups, leading to better nutrient contact with roots. Both the sol–gel method and the hydrothermal method are scalable, reproducible, and relatively cheaper for the synthesis of nano-fertilizers with superior properties. These treatment methods are becoming more common in both research and commercial fertilizer formulation, especially for micronutrient supply and abiotic stress alleviation (Kumari et al., 2022).

1.3.4 3D Printing and Precision Formulation

Three-dimensional (3D) printing, or additive manufacturing, has brought revolutionary changes to the practice of precision farming, allowing fabrication of tailor-made fertilizer matrices with complex geometries and customizable release kinetics (Estévez et al., 2025). This digital manufacturing approach can deposit materials sequentially layer by layer to create solid objects impregnated with nutrients, bio-inoculants, or even nanosensors. The 3D-printed fertilizers can be custom-made to cater to specific needs of a particular crop, soil, or growth stage, thereby improving the efficiency of nutrient use and reducing wastage (Padhiary et al., 2024). For example, the hydrogels or biodegradable polymers may include nutrients and be 3D printed as capsules or granules with different thicknesses, porosities, and release gradients. These intelligent systems may allow adjusting nutrient release in both time and space, by delivering nutrients in the medium only after moisture or root exudates have triggered the matrix. In a few sophisticated designs (e.g., M1), sensors are embedded into the print to measure pH, temperature, or nutrient content of the soil, so that feedback-based fertilization can be adopted. Precision formulation also includes AI-enabled design tools and formulation software that analyze release kinetics and optimize ingredient ratios, as well as predict the performance of the fertilizer in the field. By partnering 3D printing with these digital tools, researchers and industry can prototype and trial new fertilizer configurations rapidly, short-circuiting the cycle of innovation. There is potential for the use of 3D printing in a range of agricultural applications that are in their infancy, such as site-specific nutrient delivery, high-value horticulture,

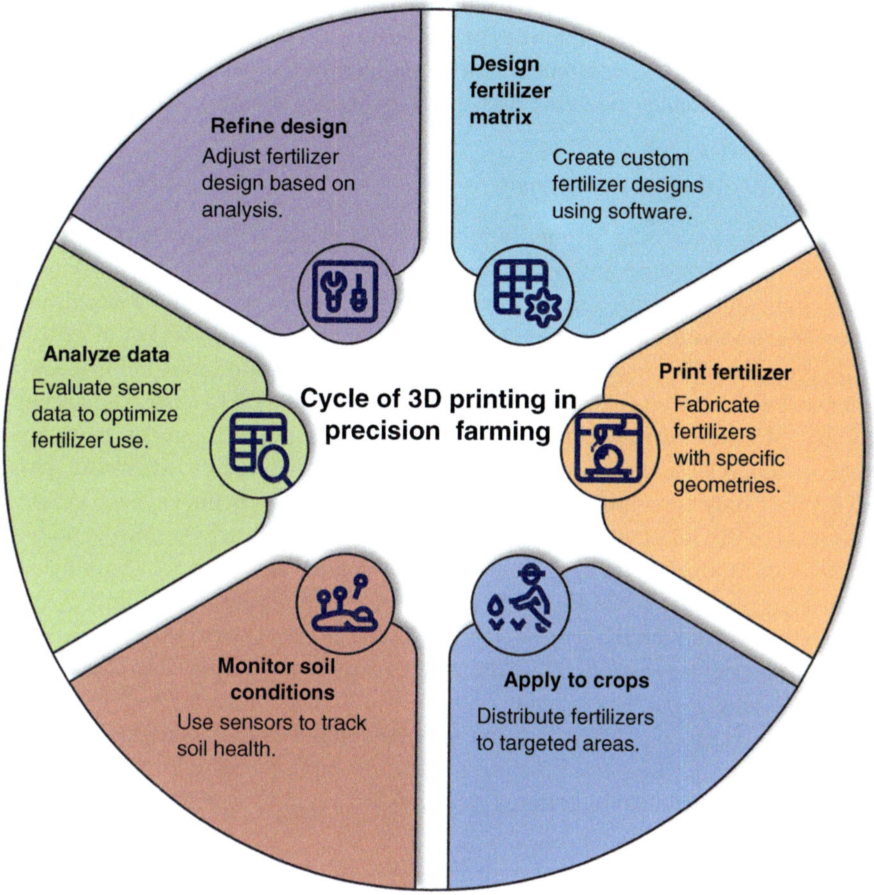

Figure 1.2 Cycle of 3D printing in precision farming.

and urban agriculture systems (Ahmad & Mahdi, 2018). This incredible versatility and precision make it a perfect fit for the development of the next generation of farm-specific, crop-specific fertilizers (Figure 1.2).

1.4 Functional Materials in Fertilizer Design

The integration of materials designed with highly advanced functions into fertilizers as a next-generation efficient, environment-friendly, and diversified option after traditional fertilizers is a great step forward. Using such functional materials helps to optimize the fertilizer application process by controlling nutrient release, increasing the efficiency of nutrient supply to plant roots as well as reducing the loss of nutrients due to leaching, volatilization, or immobilization in the soil. In this chapter, new applications of polymeric carriers, hydrogels, biodegradable coatings, stimuli-responsive molecules, and agro-industrial wastes are discussed in the context of developing highly efficient fertilizers. These materials ensure that the physicochemical properties of the fertilizer are maximized and that they are also flexible enough to be combined with other agricultural technologies (precision farming, nano-enabled delivery systems, and climate-smart agronomy).

1.4.1 Polymeric Carriers and Hydrogels

Polymeric materials have been extensively considered in the design of fertilizers for pollutants' encapsulation and any release triggered by environmental factors. Natural (chitosan, starch, and alginate) and synthetic polymers (polyvinyl alcohol and polyethylene glycol) have been used as carriers in slow-release and CRFs (Salimi et al., 2023). Of these, hydrogels can be particularly useful for both nutrient- and water-holding functions and may be defined as cross-linked polymeric materials able to absorb and retain large amounts of water. Fertilizers made in hydrogels can retain nutrients in their polymeric network and release them under the stimuli of soil moisture or temperature. This characteristic not only contributes to a steady supply of nutrients during crop growth but also improves the resistance to drought in plants. Chitosan-driven, nitrogen- and phosphorus-filled agarose gels suppress transistor leaching and enhance the nutrient uptake efficiency of plants, for example. Furthermore, superabsorbent polymers (SAPs) are being engineered for holding up to several hundred times their weight in water, which can be particularly beneficial in the arid and semiarid agricultural areas. The capability of polymeric carriers provides for the addition of other protective agents (micronutrients, PGRs, and beneficial microorganisms), forming a multifunctional delivery system. The significance of the mechanical strength, porosity, and biodegradation rates of these carriers can be tailored during fabrication for specific soil types and crop needs (Tyagi et al., 2022).

1.4.2 Biodegradable Coatings

The trend toward sustainable agriculture has resulted in the formulation of biodegradable coatings for fertilizer granules as a natural alternative to the traditional petroleum-based polymer coatings. Meanwhile, biodegradables would breakdown into innocuous constituents such as water, carbon dioxide, and biomass in natural soil under the natural weathering condition, which would solve the problem of microplastics accumulation. Starch, cellulose, lignin, and zein (from corn) are some of the most used materials for biodegradable coatings. Chemical modifications and blending with other polymers can provide desired characteristics such as moisture-proof, mechanical properties, and nutrient passage. For instance, the application of lignin-coated urea showed an obvious retardancy of nitrogen release, thus decreasing the loss of volatilization and enhancing the nitrogen use efficiency. Starch-based coatings have also been used for encapsulation of potassium fertilizers, leading to release over a longer period with the associated increase in nutrient recovery. The use of biodegradable coatings also aligns with the principles of circular agriculture, which aims to use agricultural residues and biowastes as resources. The utilization of such renewable materials also leads to less dependence on synthetics as well as waste valorization. Furthermore, the degradability of these coatings helps in preserving soil health and microorganism activity, which are usually affected by nondegradable residues of traditional fertilizers (Skrzypczak et al., 2023a).

1.4.3 Stimuli-Responsive Materials

Stimuli-responsive or "smart" materials are a frontier of fertilizer development that allows for dynamic and temporal control of nutrient release in response to environmental or biological triggers (Shen et al., 2023). These are materials with the ability to alter their structure, solubility, or permeability in response to physical or chemical stimuli (elicitants) like pH, temperature, moisture, redox potential, or even root exudates. Whether acidic or alkaline, these matrices swell or degrade in response to a change in pH, liberating encapsulated nutrients into the soil environment to fertilize plants only when the pH of the growth media is favorable to their growth. This is especially significant for control of phosphorus availability in calcareous and acid soils, where P fixation is an important issue (Adnan et al., 2025). Moisture triggering has been provided by polymers, which in the presence of moisture, such as swelling hydrogels containing (meth)acrylic acid, can swell when irrigated or following natural rainfall to release nutrients and realize the benefits of reduced waste and improved nutrient utilization. More innovative examples are enzyme-sensitive coatings, which break down in response to root-secreted enzymes to enable nutrient release in the immediate vicinity of the plant root (rhizosphere). This also became possible due to the construction of thermoresponsive polymers that expand (or contract) with the temperature of the soil, synchronizing nutrient supply to the metabolic rate of crops. The inclusion of these inputs in fertilizer formulations enables very precise nutrient management, contributing to the overall goals of precision agriculture and climate-resilient farming. While still in the research stages, these materials are opening the doors for fertilizers that are smart and self-regulating (Olasehinde et al., 2023).

1.4.4 Use of Agricultural and Industrial Waste

Valorization of agro-industrial wastes to produce fertilizers is an increasingly topic of interest in the development of sustainable fertilizers, where agro-industrial wastes are upcycled to become sources of raw materials for nutrient recovery and carrier development. It, thus, not only responds to waste problems, but also is an example of resource use efficiency and circular economy. According to Table 1.1, rice husk, sugarcane bagasse, wheat straw, and banana peels, among others, are abundant in potassium, silica, and carbon, offering the qualities to be a source of biochar, ash-based fertilizers, or nutrient carriers. For instance, ash of banana peel, which is rich in potassium, has been effectively included in slow-release potassium formulations, whereas rice husk is used as a functional carrier for micronutrients such as zinc and boron. Industrial residues also rich in nitrogen, phosphorus, and/or organic carbon are also produced by some industries, for example, fly ash, phosphonyls, distillery sludges, and food processing effluents, via methods such as composting, anaerobic digestion, pyrolysis, and/or fermentation turned into biofertilizers, organic manures, and nanoparticle precursors. An outstanding development is biochar-derived nanocomposites, and biochar is reportedly used as a medium to immobilize nanoscale nutrients or microorganisms. These

Table 1.1 Utilization of agricultural and industrial wastes in sustainable fertilizer development.

S. no.	Waste material	Processing method	Application/outcome
1	Banana peel ash	Pyrolysis	Slow-release potassium fertilizer
2	Rice husk	Thermal treatment	Carrier for zinc and boron micronutrients
3	Sugarcane bagasse	Composting	Organic carbon-rich compost
4	Wheat straw	Anaerobic digestion	Nitrogen-rich biofertilizer
5	Distillery sludge	Composting + enrichment	Phosphorus-enriched organic manure
6	Food processing waste	Fermentation	Amino acid-based foliar biofertilizer
7	Fly ash	Blending with organic manure	Potassium-rich mineral fertilizer
8	Poultry litter	Pyrolysis	Biochar for nutrient adsorption and soil health
9	Corn cob	Biochar synthesis	Carrier for NPK and micronutrients
10	Coconut shell	Gasification	Biochar-based nano-fertilizer composite
11	Cassava peels	Anaerobic digestion	Potassium-rich slurry fertilizer
12	Soybean straw	Hydrothermal carbonization	Slow-release nitrogen fertilizer
13	Brewery spent grains	Composting	Organic fertilizer enriched with phosphorus
14	Leather industry sludge	Lime stabilization + composting	Chromium-safe nutrient fertilizer

composites improve retention of nutrients and reduce emissions; ultimately, together, these composites will also contribute to soil carbon sequestration to mitigate the impacts of climate change. In addition, the utilization of waste-based materials can reduce the cost of producing fertilizers, bringing advanced fertility management technologies to smallholder farmers. State subsidies and industrial partnerships are encouraging such waste-to-fertilizer plans as part of sustainable development models (Ejedegba, 2024).

1.5 Characterization Techniques

Characterization of these NGFs is a key method in the research and the optimization of the next generation of fertilizers. It offers valuable information on the structure, composition, release behavior, degradation, and soil interaction of advanced fertilizer formulas. Characterization is required not only to achieve consistency and replicability, but also to link the physical and chemical characteristics of fertilizers to desired agronomic responses. Here, different methods of characterizing new fertilizers are reviewed with a specific focus on physicochemical characterization, nutrient release kinetics, stability, and shelf-life, as well as soil–fertilizer interactions. These are important for assessing how well fertilizers work in the field and confirming their environmental and agronomic benefits (Anas et al., 2024a).

1.5.1 Physicochemical Analysis (SEM, TEM, FTIR, XRD)

Physicochemical characterization is key to comprehend structural and morphological features of fertilizers, especially when they are doped with nanomaterials or functional coatings. These factors affect the solubility, the reactivity, and the cation release rate of the nutrient. Various sophisticated techniques are used to study the microstructure, crystallinity, chemical bonding, and composition of fertilizers. Scanning electron microscopy (SEM) is extensively used to evaluate the surface morphology and texture of fertilizer particles coated or encapsulated (Sim et al., 2021). Images from the SEM can give a surface porosity, the effectiveness of coatings, and structural integrity. Contrastingly, transmission electron microscopy (TEM) allows nanoscale imaging, which is great for nano-fertilizers or nutrient-loaded nanoparticles (Carmona et al., 2022). It is useful for the calculation of the nanoparticle size distribution, core-shell structure, and encapsulation efficiency. Functional groups within the fertilizer matrix are identified by Fourier transform infrared spectroscopy (FTIR) (Losacco et al., 2022). Analyses of FTIR spectra verify the existence of the polymers, coatings, and nutrient-binding ligands, which indicates that the modifications or encapsulation were achieved successfully. Experimental characterization, both qualitative and quantitative, of these bands may provide an indication of potential interactions with water or soil molecules (e.g., the relative depth and position of the peaks corresponding to groups —OH, $—NH_2$, —COOH, or $—SO_4$). X-ray diffraction (XRD) is used to determine whether fertilizer compounds were crystalline or amorphous (de Souza Meira et al., 2020). The diffraction patterns reflect pure phase, crystallite size, and structural phase transformation of mineral/metal fertilizers (Figure 1.3). For instance, even when applied at a similar scroll, the crystalline form of urea or phosphates has different rates of dissolution from the amorphous form, and this can affect the availability of nutrients from soils. Together, these analytical

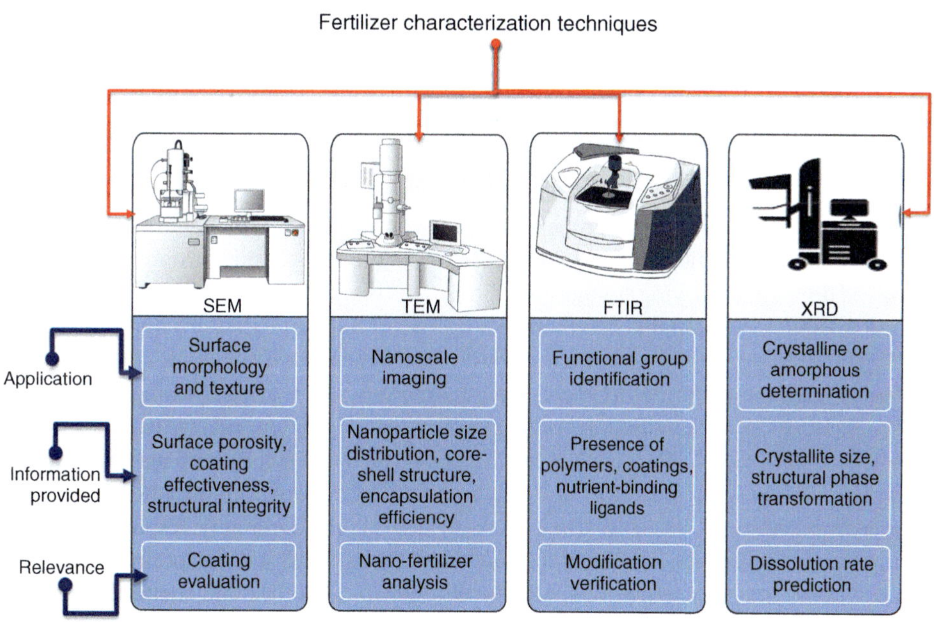

Figure 1.3 Different fertilizer characterization techniques.

tools enable scientists to optimize fertilizer formulations toward the desired physicochemical performance and product quality during scale-up processes to the commercial scale (Skrzypczak et al., 2023b).

1.5.2 Release Kinetics and Nutrient Profiling

Controlled-release characteristic is considered as one of the greatest characteristics of the future generation of fertilizers. The assessment of nutrient release kinetics is indispensable for specifying ways in which they are released over time and under various environmental conditions. Such a characterization can also lead to the development of fertilizers adapted to the plant's needs and with reduced nutrient loss. In general, in vitro leaching tests are conducted under fixed conditions, in which fertilizer samples are exposed to water or soil solutions, and the concentration of released nutrients is measured after a specific period by spectrophotometry, atomic absorption spectroscopy (AAS), or inductively coupled plasma (ICP) methods (Dimkpa et al., 2017). That would be, for example, nitrogen or phosphorus compounds released using UV–Vis spectrophotometry or micronutrients like zinc or iron with flame AAS, for instance. The release data obtained is fitted to kinetic models such as zero-order, first-order, Higuchi, and Korsmeyer–Peppas to identify whether nutrient release occurs through diffusion, dissolution, erosion, or a combination of these mechanisms. Such models offer useful parameters such as rate constants and release mechanisms, to assist in the improvement of the formulation. Nutrient profiling involves the determination of total and available nutrients by standard chemical methods of extraction. For example, the total nitrogen is determined by the Kjeldahl method and the easily available phosphorus by the Bray or Olsen method (Uygur et al., 2010). It is very important that these nutrients are presented in bioavailable form in the fertilizer for agronomic purposes. More advanced compositions might also contain multiple nutrients or bioactives, and profiling then makes sure these bioactives are stable, compatible, and deliverable in the same matrix without losing unbearable interactions (Razi & Rashidinejad, 2021).

1.5.3 Stability and Shelf-Life Evaluation

Lifetime and stability are critical properties for the commercial viability of the new fertilizers. A formulation should also be stable in physical state and chemical composition for a prolonged time under varying storage conditions. This is especially important for fertilizers that incorporate biodegradable polymers, bio-agents, or moisture-resistant nanomaterials. Several studies have employed aging tests under accelerated conditions, which expose the fertilizer to high temperatures, humidity, and light to simulate long-term storage situations (Jakhar et al., 2022). The evolution with time of moisture content, pH, particle cohesion, nutrient availability, and biological activity (for biofertilizers) is monitored to assess degradation rates. For polymer-coated or encapsulated fertilizers, barrier characteristics such as water permeability and gas diffusivity are very important. These are determined using gravimetric techniques or cells, detecting how well the coating shields the core nutrients from environmental exposure. About bioformulations, viabilities are assessed by plate counts, enzyme activity, or DNA quantification (e.g., Quantitative Polymerase Chain

Reaction (qPCR)) to ensure that microorganisms are active during storage. Thermal degradation points of materials employed in the compounds of fertilizers can be identified by means of thermal analysis techniques such as thermogravimetric analysis (TGA) and differential scanning calorimetry (DSC) (Diaz et al., 2021). A reliable and long shelf-life fertilizer product represents optimized cost performance and gains the confidence of the users, as the key driver for adoption of the product by the farmers and other stakeholders (Areo, 2024).

1.5.4 Soil–Fertilizer Interaction Studies

Most notably, to predict nutrient availability, environmental fate, and plant uptake efficiency, it is critical to know how fertilizers interact with various soil types. These interactions are regulated by soil texture, pH, organic matter, microbial community, and water retention capacity. Soil incubation experiments are used to evaluate how well the fertilizer formula behaves over time under realistic conditions. Nutrient retention, leaching potential, and microbial transformations such as nitrification or phosphate solubilization of the amended P are examined. This is useful in determining whether a fertilizer is appropriate for use in acidic, alkaline, saline, or sandy soil. Rhizobox or column leaching experiments can expose how nutrients travel in the plant–soil profile and the depth effect of fertilizer. Electrochemical sensors and Diffusive Gradients in Thin Films (DGT) techniques for in situ nutrient bioavailability monitoring in soils can be applied to provide information about nutrient fluxes in time and space. Interaction studies between soil and fertilizer also assess the adsorption–desorption behavior with isotherm-type models like Langmuir and Freundlich, to give information about the sorption strength of the nutrient to soil particles. These associations determine whether nutrients are preserved in plant-available pools or are entrapped. Furthermore, bio-agent fertilizers (e.g., mycorrhizae and nitrogen-fixing bacteria) are tested for compatibility with indigenous soil microbiota. Methods, such as metagenomics, enzyme activity determination, or microscopy-based techniques, such as fluorescence in situ hybridization (FISH), can be applied to study the responses of microbial communities and potential synergistic activity. In general, these studies are important to ascertain location-specific performance, regulatory compliance, and for guidance at the field level (Ciriaci et al., 2019).

1.6 Application Strategies and Agronomic Impacts

The shift toward novel fertilizers will call for not just new materials and characterization, but also a significant understanding of their application and their agronomic effects at a large scale. To enable these fertilizers to fulfill their claim in terms of enhancing NUE, minimizing adverse environmental impacts, and promoting sustainable agriculture, they need to be widely tested under field conditions on a range of crops and agro-ecological zones. This section covers different aspects of the performance of new fertilizers in the field, their effects on crop yield, NUE, soil health, microbial activity, and tailor-made properties depending on the typology of crop (Padhi & Agasti, 2025).

1.6.1 Field Trials and Yield Performance

If the next-generation fertilizers are really to show their utility and reliability in practical terms, then the truth will come during field trials. Field environments, in contrast to laboratory conditions, are complex and affected by many interacting factors such as soil type, climate, irrigation management, and cropping systems. The trials contribute to evaluations of whether active-release, bioavailability, or multifunction of new fertilizers contribute to real crop yield and quality enhancements. Smart fertilizers such as nanoformulation and polymer-coated nutrients were found to perform much better than traditional fertilizers for crop biomass, fruit quality, grain protein content, and harvest index in various studies. For example, application of nano-urea or nano/polymer-coated nitrogen fertilizers resulted in 10–30% yield enhancements in rice, wheat, and maize due to lowered N losses and synchronized nutrient release with crop requirements. These fertilizers are also useful in reducing lodging and synergistic overfertilization effects, which might be encountered with urea or ammonium nitrate application. Field tests also show crop tolerance of new formulations, risks of phytotoxicity, and efficacy of dosages and treatment frequency. Furthermore, through monitoring of seasonal yield responses, these trials generate information useful for refining fertilization timing, splitting fertilization doses, and integrating fertilization with other inputs such as irrigation or crop protection products (Xing & Wang, 2024).

1.6.2 Nutrient Use Efficiency (NUE)

Increasing NUE, the amount of crop biomass and/or yield own per unit of nutrient applied, is one of the key objectives of next-generation fertilizers. Enhancing NUE decreases not only the cost of fertilizers to farmers but also the pollution of the environment due to nutrient runoff, volatilization, or leaching. CRFs, nano-fertilizers, and biofertilizers have shown they could be used as efficient agents of the improved NUE of N, P, and K. For example, polymer-coated urea matches the release of nitrogen with the crop's developmental phases, cutting the potential for nutrient losses and improving NUE by as much as 50% over standard urea. In the same way, the nano-phosphorus fertilizer enhances the availability of phosphorus in alkaline soils, which are subjected to fixation. NUE is measured using isotope labeling (e.g., ^{15}N) and soil residual analysis, and by modeling nutrient uptake information derived from remote sensing. It is also affected by time and method of application (e.g., toward vs. soil) and incorporation with site-specific nutrient management. Increased NUE benefits the aspiration for sustainable intensification to get more out of less while not compromising the integrity of environment (Grzebisz & Łukowiak, 2021).

1.6.3 Soil Health and Microbial Dynamics

Although the direct use of fertilizers as plant nutrients is beneficial for short-term growth, soil stability is crucial for successful agriculture. Traditional chemical fertilizers, when used over time, can weaken the soil due to factors such as air, sunlight, rain, and wastewater. Excessive application of mineral fertilizers may lead to soil acidification, increased sodium content, and disruption of soil microbial communities. In similar contrast, the advanced fertilizers support the positive functions of the soil ecosystem. Major soil health–beneficial bioformulations include biodegradable coatings and organic nutrient carriers (such as those coated on biodegradable films), as well as bioformulations containing beneficial microbes, including nitrogen-fixing bacteria, phosphate solubilizers, and mycorrhizal fungi. These amendments amend soil structure, coefficient of water stagnation, and organic carbon content. Other issues affecting soil health include soil compaction, root compaction, and soil boring. The chemical composition and the release pattern of the fertilizer also especially affect the dynamics of the microorganisms. Slow-release fertilizers do not create nutrient toxicity that allows beneficial microbes to live right alongside consuming nutrients. Enzyme activity (dehydrogenase, urease, and phosphatase), microbial biomass, and nutrient mineralization rates were enhanced upon application of bio-stimulants and plant growth-promoting rhizobacteria (PGPR). Metagenomics, qPCR, and enzyme-linked assays are applied for the detection of soil microbial community changes. One main argument is that nano-enabled (Ray et al., 2023) and organic fertilizers enhance microbial diversity and functional abundance that are crucial for nutrient cycling and soil resistance to stress (drought and heavy metals pollution), which has been proven by recent research (Liu et al., 2024).

1.6.4 Crop-Specific Formulations

Recent advances in fertilizer chemistry include crop-specific formulations that are customized for the unique physiological and nutritional requirements of various plant species. Each crop has different nutrient uptake dynamics, morphology of the root system, and developmental stages that should be considered for optimal timing and method of nutrient application. For example, rice needs slow-release nitrogen in the vegetative growth stage for delayed lodging, and corn needs available phosphorus quickly at the root's emergence stage. Horticulture crops like tomatoes or strawberries could be more successful with nano-fertilizer enriched in micronutrients that can enhance the color and taste of fruit and the period of fruit shelf life. The next generation of fertilizers is also being tailored to high-value crops and export crops, which have exacting demands for uniformity, yield, and quality. Nano-chelated or encapsulated micronutrients for specific crops are a perfect solution for trace elements such as zinc, boron, and iron, which are generally deficient in most soils (Kapoor et al., 2022). Culture-based fertilizers also consider the soil–plant–microbe interaction. For legumes, formulations may comprise rhizobia or mycorrhizal fungi to promote nitrogen fixing and phosphorus uptake. In cereals, silicon contributes to higher disease resistance and mechanical strength. These types of personalized solutions not only increase yields but also avoid excessive nutrient application, rendering agriculture more concise and sustainable (Getahun et al., 2024).

1.7 Environmental and Economic Considerations

Given the evolution of agriculture under the pressure of environmental degradation, climate change, and the need to feed a growing population, performance properties of the next-generation fertilizers need to be assessed from technical, environmental, and economic perspectives. The sustainable and practical applications of these novel formulations will ultimately be determined by their life cycle profile, cost-effectiveness, environmental benignity, and acceptance by regulators. This part concentrates on three fundamental pillars for informed decision-making and sustainable innovations: life cycle assessment (LCA), cost–benefit analysis (CBA), and eco-toxicity and regulatory issues (Zakira, 2017).

1.7.1 Life Cycle Assessment (LCA)

LCA is an established process for determining the environmental burden of a product or production process over its life—from raw material extraction through production, packaging, and transportation to use and end use. It helps researchers, companies, and policymakers pinpoint environmental hotspots and focuses on reducing detrimental impacts (Awewomom et al., 2024). In the case of the next-generation fertilizers, LCA will be especially useful for comparing the new fertilizers with the commonly used fertilizers such as urea, ammonium nitrate, superphosphates with respect to the energy used, GHGs emitted, eutrophication potential, water use, and resource use. Modern formulations such as polymer-coated urea, nano-fertilizers, and bio-based slow-release fertilizers have also shown reduced environmental impact as compared to conventional fertilizers, including less leaching of nitrogen into groundwater, less emission of nitrous oxide, and less phosphate runoff. LCA studies also show trade-offs. For example, polymer coating manufacture or nanoparticle synthesis, if not sustainably sourced, might require more energy and emit more CO_2 (Artz et al., 2018). The implementation of greener chemistry principles, raw materials from renewable sources, as well as biodegradable matrixes are determinative for the environmental advantage of new fertilizers versus their embodied cost. Fertilizer life cycles are modeled using LCA tools, including SimaPro, GaBi, and OpenLCA under various scenarios. Comparisons are standardized by expressing them in functional units (e.g., per kg nutrient delivered or per ton crop yield). These analyses can be used to underpin policy making, eco-labeling, and decision-making support to stakeholders in the agri-food value chain (Notarnicola et al., 2015).

1.7.2 Cost–Benefit Analysis

The CBA measures the economic viability of NGF according to the ratio of total costs for production, distribution, and application to the agronomic–economic and environmental benefits. Although these new fertilizers can be more expensive to produce initially than conventional fertilizers, the potential rewards are well worth it. It is easy to see from the farmer's standpoint that less fertilizer, fewer applications, and increased crop yields lead to lower input costs and higher profit. For instance, polymer-coated urea may have a higher per-kilogram price compared with the traditional urea, but the need for less frequent applications and better NUE will increase yield and decrease N loss. Furthermore, nano-fertilizers can help in preventing waste and improve the targeted delivery of nutrients while minimizing off-site losses. These enhancements result in water and labor savings and in lower fertilizer processing and transportation energy consumption risks with reduced dosages. Some field trials and simulation models have demonstrated that smart fertilizers can reduce input cost by up to 20–40% and increase income by 15–30%, varying with the crop and region. Large-scale application is hindered by economical aspects, particularly when targeting low-economic farmer societies. Cost depends heavily on production scalability, availability of raw materials (e.g., nanomaterials or biopolymers), production infrastructure, and market access. Subsidy systems, public–private partnerships, and technology transfer strategies are necessary for making these fertilizers available and competitive in the market. CBA also includes externalities, including lower costs associated with environmental cleanup (e.g., water and soil treatments) and savings in human health costs due to reduced exposure to detrimental residues in conventional fertilizers. Factoring in these more indirect benefits makes these next-generation solutions that much more attractive and practical (Parhamfar & Zabihi, 2024).

1.7.3 Eco-toxicity and Regulatory Aspects

Since fertilizers are released into terrestrial soil ecosystems, as well as washed into aquatic ecosystems, eco-toxicological safety is a basic criterion for a nano-fertilizer. Next-generation/designed fertilizers, in contrast to conventional chemical fertilizers frequently associated with eutrophication, loss of biodiversity, and water spoilage, are frequently biodegradable, nontoxic, and eco-friendly. Nano-fertilizers and petrol-based or synthetic polymer fertilizers also pose novel risks. The eventual effects of nanomaterials like ZnO itself, silver, or silica nanoparticles on soil microbiota, plant physiology, nontarget organisms (e.g., earthworms and aquatic species), and food security have not been studied extensively (Wahab et al., 2024). Newer reports indicate that while the low concentrations of some of these compounds can increase nutrient uptake and plant resistance, higher or multiple applications can disturb beneficial populations of microorganisms or accumulate in edible tissues. To solve those problems, conventional eco-toxicity tests such as acute and chronic toxicity assays, bioaccumulation experiments, and soil respiration tests are used. International agencies, such as the OECD, EFSA, and US EPA, have established criteria for testing the toxicity and environmental behavior of new agricultural inputs. It is summarized as follows: risks are classified using metrics such as LC50, NOEC, and EC50 (Iovine et al., 2024). In addition, new fertilizers must undergo regulatory approvals prior to commercial introduction. These are associated with REACH (EU) regulations on chemical safety, biosafety regulations on bioformulations, and country-specific fertilizer registration rules, which demand extensive data on composition, efficacy, shelf life, and environmental safety. A lack of globally harmonized standards frequently inhibits innovation and slows cross-border trade. Risk-based regulation challenges of Nanomaterial(s) (NM) are addressed, and it is crucial to have a precautionary principle and standard analytic procedures and transparent reporting of nano-specific data to resolve these challenges. Promoting third-party certifications, eco-labeling schemes, and regulatory innovation hubs can facilitate the safe and responsible upscaling of next-generation fertilizers.

1.8 Conclusion

The advancement of next-gen fertilizers signifies a new era in agroeconomic nutrient management: productivity coupled with environmental stewardship. At a time when global agriculture is increasingly faced with mounting pressures—soil exhaustion, volatile climate, and the need for higher crop yields, novel fertilizer technologies prove to be viable tools to meet agronomic as well as environmental demands. This chapter has described the progress and potential of advanced fertilizers such as nano-fertilizers, polymeric-coated slow-release formulations, biodegradable carriers, and stimuli-responsive delivery systems. These next-generation fertilizers are formulated to increase NUE, minimize losses through leaching and volatilization, and deliver nutrients in a targeted manner according to the needs of the crop. Their compatibility with intelligent delivery platforms and precision agriculture systems also makes it possible to optimize the use of resources and reduce the impact on the environment. A key highlight is the use of functional materials, including hydrogels, agro waste-based carriers, and biopolymers, that enhance the fertilizers' efficiency and are consistent with circular economy logic. Moreover, a field assessment has shown increased crop yield, soil quality, and microbial diversity in the case of the abovementioned advanced inputs. LCAs and CBAs show that, despite their initial higher production expenses, some next-generation fertilizers are worth adopting due to the anticipated long-term gains in yield, environmental savings, and resource use efficiency. Impact is, however, limited, due to production scalability issues, regulatory vacuums, eco-toxicological considerations, and appropriateness in low-resource areas. Interdisciplinary cooperation among researchers, policymakers, and the industry is needed to address these barriers. Streamlined regulatory approaches, standardized field validation, and farmer-focused technology transfer mechanisms would be needed to realize the full potential of these technologies. Finally, "next-generation" fertilizers have transformational potential and can help to meet food security and environmental sustainability, twin obligations in the coming decades. When developed and applied in a responsible and data-driven way, such fertilizers have the potential to contribute to more sustainable agriculture, more climate-resilient agriculture, and to the UN-supported global work toward the SDGs, including SDG 2 (Zero Hunger), SDG 6 (Clean Water and Sanitation), 12 (Responsible Consumption and Production), and 13 (Climate Action).

References

Ahmad, L., & Mahdi, S. S. (2018). *Satellite farming*. Springer.

Ameen, A., & Raza, S. (2017). Green revolution: A review. *International Journal of Advances in Scientific Research*, 3(12), 129–137.

Anas, M., Falak, A., Khan, A., Khattak, W. A., Nisa, S. G., Aslam, Q., Khan, K. A., Saleem, M. H., & Fahad, S. (2024a). Therapeutic potential and agricultural benefits of curcumin: A comprehensive review of health and sustainability applications. *Journal of Umm Al-Qura University for Applied Sciences*, 1–16.

Anas, M., Khattak, W. A., Fahad, S., Alrawiq, N., Alrawiq, H. S., Abdelsalam, N. R., Jaremko, M., & Quraishi, U. M. (2024b). Mitigating heavy metal pollution in agriculture: A multi-omics and nanotechnology approach to safeguard global wheat production. *Journal of Hazardous Materials Advances*, 17, 100584.

Anas, M., Falak, A., Hassan, S., Khattak, W. A., Saleem, M. H., Khan, K. A., Khalid, A., & Fahad, S. (2025). Microbial interactions and bacterial responses to metal stress in plants: Mechanisms, adaptations, and applications for sustainable agriculture. *Journal of Crop Health*, *77*(1), 36.

Areo, G. (2024). Sustainable export strategies for perishable agricultural goods.

Arora, V., & Khosla, B. (2024). Synthesis of phosphorus nano-fertilisers their strategic applications and effect on plant growth. *International Journal of Environmental Science and Technology*, 22, 1–20.

Artz, J., Müller, T. E., Thenert, K., Kleinekorte, J., Meys, R., Sternberg, A., Bardow, A., & Leitner, W. (2018). Sustainable conversion of carbon dioxide: An integrated review of catalysis and life cycle assessment. *Chemical Reviews*, 118(2), 434–504.

Awewomom, J., Dzeble, F., Takyi, Y. D., Ashie, W. B., Ettey, E. N. Y. O., Afua, P. E., Sackey, L. N., Opoku, F., & Akoto, O. (2024). Addressing global environmental pollution using environmental control techniques: A focus on environmental policy and preventive environmental management. *Discover Environment*, 2(1), 8.

Carmona, F. J., Guagliardi, A., & Masciocchi, N. (2022). Nanosized calcium phosphates as novel macronutrient nano-fertilizers. *Nanomaterials*, 12(15), 2709.

Ciriaci, D., Grassano, N., & Vezzani, A. (2019). Regulations and location choices of top R&D investors worldwide. *Structural Change and Economic Dynamics*, 49, 29–42.

Dayana, D., Venkatamuni, T., Bhagyalakshmi, A., Malleswari, T. N., & Ushasukhanya, S. (2025). AI-controlled robotics in smart agricultural systems: Enhancing precision, sustainability, and productivity. In S. Mehta & F. Al-Turjman (Eds.), *Edible electronics for smart technology solutions* (pp. 351–382). IGI Global.

Dhaliwal, S., Naresh, R., Mandal, A., Walia, M. K., Gupta, R. K., Singh, R., & Dhaliwal, M. (2019). Effect of manures and fertilizers on soil physical properties, build-up of macro and micronutrients and uptake in soil under different cropping systems: A review. *Journal of Plant Nutrition*, 42(20), 2873–2900.

Diaz, M. J., Ruiz-Montoya, M., Palma, A., & De-Paz, M.-V. (2021). Thermogravimetry applicability in compost and composting research: A review. *Applied Sciences*, 11(4), 1692.

Dimkpa, C., Bindraban, P., McLean, J. E., Gatere, L., Singh, U., & Hellums, D. (2017). Methods for rapid testing of plant and soil nutrients. In E. Lichtfouse (Ed.), *Sustainable agriculture reviews* (pp. 1–43).

Eivazzadeh-Keihan, R., Chenab, K. K., Taheri-Ledari, R., Mosafer, J., Hashemi, S. M., Mokhtarzadeh, A., Maleki, A., & Hamblin, M. R. (2020). Recent advances in the application of mesoporous silica-based nanomaterials for bone tissue engineering. *Materials Science and Engineering: C*, 107, 110267.

Ejedegba, E. O. (2024). Advancing green energy transitions with eco-friendly fertilizer solutions supporting agricultural sustainability. *International Research Journal of Modernization in Engineering Technology and Science*, 06(12), 2582–5208.

Estévez, R., Quijada-Maldonado, E., Romero, J., & Abejón, R. (2025). Additive manufacturing and chemical engineering: Looking for synergies from a bibliometric study. *Applied Sciences*, 15(6), 2962.

Getahun, S., Kefale, H., & Gelaye, Y. (2024). Application of precision agriculture technologies for sustainable crop production and environmental sustainability: A systematic review. *The Scientific World Journal*, 2024(1), 2126734.

Gobinath, R. (2018). Synthesis, characterization and use of zinc and iron nanoparticles as plant nutrients.

Grzebisz, W., & Łukowiak, R. (2021). Nitrogen gap amelioration is a core for sustainable intensification of agriculture—a concept. *Agronomy*, 11(3), 419.

Heuchan, S. M., Fan, B., Kowalski, J. J., Gillies, E. R., & Henry, H. A. (2019). Development of fertilizer coatings from polyglyoxylate–polyester blends responsive to root-driven pH change. *Journal of Agricultural and Food Chemistry*, 67(46), 12720–12729.

Iovine, N., Toropova, A. P., Toropov, A. A., Roncaglioni, A., & Benfenati, E. (2024). Models for the no-observed-effect concentration (NOEC) and maximal half-effective concentration (EC50). *Toxics*, 12(6), 425.

Jakhar, A. M., Aziz, I., Kaleri, A. R., Hasnain, M., Haider, G., Ma, J., & Abideen, Z. (2022). Nano-fertilizers: A sustainable technology for improving crop nutrition and food security. *NanoImpact*, 27, 100411.

Kamal, A., Yang, J., Batool, M., Ara, U., Khattak, W. A., Touhami, D., Aljowaie, R. M., & Elshikh, M. S. (2025). Green synthesis of zinc oxide nanoparticles, characterization and its applications in inducing disease resistance against spot blotch disease in wheat by enhancing its physiological and biochemical parameters. *Journal of Crop Health*, 77(1), 1–12.

Kapoor, P., Dhaka, R. K., Sihag, P., Mehla, S., Sagwal, V., Singh, Y., Langaya, S., Balyan, P., Singh, K. P., & Xing, B. (2022). Nanotechnology-enabled biofortification strategies for micronutrients enrichment of food crops: Current understanding and future scope. *NanoImpact*, 26, 100407.

Kumari, V. V., Banerjee, P., Verma, V. C., Sukumaran, S., Chandran, M. A. S., Gopinath, K. A., Venkatesh, G., Yadav, S. K., Singh, V. K., & Awasthi, N. K. (2022). Plant nutrition: An effective way to alleviate abiotic stress in agricultural crops. *International Journal of Molecular Sciences*, 23(15), 8519.

Kundius, V. A. (2021). Development of export-oriented organic agriculture based on bio-intensive technologies. In A. V. Bogoviz (Ed.), *The challenge of sustainability in agricultural systems: Volume 1* (pp. 115–120). Springer.

Li, X., & Li, Z. (2024). Global trends and current advances in slow/controlled-release fertilizers: A bibliometric analysis from 1990 to 2023. *Agriculture*, 14(9), 1502.

Liu, Y., Lan, X., Hou, H., Ji, J., Liu, X., & Lv, Z. (2024). Multifaceted ability of organic fertilizers to improve crop productivity and abiotic stress tolerance: Review and perspectives. *Agronomy*, 14(6), 1141.

Lodhi, R. S., Das, S., & Das, P. (2023). Recent advances in polymer hydrogels for agricultural applications. Paramita Das (Indian Institute of Science Education and Research (IISER) Bhopal, India) and Subhasis Das (The Energy and Resources Institute, India) Novel polymeric materials for environmental applications, 109–177. World Scientific.

Losacco, D., Campanale, C., Tumolo, M., Ancona, V., Massarelli, C., & Uricchio, V. F. (2022). Evaluating the influence of nitrogen fertilizers and biochar on Brassica Oleracea L. var. botrytis by the use of Fourier transform infrared (FTIR) spectroscopy. *Sustainability*, 14(19), 11985.

Mahesha, K., Singh, N., Amarshettiwar, S., Singh, G., Gulaiya, S., Das, H., & Kumar, J. (2023). Entering a new agricultural era through the impact of nano-fertilizers on crop development: A review. *International Journal of Plant & Soil Science*, 35(20), 94–102.

Moradi, S., Babapoor, A., Ghanbarlou, S., Kalashgarani, M. Y., Salahshoori, I., & Seyfaee, A. (2024). Toward a new generation of fertilizers with the approach of controlled-release fertilizers: A review. *Journal of Coatings Technology and Research*, 21(1), 31–54.

Notarnicola, B., Tassielli, G., Renzulli, P. A., & Lo Giudice, A. (2015). Life cycle assessment in the agri-food sector: An overview of its key aspects, international initiatives, certification, labelling schemes and methodological issues. Bruno Notarnicola, Roberta Salomone, Luigia Petti, Pietro A. Renzulli, Rocco Roma, Alessandro K. Cerutti Life cycle assessment in the agri-food sector: Case studies, methodological issues and best practices, Springer 1–56.

Olasehinde, A. A., Blessing, A. O., Obodozie, J. C., & Chukwuemeka, S. O. (2023). Cyber-physical system integration for autonomous decision-making in sensor-rich indoor cultivation environments. *World Journal of Advanced Research and Reviews*, 20(2), 1563–1584.

Olvermann, M., Hornung, J., & Kauffeld, S. (2023). " We could be much further ahead"-multidimensional drivers and barriers for agricultural transition. *Journal of Rural Studies*, 97, 153–166.

Padhi, L. P., & Agasti, N. R. (2025). Discussion paper: Sustainable increase of crop production through improved technical strategies, breeding and adapted management–a European perspective.

Padhiary, M., Barbhuiya, J. A., Roy, D., & Roy, P. (2024). 3D printing applications in smart farming and food processing. *Smart Agricultural Technology*, 9, 100553.

Parhamfar, M., & Zabihi, A. (2024). Next-gen energy solutions: A brief study on boosting distribution efficiency with IoE technology. *Energy*, 7(3), 72–81.

Pretty, J. (2018). Intensification for redesigned and sustainable agricultural systems. *Science*, 362(6417), eaav0294.

Ray, M. K., Mishra, A. K., Mohanta, Y. K., Mahanta, S., Chakrabartty, I., Kungwani, N. A., Avula, S. K., Panda, J., & Pudake, R. N. (2023). Nanotechnology as a promising tool against phytopathogens: A futuristic approach to agriculture. *Agriculture*, 13(9), 1856.

Razi, S. M., & Rashidinejad, A. (2021). Bioactive compounds: Chemistry, structure, and functionality. In S. M. Razi & A. Rashidinejad (Eds.), *Spray drying encapsulation of bioactive materials* (pp. 1–46). CRC Press.

Sabarivasan, R., & Murali Arthanari, P. (2025). Application of nanoencapsulation technology in agriculture for effective and sustainable weed management: A critical review. *Communications in Soil Science and Plant Analysis*, 56(2), 277–291.

Sahoo, R. K., Bhardwaj, D., & Tuteja, N. (2012). Biofertilizers: A sustainable eco-friendly agricultural approach to crop improvement. In N. Tuteja & S. S. Gill (Eds.), *Plant acclimation to environmental stress* (pp. 403–432). Springer.

Salimi, M., Channab, B., El Idrissi, A., Zahouily, M., & Motamedi, E. (2023). A comprehensive review on starch: Structure, modification, and applications in slow/controlled-release fertilizers in agriculture. *Carbohydrate Polymers*, 322, 121326.

Sanghi, R., & Singh, V. (2012). *Green chemistry for environmental remediation*. John Wiley & Sons.

Sashidhar, P., Yakkala, K., Bhunia, R., Kochar, M., Brau, L., Cahill, D., & Dubey, M. (2025). Nano-biochar supported Zn delivery in plants to enhance seedling growth and ROS management in rice. *Environmental Science: Nano*, 12, 3791.

Shah, F., & Wu, W. (2019). Soil and crop management strategies to ensure higher crop productivity within sustainable environments. *Sustainability*, 11(5), 1485.

Shen, M., Liu, S., Jiang, C., Zhang, T., & Chen, W. (2023). Recent advances in stimuli-response mechanisms of nano-enabled controlled-release fertilizers and pesticides. *Eco-Environment & Health*, 2(3), 161–175.

Sim, D. H. H., Tan, I. A. W., Lim, L. L. P., & Hameed, B. H. (2021). Encapsulated biochar-based sustained release fertilizer for precision agriculture: A review. *Journal of Cleaner Production*, 303, 127018.

Singh, B. (2018). Are nitrogen fertilizers deleterious to soil health? *Agronomy*, 8(4), 48.

Skrzypczak, D., Izydorczyk, G., Taf, R., Moustakas, K., & Chojnacka, K. (2023a). Cellulose-based fertilizers for sustainable agriculture: Effective methods for increasing crop yield and soil health. *Industrial Crops and Products*, 205, 117500.

Skrzypczak, D., Trzaska, K., Mikula, K., Gil, F., Izydorczyk, G., Mironiuk, M., Polomska, X., Moustakas, K., Witek-Krowiak, A., & Chojnacka, K. (2023b). Conversion of anaerobic digestates from biogas plants: Laboratory fertilizer formulation, scale-up and demonstration of applicative properties on plants. *Renewable Energy*, 203, 506–517.

Solanki, P., Bhargava, A., Chhipa, H., Jain, N., & Panwar, J. (2015). Nano-fertilizers and their smart delivery system. In M. Rai, C. Ribeiro, L. Mattoso, & N. Duran (Eds.), *Nanotechnologies in food and agriculture* (pp. 81–101). Springer.

de Souza Meira, R. C., da Paz, S. P. A., & Corrêa, J. A. M. (2020). XRD-Rietveld analysis as a tool for monitoring struvite analog precipitation from wastewater: P, Mg, N and K recovery for fertilizer production. *Journal of Materials Research and Technology*, 9(6), 15202–15213.

Tyagi, P., Agate, S., Velev, O. D., Lucia, L., & Pal, L. (2022). A critical review of the performance and soil biodegradability profiles of biobased natural and chemically synthesized polymers in industrial applications. *Environmental Science & Technology*, 56(4), 2071–2095.

Uygur, V., Irvem, A., Karanlik, S., & Akis, R. (2010). Mapping of total nitrogen, available phosphorous and potassium in Amik Plain, Turkey. *Environmental Earth Sciences*, 59, 1129–1138.

Wahab, A., Muhammad, M., Ullah, S., Abdi, G., Shah, G. M., Zaman, W., & Ayaz, A. (2024). Agriculture and environmental management through nanotechnology: Eco-friendly nanomaterial synthesis for soil-plant systems, food safety, and sustainability. *Science of the Total Environment*, 926, 171862.

Xing, Y., & Wang, X. (2024). Precise application of water and fertilizer to crops: Challenges and opportunities. *Frontiers in Plant Science*, 15, 1444560.

Zakira, U. (2017). *Life cycle assessment and cost-benefit analysis of landfill mining at the City of Denton Landfill*. The University of Texas at Arlington.

2

Revolutionizing Farming: The Rise of Next-Generation Fertilizers

Sharjeel Haider[1,2], Nosheen Mirza[3], Guangda Ding[1], Muhammad Anwar-ul-Haq[2], Khadija Bibi[4], Venuste Munyaneza[1], Iftikhar Ali Ahmad[1], Muhammad Mehran[1], Sidra Sohail[5], Asma Zafar[6], Dua e Zainab[7], and Bushra Huma[8]

[1] *College of Resources and Environment/Microelement Research Center/Key Laboratory of Arable Land Conservation (Middle and Lower Reaches of Yangtze River), Ministry of Agriculture and Rural Affairs, Huazhong Agricultural University, Wuhan, China*
[2] *Institute of Soil and Environmental Sciences, University of Agriculture Faisalabad, Faisalabad, Pakistan*
[3] *Department of Soil and Environmental Sciences, Ghazi University, Dera Ghazi Khan, Pakistan*
[4] *Department of Botany, Faculty of Sciences, Ghazi University, Dera Ghazi Khan, Pakistan*
[5] *State Key Lab of Soil Erosion and Dryland Farming on the Loess Plateau, Institute of Soil and Water Conservation, Northwest A & F University, Xianyang, China*
[6] *College of Natural Resources and Environment, Northwest A&F University, Yangling, China*
[7] *Key Laboratory of Green and Low Carbon Agriculture on Dryland in Northwest China, Ministry of Agriculture and Rural Affairs, College of Natural Resource and Environment, Northwest A & F University, Xianyang, China*
[8] *Hubei Key Laboratory of Soil Environment and Pollution Remediation, College of Resources and Environment, Huazhong Agricultural University, Wuhan, China*

2.1 Introduction: The Future of Farming

Nutrients or plant fertilizers are substances that cause development and growth of plants in the presence of elements or nutrients. The fertility of the soil has been reduced by killing the useful microbes due to heavy usage of fertilizers causing a great requirement of replacement and eco-friendly nanofertilizers (NFs), which is the most important use of nanotechnology in the field of the agriculture sector (Abobatta, 2018). This is because the nutrient carriers or nutrient transporters taking substrates that are of nano dimension, that is, 1–100 nm, are coming up. The conventional fertilizers or bulk fertilizer materials can be processed as NFs or extracted in other plants by the use of encapsulation/coating of nanomaterials (NMs). Along with the big surface area, nanoparticles have the ability to carry a lot of nutrients and release them at a gradual and consistent rate. At this point, the crop is able to absorb elements based on its requirement without encountering any negative factors that are experienced with conventional fertilizers (Igiehon & Babalola, 2018). The need to use fertilizers to increase crop production through increased quantity has become one of the vital requirements to sustain the continuously growing world population (Brown et al., 2009; Karp & Richter, 2011; Prosekov & Ivanova, 2018). Urea is one such commonly used fertilizers, majorly due to its high nitrogen concentration (47%) and low costs and economical availability.Its vulnerability to losses through leaching, volatilization and denitrification, however, leads to environmental, health and ecological challenges, which have seen the escalation of these concerns and, consequently, the introduction of Next Generation Fertilizers (NGF) (also known as Advanced or Controlled Release Fertilizers (CRF)), which are innovative products aimed at increasing Nutrient Use Efficiency (NUE) and reducing environmental effects. (Plimmer & Gammon, 2003; Shaviv, 2001). These fertilizers are discussed and divided into two categories: (i) CRFs, where the nutrients are coated with polymers or entangled in matrix substances in order to control nutrient release over time; and (ii) stabilized fertilizers, in which chemical or biological additives prevent a larger degree volatilization, leaching, or other processes to the loss of nutrients. The use of controlled-release (CR) formulations has been a very practical method of quasi-predicting and predictably attained desired rates of nutrient release. CRF development is a very good example of a sustainable technology that would help lower dissolution rates of nutrients, improve need-based use of the nutrients, and reduce the negative ecological effects. Therefore, in order to attain controlled nutrient release, the urea fertilizer may be coated or dissolved in a blend with water-insoluble polymers in order to prolong the release period and inhibit the nitrogen accessibility rate to plants (Azeem et al., 2016; Beig et al., 2020; Calabi-Floody et al., 2018; Hauck, 1985; ME Trenkel, 2021; Sharma & Singh, 2019).

This chapter introduces a step-by-step analysis of NGF technologies: how they were developed, their mechanisms of action, and some of the many advantages they have. It delves into the manner in which these innovations overcome the shortcomings of the traditional fertilizers, in addition to promoting sustainable intensification of agriculture. Also, this chapter addresses issues of the economy, regulation, and adoption of these technologies as well as some case studies of their use in various agroecological zones. This chapter will attempt to give a direction to the stakeholders in the field, researchers, policymakers, and farmers on how to tap into the power of NGFs to revolutionize farming into a more viable and productive future.

2.2 The Science Behind Next-Generation Fertilizers

Nowadays, food insecurity and climate change are persistent challenges faced by human beings. Despite tremendous advancements in public health, nutrition, personal cleanliness, and medicine, the growing world population has caused a population explosion (Mueller et al., 2012). According to the Food and Agriculture Organization of the United Nations (FAO), globally the population is expected to reach 11 billion by 2100, and food production must expand by 60–70% to meet the growing demand (El Idrissi et al., 2023; Mariyam et al., 2024). Conventional strategies to meet the challenges of food production have heavily relied on pesticides, synthetic fertilizers, and genetically engineered seeds (Channab et al., 2023; Mehran et al., 2023).

Fertilizers, including primary macronutrients (nitrogen (N), phosphorus (P), and potassium (K)), secondary macronutrients (sulfur (S), magnesium (Mg), and calcium (Ca)), and micronutrients (iron (Fe), copper (Cu), zinc (Zn), silicon (Si), manganese (Mn), boron (B), molybdenum (Mo), nickel (Ni), sodium (Na), and chloride (Cl)), play an important role in the availability of nutrients for improving the yield of crops and for producing more healthy food (Kopittke et al., 2019). Delivering nutrients in precise dose, form, place, and time while keeping costs competitive still proves elusive. Studies revealed that 40–70% of applied nitrogen, 80–90% of phosphorus, and 50–90% of potassium in soil are lost to the environment, washed into waterways, volatilized into the air, or locked into plants in unavailable forms (Zulfiqar et al., 2019). Poor NUE forces the farmers to excessive use of chemical fertilizers just to sustain cereal yields (Ikram et al., 2024a). Yet, this strategy has barely improved cereal yields while squandering resources and increasing nutrient pollution, acidifying soil, causing water eutrophication, releasing greenhouse gas, degrading soil health, diminishing food quality, and creating other environmental problems.

A considerable amount of these chemical fertilizers is underutilized and might contribute to difficulties, including leaching, mineralization, and bioconservation. Chemical fertilizer accumulation has larger environmental repercussions, influencing ecosystems like disrupting soil microflora, marine habitats, and even parasite populations (Figure 2.1) (El Gharrak et al., 2022). Concerns over declining soil fertility, increasing environmental harms, and the poor efficiency of traditional fertilizers are major factors for a transformative innovation in farming and agriculture (Aziz et al., 2015; Chhipa, 2017; Goyal et al., 2023; Mehran et al., n.d.).

However, one proposed solution is the adoption of next- or new-generation fertilizer technology for achieving sustainable agriculture and boosting crop yield (Usman et al., 2020). New-generation fertilizers are advanced formulations designed to improve agriculture yield

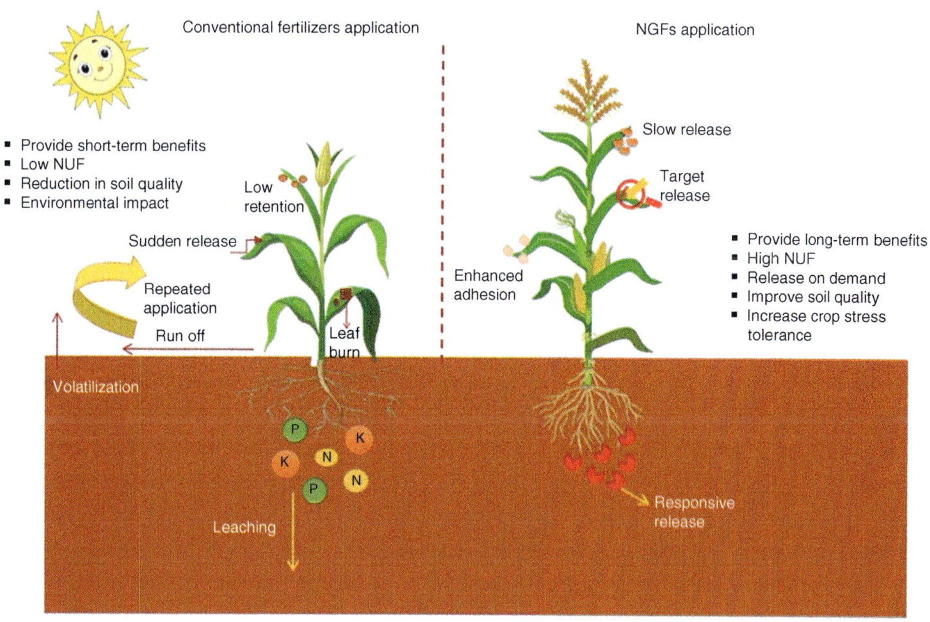

Figure 2.1 Difference and efficacy of next-generation fertilizers compared to conventional fertilizers.

while minimizing environmental stewardship. These fertilizers characterize a modern agricultural approach that is rooted in precision, sustainability, and environmental stewardship. These advancements are critical to meeting the world's growing food demand while minimizing farming's environmental impact. According to current studies, sustainable agriculture methods are increasingly required to preserve food security while lowering environmental effects. The development and acceptance of NGFs is an important part of this transition to sustainability. These fertilizers aim to address the shortcomings of conventional fertilizers and promote environmentally accountable and more productive agriculture.

2.2.1 Nanotechnology in Fertilizers

Achieving sustainable agricultural hinges on merging multidisciplinary insights with cutting-edge technologies to revolutionize traditional fertilizers and increase NUE. Among the most promising tools are nano-fertilizers, which are NMs that function either as a direct source of micro- or macronutrients or as carriers/additives (delivering nutrients to the crop) (Kah et al., 2018). Nano-fertilizers can also be developed by enclosing nutrients within the NMs (DeRosa et al., 2010).

By delivering nutrients to crops more efficiently, nano-fertilizers, including nanoparticles, increase both yield and quality, reduce the production cost, and alleviate both biotic and abiotic stresses in plants, thus contributing toward agricultural sustainability (Haider et al., 2025; Munyaneza et al., 2025a,b). Kah et al. (2018) conducted a comprehensive meta-analysis of 78 studies and found that nano-fertilizers provide a median efficacy gain of 18–29% over the conventional fertilizers (Kah et al., 2018). When broken down by types, nanocarrier-based products showed substantial improvement with a median yield increase of 29%, but nanoformulated micro- and macronutrients registered gain of 19% and 18%, respectively. The use of phosphatic nano-fertilizers has also been associated with an improved growth rate (by 32%) and seed production (by 20%) of soybean (*Glycine max* L.) when compared to those treated with traditional fertilizers (Liu & Lal, 2014). Nano-fertilizers also advance the plant metabolism and the nutrient uptake by accessing nanoscale pores facilitated by molecular transporters or exploiting nanostructure cuticle channels (Ikram et al., 2024b; Rico et al., 2011). Integrating nanotechnology in plant nutrition enables the generation of slow/controlled-release fertilizers that rises the fertilizer use efficiency and curb the losses of nutrients, offering a greener alternative (Figure 2.1) (Liu & Lal, 2014). Overall, employing slow-release solutions reduces the amount of fertilizer required. Ideally, nano-fertilizers should be able to dispense nutrients when and where needed by crops, preventing volatilization and downstream leaching.

This goal can be achieved if the nutrient release is triggered by plant signals as first proposed by DeRosa (2010), enabling "intelligent" fertilizers that respond to communicational signals between plant roots and soil microbes. In short, NFs improve the NUE of traditional inorganic, organic, and biological fertilizers, curbing environmental loss and economic costs while sustaining higher crop yields. Nanotechnology in fertilizers includes next-generation nano-fertilizers, inorganic nano-fertilizers, organic nano-fertilizers, and nano-biofertilizers (Liao et al., 2023).

2.2.2 Microbial Innovations in Next-Generation Fertilizers

Microbial technologies are central to NGFs, providing sustainable and efficient alternatives to conventional chemical fertilizers. Among the key players, endophytes and cyanobacteria excel at fixing and delivering nutrients, thereby rejuvenating soils and improving crop yield.

2.2.2.1 Endophyte

Endophytes are bacteria and fungi that live inside plant tissues for part or all of their life span, without causing any apparent damage to the host plant. Endophytes produce plant hormones such as indole-3-acetic acid, solubilize phosphate, release siderophores, and enhance plant resilience to biotic and abiotic stresses (Gaiero et al., 2013; Haider et al., 2024; Lebeis, 2014; Rosenblueth & Martínez-Romero, 2006). Furthermore, some bacterial endophytes like *Azoarcus*, *Azospirillum*, and *Herbaspirillum* harbor genetic material for biological nitrogen fixation (BNF), allowing them to transform dinitrogen gas (N_2) into plant-useable forms of ammonium and nitrate directly inside the host plant (Ashraf et al., 2022; Bhattacharjee et al., 2008; Santi et al., 2013). Microbial activity can also transform excessive ammonium or nitrate in the soil into nitrous oxide, which is a powerful greenhouse gas (Kandel et al., 2017). This process of using bacterial endophytes in agriculture offers a powerful way to curb the environmental impacts such as greenhouse gas emission and pollution because of using chemical fertilizers, especially N fertilizers. Therefore, by relying on these natural symbionts such as bacterial endophytes, farmers can cut fertilizer use and move toward more sustainable crop production in the future.

2.2.2.2 Cyanobacteria

Cyanobacteria are among the earth's oldest life forms, making up about 33% of the total algal population nationwide and dominating at 50% in some southern and eastern states. These free-living, photosynthetic diazotrophs include Anabaena, Nostoc, Calothrix, and Aulosira genera, which are the dominant nitrogen fixers (Sahu et al., 2012). Cyanobacteria contribute significantly to soil fertility, thus enhancing the yield. They can thrive in different soil types, resulting in cosmopolitan spread. Blue-green algae excrete a variety of growth-promoting chemicals, such as phytohormones, various vitamins, and amino acids (Bhuyan et al., 2023), that enhance the water-holding capacity, elevate the soil salinity, avert the growth of the weeds, and improve the availability of soil P by producing various organic acids (Ammar et al., 2022). Furthermore, cyanobacteria have been shown to be able to detoxify a wide range of pollutants, positioning them as a promising agent for bioremediation. Cyanobacteria can function as biofertilizers, either as free-living organisms in the soil or as symbionts in plant root nodules. Whether living freely in the soil or as symbionts within root nodules, strains such as *Nostoc* and *Anabaena* colonize rocks and soil and can fix about 20–25 kg of N/ha, making them an attractive alternative to chemical nitrogen fertilizers, as they can reduce the need for chemical inputs while maintaining or even improving crop productivity.

2.2.3 Enhanced Efficiency Fertilizers (EEFs)

These refer to fertilizer engineered to restrict nutrient losses, whether by gaseous losses, leaching, or runoff, relative to standard reference, thereby increasing uptake by plants (Prasad & Shivay, 2021). According to Asadu et al., (n.d.) (2014), the baseline for comparison is that any soluble fertilizer products are left unmodified, with no coating, encapsulation, compaction, inhibitors, and other treatments. Against this benchmark, enhanced efficiency fertilizers (EEFs), whether stabilized, CR, and slow-release fertilizers, modulate nutrient availability so that supply aligns with plants' demand to absorb nutrients, thereby reducing nutrient losses either by volatilization or by leaching into groundwater (Anwar-ul-Haq et al., 2023; Chalk et al., 2015; Timilsena et al., 2015). By reducing nutrient losses, EEFs help maintain a balanced nutrient profile and prevent soil acidification. Though their initial price is higher, the long-term benefits of reduced nutrient losses and improved crop yields translate them into long-term savings for growers (Goyal et al., 2023). A prominent most illustration of an EEF is EuroChem's ENTEC® product line. ENTEC® formulations deploy a proprietary technology that keeps applied fertilizer stabilized in the ammonium form for an extended duration. By reducing nitrification, in which ammonium is converted into nitrate, this prolongs ammonium retention and reduces subsequent denitrification, the microbial reduction of nitrate to nitrogen gas, which can cause significant nitrogen losses from the soil (Rakhimol et al., 2021). Another illustration involves the use of CRFs whose polymer coatings are designed to regulate the release of nutrients at predetermined rates, ensuring that nutrients are released in a manner that matches the plant's uptake demand (Fufaeva et al., 2024).

In short, the science behind NGFs is multifaceted, which waves together nanotechnology, microbial innovations, and enhanced efficiency formulations. Together these advancements promise to transform agriculture by raising the efficiency while reducing environmental impacts (Usman et al., 2020). As further research confirms their performance and economies of scale drive down prices, the widespread adoption of these fertilizers is expected to rise significantly.

2.3 Control-Released Fertilizers: Maximizing Efficacy

Conventional fertilizers, while essential for higher yield, often result in low NUE. Where a significant portion of the fertilizer is not effectively taken up by the plants and lost through several pathways like leaching, volatilization, immobilization, and runoff (Vejan et al., 2021). These inefficiencies impose a dual cost: direct economic losses for farmers but also imposes environmental consequences, including nitrate leaching, nitrous oxide emissions, eutrophication of water bodies, and significantly degrading soil and air quality (Govil et al., 2024).

CRFs are an emerging technology efficiently designed to synchronize nutrient release precisely with crop demand. CRFs are smartly engineered so that the required concentration of targeted nutrients is stored, which minimizes chances of nutrient loss and improves the efficiency of required nutrient utilization by plants (Maaz et al., 2025).

2.3.1 Main Types and Mechanisms of CRFs

There are many types of CRFs available; each has different mechanisms and releasing pattern under different factors.

2.3.1.1 Polymer-Coated Fertilizers (PCFs)

PCFs contain a nutrient core, such as nitrogen, phosphorus, potassium granules or urea, encapsulated in a semipermeable polymer membrane. These nutrients are released through this membrane via diffusion that is mainly driven by the concentration gradient between the inside of the granule and the soil solution (Lakshani et al., 2023). The whole process of PCF release is based on these phases: **(i) Lag-phase**: water slowly penetrates into the polymer coating through microscopic cracks or pores. During this process, only a small portion of the fertilizer core starts to dissolve, yet no nutrients are released; this stage is called "lag phase" (Chandra et al., 2019). During this phase, osmotic pressure gradually builds as the internal solution becomes increasingly concentrated, eventually reaching the critical concentration required to release nutrients (Tong et al., 2018). **(ii) Constant release phase**: once the internal osmotic pressure exceeds the resistive strength of the polymer coating, a steady diffusion of nutrients starts through the membrane into the soil (Mansouri et al., 2023). This phase drives a near-linear, controlled release of nutrients that is synchronized with nutrient uptake demand. **(iii) Decay phase**: once the nutrient concentration within the granule begins to decline, the concentration gradient across the membrane weakens, resulting in a gradual decline in release rate until total depletion of the core is achieved (Du et al., 2019). The main examples of such fertilizers are polymer-coated urea (PCU) and NPK blends. In these fertilizers, urea granules are coated with polymers such as polyurethane or polyethylene. The polymer coating controls nitrogen release that reduces the chance of leaching and volatilization. Polymer-coated NPK blends are complex fertilizers, where nitrogen (N), phosphorus (P), and potassium (K) are coated in order to release a balanced amount of nutrient (Yaseen et al., 2017).

Nutrient release from these fertilizers is mainly controlled by temperature, for example, higher temperature results in faster nutrient release as it increases solubility of the fertilizer core and ultimately accelerates the diffusion process through the polymer membrane. Such temperature sensitivity allows these fertilizers to synchronize nutrient release with plant demand, as nutrient uptake by crops generally increases with temperature during active growth periods (Adams et al., 2013; Kassem et al., 2024).

2.3.1.2 Sulfur-Coated Fertilizers (and Polymer-Sulfur-Coated)

Sulfur-coated fertilizers are a widely known practice as CRFs. These sulfur coatings are mainly used for soluble fertilizer cores as they are cost-effective and have higher nutrient value. These coatings are mainly used for urea fertilizers, and one of the most useful innovations is polymer-sulfur-coated urea (PSCU), where a thin sulfur-coated polymer layer is incorporated on urea for a controlled release of fertilizer and reduces the burst effect, which is the initial rapid release observed in SCU.

The mechanism of PSCU is governed by the breakdown of both the polymer and sulfur layers, where physical and microbial factors play key roles (Ying-hui et al., 2021).

The nutrient release mechanisms are based on a two-stage process. Initially, the fertilizer release mechanism is driven by the physical factor of the sulfur coating. Urea granules are coated with molten sulfur to create a barrier that is not properly uniformed; this imperfection results in microscopic cracks and pores. Once fertilizer is applied, these microscopic cracks and pores allow water to enter and dissolve the soluble core of urea fertilizer. Then nutrient solution reaches the soil via diffusion (Essa et al., 2021). Once the initial mechanism is completed, the remaining sulfur shell is degraded by soil microbes that significantly control the nutrient release process. In some SCUs, a wax-type sealant is applied on sulfur to further resist water entry, activating soil microbes to break down the wax layer to reach sulfur. This gradual degradation of the shell sustains the release of nutrients over time (Eghbali Babadi et al., 2021).

The rate of nutrient release from these fertilizers is not mainly influenced by factors related to the fertilizer granule and the soil environment. Thus, it is crucial to understand the following factors to maximize SCU and PSCU. **(i) Coating thickness and integrity**: the thickness of the sulfur and polymer coating plays an important role in the release rate. Thick coating ensures slower release of nutrients by creating a substantial barrier for water and microbial degradation (Figure 2.2) (Sun et al., 2023). Also, the integrity of the coating plays a critical role; if the coating is damaged or has any cracks, it can lead to a rapid and uncontrolled release of fertilizers. **(ii) Soil temperature**: higher temperatures accelerate the process of nutrient release by increasing soil microbial activity and diffusion. **(iii) Soil moisture**: adequate moisture in soil is necessary to drive the nutrient release process in two ways. First, water penetrates into the fertilizer coating to dissolve the urea core. Second, it is essential to support soil microbes for degradation of sulfur coating (Mustafa et al., 2022). **(iv) Microbial activity**: soil microorganisms are directly involved in the breakdown of the

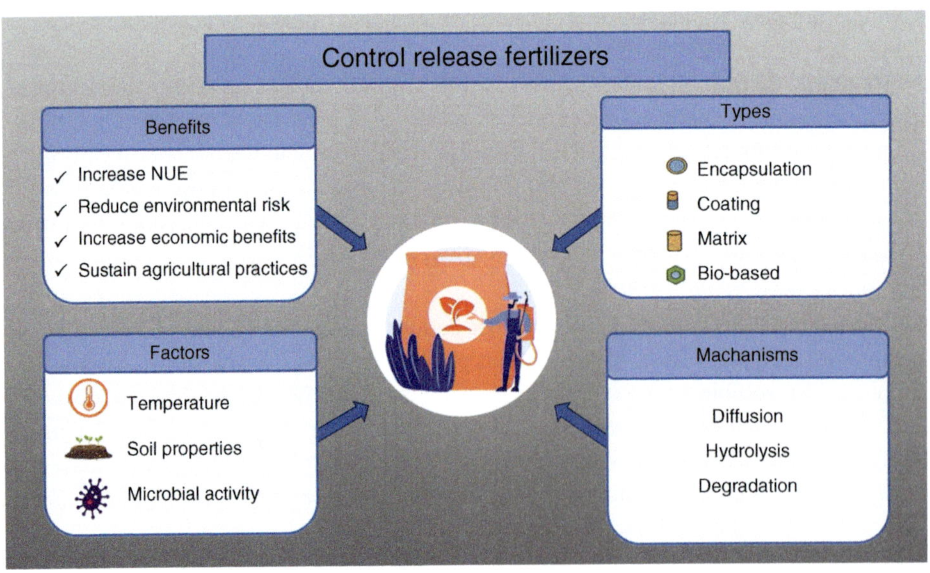

Figure 2.2 Types, benefits, and mechanisms of controlled-release fertilizers to maximize nutrient efficiency.

sulfur coating (Guo et al., 2025). Their activities are mainly influenced by soil pH, soil organic matter, moisture, and aeration. Thus, the process of soil fumigation can delay the release of nutrients, as it reduces the population of soil microbes (Figure 2.2) (Dincă et al., 2022).

2.3.1.3 Matrix-Based Fertilizers

Matrix-based fertilizers are physically incorporated into a solid matrix. These matrix materials are important to integrate the fertilizer granules within the matrix. Generally, hydrophilic and hydrophobic materials are used to form a barrier around the fertilizer and create a matrix, that is, hydrogels, polyolefins, polysulfones, and rubber (Ravindran et al., 2025).

These matrix-based systems are mainly based on the uniform dispersion of soluble granular fertilizers like urea, with the help of a carrier. There are two main mechanisms involved in the release of these nutrients to the soil. **(i) Degradation-based release**: in this type, the matrix is based on biodegradable polymers susceptible to breakdown in the soil system, that is, polyhydroxyalkanoate (PHA), polybutylene adipate-co-terephthalate (PBAT), polylactic acid (PLA), and starch. These biodegradable polymers are metabolized by microorganisms in soil, as these microbes feed on the polymer matrix that gradually releases embedded nutrients. Thus, the rate of release is directly linked to biodegradation rate (Figure 2.2) (Firmanda et al., 2023; Majeed et al., 2015; Yuste et al., 2014). **(ii) Diffusion-based release**: in diffusion-based release, porous matrix is used, such as zeolites, that allows entry of water and dissolves the embedded nutrient based on the concentration gradient between the matrix and soil solution. Here, the porosity of the matrix is critical in governing the diffusion rate (Rani et al., 2024).

Matrix-based fertilizers are influenced by environmental and product-specific factors. For example, soil temperature, soil moisture, microbial activities, soil pH, soil texture, granule surface area, and type/nature of polymer (Entry & Sojka, 2008, 2009).

2.3.1.4 Chemically Bound/Slowly Soluble Compounds

These are slow-release fertilizers that contain chemical compounds having low water solubility. In these chemically bound and/or slowly soluble compounds, nutrient release is dependent on chemical or biological transformation (Velázquez-Antunez et al., 2023). The nutrient release mechanism of these fertilizers is based on slow hydrolysis and/or microbial breakdown (Al-Rawajfeh et al., 2021).

The most common examples of this class are urea-formaldehyde (UF) and isobutylidene di-urea (IBDU). **(i) UF**: they are produced by the reaction of urea with formaldehyde; this reaction creates methylene urea polymers that have varying chain lengths. In these fertilizers, release of nitrogen is entirely dependent on soil microbial activity. As microbes produce some enzymes, like methylenediurease, these enzymes break down polymer chains and mineralize nitrogen into plant-available forms such as ammonium. In this type, the length of the polymer chains is a key factor because shorter chains can break down more easily, meanwhile longer chains extend the nutrient release period. Consequently, nitrogen release rate is controlled by factors that influence soil microbial life, including aeration, moisture, temperature, and pH of soil (Almutari, 2023; Nardi et al., 2018) **(ii) IBDU**: it is a chemical compound formed after the reaction of urea and isobutyraldehyde. It is a crystalline solid that contains about 31–32% of nitrogen, with most of the compound being water-insoluble. In contrast to UF, the main mechanism for nitrogen release is slow hydrolysis. In contact with soil water, it slowly dissolves and hydrolyzes into urea and isobutyraldehyde. IBDU does not depend on soil temperature and microbial activities. The release rate is mainly influenced by particle size and low water solubility; smaller particles with larger surface area dissolve faster in soil moisture. It is also observed that lower soil pH and higher temperatures accelerate the process of hydrolysis (Al-Rawajfeh et al., 2021; Prasad & Shivay, 2021).

2.3.2 Challenges and Limitations for Current CRFs

Despite the significant results of CRFs, there are several challenges that hinder their widespread adoption. The most significant challenge is the cost of CRFs, as they are 1.3–10 times more expensive than conventional fertilizers (Kandulu et al., 2017). The premium cost of CRFs is driven by the application of specialized polymer coatings and energy-intensive manufacturing processes. For micro-farmers, these prices are out of budget to fulfill the fertilizer requirement of crops (Liu & Lal, 2014). Beyond cost, environmental factors are a great challenge for CRFs. Performance of CRFs under field conditions can vary and is frustratingly unpredictable. As discussed above, nutrient release mechanisms of CRFs are highly sensitive to several factors like soil temperature, moisture, pH, and microbial activity that can drastically affect nutrient release rates.

Material stability is also a critical factor. Many CRFs face two problems, such as "initial burst effect" and "final tailing effect." Where, right after application, a sudden and uncontrolled release of nutrients occurs (ME Trenkel, 2021). This happens if the coating stability is compromised, which defeats the purpose of controlled delivery (Davidson et al., 2013). Conversely, the tailing effect results in prolonged and slow release of nutrients even after the crop requirement is completed, resulting in risk of environmental loss (Lawrencia et al., 2021). Also, the most pressing concern is the environmental footprint of the coatings themselves. Most of the CRFs only rely on nonbiodegradable, petroleum-based polymers such as polyolefins and polyurethane. In such fertilizers, once nutrients are delivered, empty shells remain in the soil and degrade into microplastics with time. These remaining polymers result in the accumulation of microplastics in agricultural lands. According to a rough estimate, each year around 50 kg/hectare of coating material is accumulated in soil as microplastics (Lubkowski, 2016). Meanwhile, biodegradable polymer coatings are promising but currently face the problem of higher production costs, plus they face manufacturing difficulties in achieving consistent and durable coatings (Ezema et al., 2022).

2.3.3 Next-Generation Innovations

The evolution of CRFs is entering a new phase of advancements in materials science, biotechnology, and nanotechnology. The current goal of CRFs is no longer to reduce the rate of nutrient release but to synchronize it with unprecedented precision, with a major focus on eliminating their environmental footprints. These next-generation innovations in CRFs are discussed below.

2.3.3.1 Biodegradable Coatings and Bio-based Matrices

The most critical area of innovation is to develop bio-degradable coatings that can be degraded completely; to replace the nondegradable coatings, these synthetically developed coatings are responsible for microplastic pollution. Keeping the main focus on natural, renewable, and environmentally friendly materials, starch, cellulose, lignin, and PHAs are emerging as biodegradable coatings.

Starch and cellulose are plant-based polysaccharides having low cost with inherited biodegradability (Arora et al., 2021). Researchers are modifying these two wonderful biodegradable materials through the technique of esterification or blending them with other biopolymers in order to improve their water resistance and mechanical strength to create effective barriers for fertilizer. Lignin is a by-product of the biorefinery and paper industries. It is a low-cost phenolic polymer with natural hydrophobic and antimicrobial properties; these qualities make it an excellent source for coating materials to slow down dissolution and provide protection against premature microbial breakdown. Researches are developing lignin for both purposes: as a coating material and as a component to blend in matrix-based fertilizers (Shu et al., 2021; Vasile & Baican, 2023). PHAs are natural polyesters produced by microorganisms. PHAs are highly biodegradable; upon degradation, they release water and carbon dioxide in soil. They have special properties called thermoplastic, making them similar to conventional plastics, these properties allow them to be processed into durable coatings (Mathuriya & Yakhmi, 2017).

2.3.3.2 Smart/Stimuli-Responsive Fertilizers

Perhaps the most advanced concept is the "smart" fertilizer. This fertilizer remains stable in the soil without changing until there is any specific signal that allows it to release nutrients. Such demand-driven fertilization controlled by the plant itself is considered as the most efficient way to fertilize. These types of fertilizers include pH-responsive, enzyme-triggered, and root exudate-responsive fertilizer. pH-responsive fertilizers have hydrogel coatings/matrices that respond in swelling or shrinking to changes in soil pH. They are designed to release nutrients when soil pH near the root zone drops; these phenomena help in nutrient uptake by the plant (Shaghaleh et al., 2022).

While the concept of enzyme-triggered nutrient release involves developing a coating that can only be broken down by specific enzymes released from roots. When a plant demands a nutrient, a certain enzyme is released; then a smart fertilizer could detect and respond to this specific signal and release nutrients exactly at the time and place of need (Calabi-Floody et al., 2018; Symanowicz et al., 2022). Plant root exudate-responsive fertilizer is a more sophisticated version of enzyme-triggering fertilizers; the release mechanism of such fertilizers is based on root exudates (Phong et al., 2025). Here, the fertilizer will respond to a plant-specific chemical, in order to perfectly synchronize plant demand.

2.3.3.3 Nano-fertilizers and Nano-encapsulation

Nanotechnology is a revolutionary science that involves engineering materials at the atomic and molecular scale. For fertilizer delivery, "nano-encapsulation" techniques are a well-known approach that encapsulates nutrients in nano-sized shells or embeds them within nanoporous materials. Due to a higher surface-area-to-volume ratio, nutrients are released with a more precision. These "nanocarriers" protect the nutrient from premature interactions with the soil and deliver it more precisely to the roots. Researchers are exploring several materials (chitosan, clays, and carbon nanotubes) to create the nanoscale delivery vehicles (Rajani & Meena, 2025; Singh et al., 2024).

There is another approach called "Targeted Delivery"; in the near future, it may be possible to use nanoparticle surface to target specific sites on plant roots. This approach will be similar to targeted drug delivery in medicine, reducing fertilizer application rates and environmental risks. These innovative technologies are still in the research and development phase; they show a profound shift in our crop nutrition approach. By moving forward from slow-release and passive to active systems, we can make one step forward to the ultimate goal of a truly sustainable and productive agricultural system.

2.4 Bio-based and Organic Fertilizers: Nature Meets Innovation

The term bio-based fertilizer, or BBF, has recently increased in academic journals to describe a broad range of biologically derived substances used to feed nutrients to crops (Chojnacka et al., 2020; Kurniawati et al., 2023; Wester-Larsen et al., 2022). However, separating BBFs from organic fertilizers and biowaste is a hurdle. The word "biowaste" refers to biological materials that have been dumped as garbage, such as byproducts of agriculture, food production, animal husbandry, and human life. Nutrient-rich biowastes, when applied to soils to supply nutrients for plants, are considered organic fertilizers (Naz et al., 2025; Pain & Menzi, 2011). In daily parlance, the distinction between a generic biowaste and an organic fertilizer is dependent solely on their real use. The word "animal manure" refers to a fertilizer made from animal excrement, urine, and bedding material. Animal manures are frequently processed, such as by anaerobic storage, water addition, or homogenization, before being applied to the land. However, the term BBF is used in literature supporting alternatives to mineral fertilizers or describing materials with enhanced fertilizer value, signifying a more technologically advanced substance than a basic, raw organic fertilizer such as animal dung (Egas et al., 2023; Garmendia-Lemus et al., 2024; Moshkin et al., 2023). Therefore, it is arguable that animal manure is not a BBF, since it had no treatment specifically to improve its fertilizing qualities.

The core value of a BBF lies in its circularity (Chojnacka et al., 2020). BBFs are intended to allow for the reintroduction of nutrients that would otherwise be wasted, to improve NUE when compared to less processed organic fertilizers, or to reduce potential negative environmental impacts when compared to organic or mineral fertilizer alternatives. Furthermore, organic fertilizers are related to the presence of organic matter and the ability to improve soil structure or serve as an organic carbon source (Mehran et al., 2025). However, in the absence of organic matter, a BBF can contain nutrients produced from biomass, which resemble mineral fertilizers in their chemical and physical forms, as long as the nutrients are biological in origin.

The clear definition of the term BBF is difficult to determine and is still being debated. One proposed definition is that BBFs are materials or products created from biological materials, with a content of bioavailable nutrients high enough to be employed as fertilizer (Wester-Larsen et al., 2022). On the other hand, Rodriguez and Miller (2023) proposed that waste streams, that is, manure, sewage sludge, and food processing wastes, which can be valorized, through processing, into a fertilizer product, should be termed BBFs. Both definitions denote that a biowaste or an organic fertilizer requires a certain degree of processing in order to attain BBF status. In addition, bio-based status has also been put on the fertilizers channel whereby recovery of the nutrients (e.g., using ammonia stripping or P precipitation) from biowastes occurs, as well as in the one in which the use of fertilizers involves new products in fertilizer production consisting of a combination of synthetic nutrients with nutrients of biological origin. We do not want to substitute the phrase organic fertilizer with BBF. It is more so to give us an impression of the prospects and challenges of reincorporating nutrients within bio-wastes into the food system and enhancing the agronomic efficiency of already existing organic and BBF, hence, the development of newer and better BBFs. As applied to the current review, a BBF can be defined as any material of biological source (e.g., a biowaste) that has been transformed with the aim of further increasing fertilizing characteristics (e.g., an acidified slurry), as well as all fertilizing products where a large part of nutrients are of biological nature, for example, a fertilizer of organo-mineral origin consisting of a combination of rock phosphate extracted by mining and ammonium nitrate formed through ammonia stripping of manure (Wester-Larsen et al., 2022).

In the last 20 years, various new BBFs have arisen and are now being investigated for commercial use in agriculture. Ideally, the use of these BBFs in agriculture will improve soil quality and help to mitigate climate change. Aside from the good impacts, the use of BBFs may increase the risk of spreading organic and inorganic contaminants and infections.

The increasing demand for food production, combined with growing worries about the environmental impact of chemical fertilizers, has prompted the investigation of alternate techniques to improve soil fertility. BBFs, which are made utilizing biotechnological processes, have emerged as a possible answer to these problems. These fertilizers are made from natural materials such as organic waste, microbes, and plant residues, and they have various benefits, including enhanced soil health, less pollution, and greater nutrient availability.

2.4.1 Nitrogen-Fixing Bacteria

Microbial inoculants, which include nitrogen-fixing bacteria like Rhizobium, Azotobacter, and Azospirillum, are commonly employed in BBFs to improve soil fertility. These bacteria create symbiotic interactions with plants, particularly legumes, and help transform atmospheric nitrogen into a form that plants can absorb and use. The use of nitrogen-fixing bacteria as BBFs eliminates the need for synthetic nitrogen fertilizers, which are a major source of environmental contamination.

2.4.2 Phosphate-Solubilizing Microorganisms

Phosphate-solubilizing microorganisms, such as some bacteria and fungi, are critical to the bioavailability of phosphorus in soils. These microorganisms transfer phosphorus from insoluble to soluble forms that plants may absorb. BBFs can improve phosphorus availability by inoculating soils with phosphate-solubilizing microorganisms, resulting in increased plant growth and agricultural yields.

2.4.3 Bioreactors for Bio-based Fertilizer Production

1) Industrial production of BBFs. Bioreactors are used to manufacture BBFs on a large scale, providing regulated settings for the development of microorganisms capable of producing useful substances. Bioreactors can be used to develop microorganisms that create nitrogen-fixing bacteria, phosphates, or enzymes that degrade organic materials by adjusting temperature, pH, and nutrient availability. These technologies enable the efficient and scalable production of microbial inoculants and other BBFs suitable for agricultural use.
2) Processing organic waste. Bioreactors can also be used to convert organic waste like animal manure, agricultural leftovers, and food waste into BBFs. Bioreactors aid in the breakdown of organic matter into useful nutrients that can be used as soil amendments via processes such as composting and anaerobic digestion. This not only reduces trash but also produces long-lasting and nutrient-dense fertilizers.

2.4.4 Nutrient Cycling and Bio-based Fertilizers

Microorganisms are also important in nutrient cycling in soil and degrading organic matter to release some key nutrients like nitrogen, phosphorus, and potassium. How BBFs work is to activate these natural processes and increase the soil microbial activity and nutrient availability to the plants. This positively affects healthier soil ecosystems and sustainable agriculture and lessens the use of chemical fertilizers. Composting and vermicomposting: composting is a process of making BBFs out of organic waste and vermicomposting is making BBFs with the use of worms. In the two processes, microorganisms decompose organic matter, including plant remains and food waste, into compost material that is humus-rich and can be used as natural fertilizer. Vermicomposting, which entails earthworm use, maximizes the content and structure of the compost, making it better to use in the improvement of soil health and fertility.

2.4.5 Future Directions for Bio-based Fertilizers

2.4.5.1 Integration with Precision Agriculture

BBFs can be combined with precision agriculture technologies in order to streamline their use and adaptation. Integration of BBFs, sensors, GPS, and data analytics provides an opportunity to help farmers apply appropriate levels of fertilizer at the optimal time, resulting in minimized wasted fertilizer and nutrient efficiency. Biotechnology has also been instrumental in the production of bio-based fertilizers, which provide substitutes for chemical fertilizers and provide up-to-par sustainable solutions. BBFs have the capacity to enhance fertility of soil through microbial inoculants, bioreactors, and nutrient cycling facilities that lead to achievement of healthier ecosystems and reduced pollution of the environment (Table 2.1). Despite these limiting factors, like scalability, cost, and regulatory challenges, there is a lot of potential regarding the future of BBFs in sustainable agriculture because research and innovation continue in the field of biotechnology.

Table 2.1 Types of potential main input materials for new bio-based fertilizers.

Material	Description	Examples of transformation methods into BBF
Animal by-products	E.g., Raw animal excreta, meat, and bone meal	Acidification, alkalinization
Compost	Biowastes that went through a composting process	Incorporation into organo-mineral fertilizers
Digestates	Solid or liquid output of the anaerobic digestion process of biowastes	Biomass pretreatments Electrokinetic pretreatment Ensiling pretreatment Redigestion after solid–liquid separation
Mineral precipitates or concentrates	Mineral compounds extracted, separated, or precipitated from biowastes (e.g., struvite, ammonium sulfate)	Incorporation into organo-mineral fertilizers and direct application
Plant-based materials	Plant juices, extracts, and solids, olive, pomace	Composting, anaerobic digestion, solid–liquid separation
Sewage sludge	By-product of wastewater treatment plants that concentrates solids of sewage.	Acidification, alkalinization Incorporation into organo-mineral fertilizers
Thermally treated biowastes	Biowastes converted into ash or biochars via incineration or pyrolysis, respectively.	Pretreatments

2.5 Nano-enhanced Fertilizers: Tiny Particles, Big Impact

Agriculture, being one of the keystones of the civilization of man, has always continued to develop with the change of demands of the growing world population and confrontations of a changing environment. One of the latest evolutions, which has the potential to make a revolution in how we feed our crops, is the introduction of nano-fertilizers. However, in order to understand their capability and how to deal with related issues, some background information would be required about what nano-fertilizers are and how they should be classified. Finally, a nano-fertilizer is not significantly different from a standard fertilizer. They both feed the plants with the necessary nutrients to enhance growth, yield, and general health of the plant (Figure 2.3). The difference is in their size and mode of delivery. Nano-fertilizers effectively constitute fertilizers that have been reduced to the nanometer scale, more often than not between 1 and 100 nm in at least one dimension (AS, 2023; Manjunatha et al., 2019). Such nanoscale restructuring provides sizeable surface area-to-volume ratios and thus a high level of reactivity and the ability to provide interactions with plant tissues that otherwise could not be attained with bulkier, conventional fertilizer particles. Science and nature intermingle in the form of the composition of nano-fertilizers. They are composed of important nutrients such as

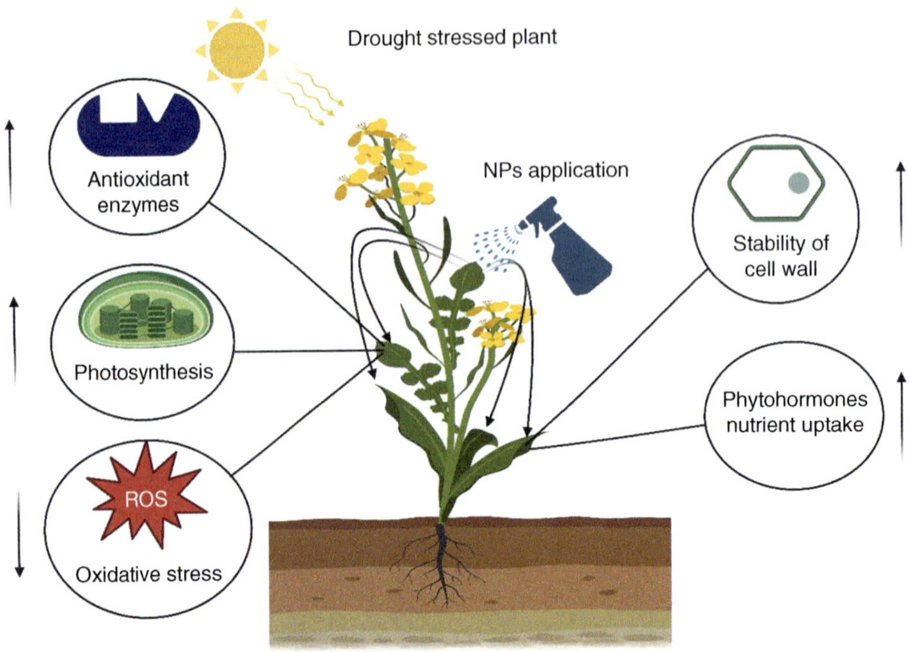

Figure 2.3 Effect of nanoparticle-based nano-enhanced fertilizers on crop growth.

nitrogen, phosphorus, and potassium and other micronutrients. Nevertheless, they are encapsulated, embedded, or coated in nanoparticles, which may be characterized as of different origin: lipid-based, polymer-based, or even inorganic, such as those based on silica or metallic nanoparticles. Nanoscale delivery system promises a more controlled release of nutrients so that plants have a more efficient time picking it up, hence, less wastage and likelihood of pollution of the environment, and enhance tolerance of plants to various stresses (Figure 2.3) (Ahmad et al., 2025; Harish et al., 2022; Munyaneza et al., 2025b). It is also significant to add that the smaller size does not only modify the physical characteristics of the fertilizers but may also modify their chemical and biological characteristics. It would be easier to penetrate plant cells and tissues as nanoparticles, which can make delivery of nutrients to reach the cell site where they are most required (Avellan et al., 2021). Taking a closer look at the realm of nano-fertilizers, one will discover a universe of diversity. Such diversity is not only necessary to serve the various crops and soils but also plays a characteristic role in maximizing the potential level of nanotechnology in agricultural aspects. **Organic nano-fertilizers**: These are the ones created based on the biological substances such as proteins or lipids to release nutrients in the organic form of a matrix. As an example, the nutrients can be incorporated into a lipid-based nanoparticle to allow a gradual release. **Inorganic nano-fertilizers**: They are composed of inorganic nanoparticle such as zinc oxide and even gold nanoparticles. They can be employed to provide particular micronutrients to the plants, utilizing their particular features, such as great solidness or peculiar communication with the plant cells

(Tao et al., 2021). **Bio-derived nano-fertilizers**: They are produced through biological mechanisms. As an example, some fungi and bacteria would produce nanoparticles, having provided the fungus with the metal ions. It can then load such bio-derived nanoparticles with nutrients and utilize them as fertilizers. **Chemically produced nano-fertilizers**: Their production is based on chemical reactions and processes, often in laboratories. The benefit in this case is its precision because the chemists have the ability to determine the size, shape, and properties of the nanoparticles very precisely. **Slow release nano-fertilizers**: These will sustain the release of nutrients in the plant over a long interval, which is why they will not require regular fertilizing of the plant to provide an adequate supply of nutrition to the plant. **Targeted nano-fertilizers**: The advantage of such fertilizers is that by using the particular characteristics of nanoparticles, they are able to provide nutrients to specific organs of a plant, such as roots, or to a particular organelle within the cells of the plants. It can be especially effective in the case of certain nutrient deficiencies or in the case of the desire to activate certain physiological processes. **Multifunctional nano-fertilizers**: These are the ultimate high-tech works to be realized in agriculture using nanotechnology. The nutrients provided by them are not the only elements of use but can also prevent the pests, provide better soil quality, or even track the condition of the plant by containing the elements of sensing included in it.

2.6 Precision Farming: Fertilizer and Digital Innovation

Precision agriculture is a transformative modern farming approach that uses advanced technologies and data-driven methods that are totally different from traditional farming methods. The central goal of precision agriculture is to apply the right input at the right place, in the right amount, at the right time (Linaza et al., 2021). This concept is growing exponentially and being practically applied. The cornerstone of precision fertilizer lies in data-driven decision-making, where various technologies are used to collect digital data (Table 2.2). These technologies allow targeted application of fertilizers for specific crops and soil types to maximize nutrient uptake and reduce environmental losses. The real-time data enable farmers to directly monitor field conditions like nutrient deficiencies, pest infestations, or disease outbreaks. This proactive approach for farm management, powered by real-time data, is reshaping the future of agriculture by optimizing inputs and enhancing overall profitability and sustainability (Khan & Babar, 2024).

2.6.1 Key Technologies in Precision Fertilization

There are several advanced technologies that aim precision fertilization. These emerging technologies enable farmers to make site-specific decisions. These innovations technologies provide a comprehensive understanding of field conditions to farmers for efficient and effective nutrient management. Technologies with significant positive impact on fertilization systems are discussed in Table 2.2.

Table 2.2 Key technologies in precision fertilization.

Technology	Classifications	Description	Application in precision fertilization	Key benefits	References
Remote sensing	Drones (UAVs)	Unmanned aerial vehicles (UAVs) are usually equipped with various sensors and cameras to capture high-resolution imagery from low altitudes	Provides ultra-high-resolution imagery details about field conditions to identifying variability across small areas. This is often used for targeted spot-spraying of fertilizers and helpful in data analysis for fertilizer dose management	Captures extremely detailed data for micromanagement and is not bothered by cloud cover	Xia et al., (2025)
	Satellites imaging	Earth-orbiting platforms that capture imagery of area of concern	Monitors crop growth data that can be used to generate vegetation indices (NDVI) to correlate with nitrogen status of specific area	Cost-effective to monitor on a large scale. Provides historical data for long-term analysis	Schut et al. (2018)
	Multispectral imaging	These are sensors used to capture data from a few specific bands of light relevant to plant health (e.g., red, green, blue, near-infrared)	Calculate vegetation indices (NDVI) to assess nutrients and crop biomass to further compute fertilizers need	Well-established methodology that can differentiate between healthy and stressed vegetation	Cardim Ferreira Lima et al. (2020)
	Thermal imaging	Sensors based on variation in temperature across a crop canopy	Identifies plant stress linked to nutrient deficiencies that impair physiological functions Delineates stressed zones for investigation	Physiologically measure plant stress even before visual symptoms appear	Guo et al. (2016), Moon et al. (2022)

Internet of Things (IoT) sensors	Soil moisture sensors	Moistures sensor probes are placed at various soil depths to measure soil volumetric water content in real time	These sensors optimize process of fertigation and ensure nutrients are applied when needed by plant	Prevents nutrient loss from leaching, reduces water usage, and ensures nutrients are in a plant-available solution	Coelho et al. (2020)
	Nutrient sensors (ion-selective electrodes)	These are in situ sensors, measure specific nutrients like nitrate, phosphate, and potassium in the soil solution	Provides real-time measurement of available nutrients, and allow "just-in-time" applications of fertilizer when needed	Enables highly responsive nutrient management and reduce risk of over-application by measuring already present nutrients	Kumari et al. (2024)
	pH sensors	Based on electrodes that measure the acidity or alkalinity of the soil	As soil pH controls the availability of most nutrients. This data helps in making decisions on soil amendments (e.g., lime) to optimize nutrient uptake	Ensures applied fertilizers are actually available to the plant by maintaining optimal pH. Maximizes fertilizer efficiency	Kumar and Kumar (2019)
	EC sensors	Sensors that measure soil electrical conductance that correlates with soil texture, salinity, and nutrient levels	Raed soil texture variability and helps to delineate management zones for VRA of fertilizers	A rapid method to map soil texture and salinity, which are key factors in fertilization planning	Othaman et al. (2021)
AI and machine learning	Predictive analytics models	These models are based on algorithms that use historical data related to yield, soil and weather. and use real-time data to forecast future outcomes	Provide information about crop nutrient requirements, predicts potential yield for different zones, and estimates nutrient loss risk under various weather scenarios	Enables proactive fertilizer management by allowing effective planning for input purchasing and application timing	Dave and Verma (2025)
	Recommendation engines	These are AI-driven systems used to synthesize data obtained from different AI-sources to advice the farmers	Provide real-time recommendations for fertilizer type, rate, timing, and placement by using a software platform or mobile app	Simplifies complex data and improves accuracy of decision-making	Musanase et al. (2023)

(Continued)

Table 2.2 (Continued)

Technology	Classifications	Description	Application in precision fertilization	Key benefits	References
	Computer vision	This is field of AI used computers to interpret information obtained from digital images and videos	Identify and quantify areas of nutrient stress with high accuracy	Provides quantitative assessments of field conditions and reduce human error	Lippel et al. (2025)
Positioning and application systems	GPS/RTK	Global positioning systems (GPS) provide location data while real-time kinematic (RTK) provides sub-inch accuracy by correcting the GPS signal in real time	These technologies provide auto-guide to machines for precise mapping and soil sampling. Here, accuracy of RTK is a critical factor for precise fertilizer banding	Provide accurate field maps. Allows perfect guidance and precise nutrient placement by eliminating overlaps/skips	Michael (2025), Radicioni et al. (2020)
	Variable rate controllers	These are electronic devices placed on tractors to interpret digital prescription maps (VPA) and control fertilizer spreader	It is called as "brain" of VRT because it adjusts flow of fertilizer according to the command given by VRA map as the vehicle moves	Ensures that fertilizer is applied at right rate at the right place	Pawase et al. (2024)
	Section control	This feature turns individual sections of a fertilizer sprayer at field boundaries or area that needs be fertilized	Reduce over application of fertilizer in lands and avoids application outside a specific boundary	Lowers input costs and reduces fertilizer losses into the environmental	Wang et al. (2022)

2.6.2 Digital Innovations in Fertilizer Management

The digital revolution continues to create innovative solutions to further refine and enhance fertilizer management to sustain the idea of precision agriculture. Most useful digital innovations for the management of fertilizers are AI-powered systems, blockchain technology, and smart-farming software (Table 2.2).

2.6.2.1 AI-Powered Systems

This system provides real-time nutrient advice. This technology develops sophisticated recommendation systems that provide tailored advice to farmers in real-time about the application of nutrients (Pundhir et al., 2026). These systems analyze continuous data streams from different sources, such as soil sensors, satellite imagery, and weather forecasts, to provide precise recommendations about the timing, rate, and placement of fertilizer (Table 2.2) (Hedley, 2015; Ukhurebor et al., 2022). Such technologies empower farmers to make well-informed and immediate decisions with the aim of optimizing the use of nutrients and responding to changing conditions in the field.

2.6.2.2 Blockchain Technology

This technology is known as transparent supply chain tracking. This is an emerging and powerful tool used to enhance transparency and traceability of agricultural supply chains (Bosona & Gebresenbet, 2023). In the context of fertilizer management, blockchain system creates a secure record of a fertilizer's journey; starting from the manufacturer and leading to the farm. This system ensures the authenticity of fertilizers and prevents the distribution of substandard products (Jangid et al., 2023).

2.6.2.3 Smart-Farming System

The proliferation of digital tools has led to the development of integrated smart platforms for fertilizer management. These smart-farming platforms are known to be central hubs that bring data from various sources together into a single farm management software system. For example, IoT sensors, drones, and machinery. This integration provides a holistic view of the farm operation that facilitates comprehensive data analysis across different farm management aspects, including fertilization, irrigation, and pest control (Toselli et al., 2023).

2.6.3 Challenges in Adoption

Despite the widespread adoption of precision fertilization technologies, there are several significant hurdles for the farming community. The major challenges are discussed later.

2.6.3.1 High Initial Investment and Lack of Technical Expertise

A major barrier for farmers with limited financial resources to adopt these technologies is the cost of precision agriculture equipment, software, and training (Galati et al., 2025; Pandeya et al., 2025). Furthermore, farmers need a certain level of technical expertise to effectively utilize advanced technologies. Aside from operating the equipment, interpretation of data generated in vast amounts is another change for farmers that are directly involved in making informed management decisions.

2.6.3.2 Data Privacy and Connectivity Issues in Rural Areas

The data-based precision agriculture has concerns about the privacy of data (Pandeya et al., 2025). Sometimes farmers are not comfortable to share data of their farms because they have fear of misuse or it being used against their interests. Another problem is the lack of reliable internet connectivity in many rural areas that is not easy to assess for many farmers. These issues limit the functionality of many precision agriculture tools that are based on real-time transfer of data and cloud computing.

2.6.3.3 Smallholder Farmer Accessibility

Most of these technologies are of higher cost, challenging for smallholder farmers, who make up a significant portion of the agricultural sector globally (Van Loon et al., 2018). Fragmented land holdings and limited access to agriculture extension services further create hurdles for farmers to adopt and get benefits from advanced precision agriculture technologies. Meanwhile, the development of mobile applications is emerging as a powerful digital solution to make these technologies more accessible for small-scale farmers.

2.6.4 Future Trends

The field of precision fertilization is making progress in generating new innovative idea; researchers are working to make advancements in the fields of robotics, data integration, and supportive policies. **(i) Robotics fertilizer application**: the future of precision farming is increasingly autonomous. For example, robotic machineries and self-driving tractors are being developed to perform several tasks, including sowing, weeding, irrigation, and fertilization, that require minimal human intervention (Ghafar et al., 2023). These AI-driven robotic machines are generally guided by GPS and well-equipped with advanced sensors. These sensors are used to apply nutrients with high precision to enhance efficiency and reduce labor costs (Khan et al., 2018). These are promising innovations that ensure scalable and consistent farming operations. **(ii) Integration of precision application technologies with CRFs**: this is a remarkable future trend that involves integration of precision application technologies with CRFs for highly efficient closed-loop systems. This system deals with sensors placed in the field to detect the requirement of crops for a specific nutrient, which triggers a precisely located CRF. This smart technology will be able to release nutrients in a controlled manner according to the plant's uptake pattern. This is helpful in minimizing losses of nutrients into the soil and environment by being the ultimate demand-driven fertilization. **(iii) Policy and infrastructure support for digital agriculture**: the emerging precision agriculture system will require a supportive ecosystem, including government policies that will be helpful in improving adoption of sustainable farming practices and in providing subsidies for small-scale farmers in order to overcome the initial investment costs (Mohan et al., 2021). Development of rural infrastructure and having reliable and affordable broadband internet access are needed to bridge the digital gaps. Apart from this, establishment of data privacy is essential to build trust and encourage data sharing among farmers. Thus, collaborative efforts between governments, the private sector, farmers, and research institutions will be important to create a healthy environment for sustainable digital agriculture.

2.7 Challenges to Adoption: Bridging the Gap

Despite their promises, NGFs are facing a lot of challenges in adoption, rooting in economics, technology, education, regulations, and on ground logistics. Tackling each of these challenges is crucial for the successful integration of NGFs into mainstream agricultural practices.

2.7.1 Economic Factors

Initially, next-generation fertilizers remain significantly more expensive to produce than standard chemical fertilizers, limiting their widespread use in agriculture. This increased cost is due to the complicated manufacturing techniques required to coat or encapsulate the nutrients (Lawrencia et al., 2021). Despite the long-term benefits of higher productivity and less environmental effect, the initial investment can be a considerable barrier for farmers, especially in areas with low financial resources. This economic hurdle is exacerbated by the necessity for specialized application equipment and the possibility of increasing production expenses. So, concerns about the affordability of NGFs can be reduced by choosing cost-effective raw materials, prioritizing apatite, silica NMs, and bio-based alternatives, and integrating biofertilizers and enhanced efficiency fertilizers, particularly for high-value agricultural applications.

2.7.2 Education and Awareness

Public acceptance, influenced by risk–benefit perceptions, plays a key role in the introduction of new NGFs to agriculture (Alayón et al., 2022). According to recent reviews, farmers are not well-informed about the advantages and appropriate application of NGFs, and public perceptions of contemporary technology in the agri-food industries are generally moderate. Knowledge gaps about the functionality, advantages, and resistance to change of this technology are cited as obstacles to the adoption of NGFs (Frewer et al., 2014). Practitioners face significant issues in grasping NF properties, operational precautions, and efficacy determinants. Policymakers should provide specialized guidance for certain crop nutritional needs, soil conditions, and environmental factors in order to address these problems. Comprehensive educational programs and awareness campaigns that educate farmers about these innovations and their potential effects on sustainability and productivity can also help address this knowledge gap by empowering them to make well-informed decisions about incorporating these cutting-edge products into their farming operations (Goyal et al., 2023).

2.7.3 Technological Barriers and Regulatory Hurdles

NGFs often require specific application methods that vary from conventional fertilizers. For example, farmers used to more straightforward application procedures may find the complexity of CRFs (which require a specific application technique to ensure the controlled release of nutrients) and nano-enhanced fertilizers (which require precise calibration to ensure the nanoparticles are delivered to the plant roots effectively without causing harm) off-putting. This barrier can be lessened by streamlining application procedures and offering easily accessible training on the use of new technologies (Du et al., 2019).

Furthermore, compatibility with current farming equipment and methods is also essential for adoption. Certain NGFs may require the use of special equipment during application. For instance, equipment that can handle and disperse nanoparticles without contaminating the environment or causing danger to the operator may be necessary for the accurate application of fertilizers enriched with nanotechnology. Similarly, some dispensers that can adjust the release rate based on environmental circumstances may be required for CRFs (Chhipa, 2017). Such equipment is very expensive and requires a lot of care, which can be a significant barrier, especially for small-scale farmers. Another technological challenge is integrating NGFs with precision farming technologies, which optimize nutrient application by using sensors, drones, and data analytics. Farmers might not have the infrastructure or technical know-how to successfully use these technologies (Goyal et al., 2023). Farmers may resist adopting new technologies because of a preference for familiar practices or disbelief about the benefits of new technologies. This confrontation to alteration can be particularly pronounced when the new technologies need significant changes to established farming routines.

To overcome these barriers, a multifaceted approach is required. For example,

a) Develop fertilizer formulations and application techniques that are easy to use and work with current agricultural practices. This could entail producing fertilizers that can be applied using already existing equipment or devising application methods that are simple to learn and execute.
b) Make training courses that teach farmers about the advantages and correct application of NGFs easily accessible. These strategies must be customized to the local environment, taking into account particular soil types and crops.
c) Promote the creation of long-lasting, reasonably priced equipment that works with NGFs. This could entail collaborations between government organizations, agriculture cooperatives, and manufacturers.
d) The application of NGFs technology may be improved by integrating it with other smart agriculture practices, such as precision agriculture. Precision agriculture employs contemporary technologies, like satellite images and field mapping, to increase crop quality and profitability (Padhiary et al., 2024). Additionally, it improves the utilization of traditional resources, facilitating the advancement of sustainable agriculture and addressing progressively pressing economic and ecological challenges. Integration of NGFs with precision agriculture technology will assist the growers in terms of cost and also management of time.
e) Implement demonstration projects that showcase the comfort of use and efficiency of NGFs. These projects can deliver hands-on experience and tangible results that can help convince unconvinced farmers of the benefits of new technologies.
f) Offer incentives such as attractive subsidies or tax breaks for farmers who invest in the necessary equipment or training to use NGFs.
g) Improving awareness and farmer understanding is crucial for promoting the widespread use of this technology. Strategies such as enhancing the management of the currently available information, careful design of economic assistance programs, and substantial efforts expected to deliver information regarding NGFs might assist in overcome the social barriers in the adoption of this technology.

2.8 The Future of Fertilizers: Innovation on the Horizon

The field of precision fertilization is making progress in generating new innovative idea; researchers are working to make advancements in the fields of robotics, data integration, and supportive policies. **(i) Robotics fertilizer application**: the future of precision farming is increasingly autonomous. For example, robotic machineries and self-driving tractors are being developed to perform several tasks, including sowing, weeding, irrigation, and fertilization, that require minimal human intervention. These AI-driven robotic machines are generally guided by GPS and well-equipped with advanced sensors. These sensors are used to apply nutrients with high precision to enhance efficiency and reduce labor costs (Khan et al., 2018). These are promising innovations that ensure scalable and consistent farming operations. **(ii) Integration of precision application technologies with CRFs**: this is a remarkable future trend that involves integration of precision application technologies with CRFs for highly efficient closed-loop systems. This system deals with sensors placed in the field to detect the requirement of crops for a specific nutrient, which triggers a precisely located CRF. This smart technology will be able to release nutrients in a controlled manner according to the plant's uptake pattern. This is helpful in minimizing losses of nutrients into the soil and environment by being the ultimate demand-driven fertilization. **(iii) Policy and infrastructure support for digital agriculture**: The emerging precision agriculture system will require a supportive ecosystem, including government policies that will be helpful in improving adoption of sustainable farming practices and in providing subsidies for small-scale farmers in order to overcome the initial investment costs. Development of rural infrastructure and having reliable and affordable broadband internet access are needed to bridge the digital gaps. Apart from this, establishment of data privacy is essential to build trust and encourage data sharing among farmers. Thus, collaborative efforts between governments, the private sector, farmers, and research institutions will be important to create a healthy environment for sustainable digital agriculture.

2.9 Conclusion

The NGF will be a breakthrough in contemporary farming, as they will remedy the inefficiency of the established methods as well as the critical necessity of environmental sustainability. In this entire chapter, we have discussed various novel fertilizers, including nanotechnology-improved carriers and microbial-based inoculants, enhanced-efficiency fertilizers, and BBFs, which are discussed to ultimately enhance the efficiency of nutrient uptake, decrease environmental impacts, and favor sustainable soil health. These fertilizers are not merely foods for crops, but they are redefining the basis of agricultural inputs management due to their ability to create a nutrient supply to meet the crop requirements, environmental factors, and growth needs. Though the innovations show positive outcomes, there are still major problems in scaling them. Its widespread adoption is hindered by high production costs, absence of technical knowledge on the part of the farmers, poor market infrastructure, and regulatory uncertainty. In addition, the problems of the type of environmental impact that some of the nonbiodegradable coatings have and field performance variability should be passed on to ascertain the sustainability in the long term.

In the future, a combination of NGFs with precision agricultural technologies will allow them to become particularly useful and effective, which includes decision-making based on artificial intelligence, the use of robots to apply fertilizers, and connected real-time monitoring of fields. Also, the emergent research on the biodegradable materials, smart responsive delivery, and the circular bioeconomy models will further prove its adoption. The best and efficient use of these technologies can be applied with the assistance of collaborations among policymakers, researchers, manufacturers, and farmers. Educational outreach, financial incentives, and conducive policy framework will gear up the transition. With a shift to a climate- and resource-limited future, NGFs provide a much-needed technology in the creation of a sustainable, productive, and resilient agricultural system.

References

Abobatta, W. F. (2018). Over view of nano-fertilizers. *Asian Journal of Ethnopharmacology and Medicinal Foods*, 4(4), 17–20.

Adams, C., Frantz, J., & Bugbee, B. (2013). Macro-and micronutrient-release characteristics of three polymer-coated fertilizers: Theory and measurements. *Journal of Plant Nutrition and Soil Science*, 176(1), 76–88.

Ahmad, I. A., Hu, H., Islam, M. S., Fu, Q., Zhu, J., Miao, F., Mehran, M., Haider, S., Murad, Z., & Ali, A. (2025). Simultaneous adsorption of cadmium and arsenic by goethite-modified rice straw-derived biochar in water and soil: Interactive ion effects and co-adsorption mechanism. *Environmental Monitoring and Assessment*, 197(6), 648.

Alayón, C. L., Säfsten, K., & Johansson, G. (2022). Barriers and enablers for the adoption of sustainable manufacturing by manufacturing SMEs. *Sustainability*, 14(4), 2364.

Almutari, M. M. (2023). Synthesis and modification of slow-release fertilizers for sustainable agriculture and environment: A review. *Arabian Journal of Geosciences*, 16(9), 518.

Al-Rawajfeh, A. E., Alrbaihat, M. R., & AlShamaileh, E. M. (2021). Characteristics and types of slow-and controlled-release fertilizers. In F. B. Lewu, T. Volova, S. Thomas, & K. R. Rakhimol (Eds.), *Controlled release fertilizers for sustainable agriculture* (pp. 57–78). Elsevier.

Ammar, E. E., Aioub, A. A. A., Elesawy, A. E., Karkour, A. M., Mouhamed, M. S., Amer, A. A., & El-Shershaby, N. A. (2022). Algae as bio-fertilizers: Between current situation and future prospective. *Saudi Journal of Biological Sciences*, 29(5), 3083–3096.

Anwar-ul-Haq, M., Mehmood, M. T., Alisha, S., Haider, S., Awais, M., Nadeem, M., Mubeen, M., & Iftikhar, I. (2023). Nutrient management under changing climate. In W. N. Jatoi, M. Mubeen, M. Z. Hashmi, S. Ali, S. Fahad, & K. Mahmood (Eds.), *Climate change impacts on agriculture: Concepts, issues and policies for developing countries* (pp. 281–297). Springer.

Arora, S., Singh, D., Rajput, A., Bhatia, A., Kumar, A., Kaur, H., Sharma, P., Kaur, P., Singh, S., & Attri, S. (2021). Plant-based polysaccharides and their health functions. *Functional Foods in Health and Disease*, 11(4), 179–200.

As, Q. (2023). Biosynthesis and evaluation of cinnamomum zeylanicum nanomaterials for the treatment of polycystic ovary syndrome in mice. *Pakistan Veterinary Journal*, 43(1), 118–124.

Asadu, C. O., Ezema, C. A., Ekwueme, B. N., Onu, C. E., Onoh, I. M., Adejoh, T., Ezeorba, T. P. C., Ogbonna, C. C., Otuh, P. I., & Okoye, J. O. (n.d.). Environmental pollution and management.

Ashraf, M., Shahzad, S. M., Abid, M., Mehmood, K., Aziz, A., Sarwar, A., Akhtar, N., & Mehran, M. (2022). Ionic homeostasis and growth characteristics of tomato (Solanum lycopersicum L.) grown with municipal wastewater by supplying silicon, farmyard manure and plant growth promoting rhizobacteria. *Silicon*, 14(18), 12855–12867.

Avellan, A., Yun, J., Morais, B. P., Clement, E. T., Rodrigues, S. M., & Lowry, G. V. (2021). Critical review: Role of inorganic nanoparticle properties on their foliar uptake and in planta translocation. *Environmental Science and Technology*, 55(20), 13417–13431. https://doi.org/10.1021/acs.est.1c00178

Azeem, B., KuShaari, K., & Man, Z. (2016). Effect of coating thickness on release characteristics of controlled release urea produced in fluidized bed using waterborne starch biopolymer as coating material. *Procedia Engineering*, 148, 282–289.

Aziz, T., Maqsood, M. A., Kanwal, S., Hussain, S., Ahmad, H. R., & Sabir, M. 2015. Fertilizers and environment: Issues and challenges. Hakeem, K. (eds) Crop production and global environmental issues, 575–598. Springer

Beig, B., Niazi, M. B. K., Jahan, Z., Hussain, A., Zia, M. H., & Mehran, M. T. (2020). Coating materials for slow release of nitrogen from urea fertilizer: A review. *Journal of Plant Nutrition*, 43(10), 1510–1533.

Bhattacharjee, R. B., Singh, A., & Mukhopadhyay, S. N. (2008). Use of nitrogen-fixing bacteria as biofertiliser for non-legumes: Prospects and challenges. *Applied Microbiology and Biotechnology*, 80(2), 199–209.

Bhuyan, P. P., Nayak, R., Jena, M., & Pradhan, B. (2023). Convoluted role of cyanobacteria as biofertilizer: An insight of sustainable agriculture. *Vegetos*, 36(2), 309–321.

Bosona, T., & Gebresenbet, G. (2023). The role of blockchain technology in promoting traceability systems in agri-food production and supply chains. *Sensors*, 23(11), 5342.

Brown, M. E., Hintermann, B., & Higgins, N. (2009). *Markets, climate change, and food security in West Africa*. ACS Publications.

Calabi-Floody, M., Medina, J., Rumpel, C., Condron, L. M., Hernandez, M., Dumont, M., & de La Luz Mora, M. (2018). Smart fertilizers as a strategy for sustainable agriculture. *Advances in Agronomy*, 147, 119–157.

Cardim Ferreira Lima, M., Krus, A., Valero, C., Barrientos, A., Del Cerro, J., & Roldán-Gómez, J. J. (2020). Monitoring plant status and fertilization strategy through multispectral images. *Sensors*, 20(2), 435.

Chalk, P. M., Craswell, E. T., Polidoro, J. C., & Chen, D. (2015). Fate and efficiency of 15N-labelled slow-and controlled-release fertilizers. *Nutrient Cycling in Agroecosystems*, 102(2), 167–178.

Chandra, M. S., Lal, M., Naresh, R. K., Yadav, S., Kumar, R., Kumar, R., & Lavanya, N. (2019). Role of polymer coated fertilizers (PCFS) an advance technology for improving nutrient use efficiency and crop productivity: A review. *International Journal of Chemical Studies*, 7(6), 2667–2679.

Channab, B.-E., El Idrissi, A., Zahouily, M., Essamlali, Y., & White, J. C. (2023). Starch-based controlled release fertilizers: A review. *International Journal of Biological Macromolecules*, 238, 124075.

Chhipa, H. (2017). Nanofertilizers and nanopesticides for agriculture. *Environmental Chemistry Letters*, 15(1), 15–22.

Chojnacka, K., Moustakas, K., & Witek-Krowiak, A. (2020). Bio-based fertilizers: A practical approach towards circular economy. *Bioresource Technology*, 295, 122223.

Coelho, A. D., Dias, B. G., de Oliveira Assis, W., de Almeida Martins, F., & Pires, R. C. (2020). Monitoring of soil moisture and atmospheric sensors with internet of things (IoT) applied in precision agriculture. *2020 XIV Technologies Applied to Electronics Teaching Conference (TAEE)*, 1–8.

Dave, H., & Verma, J. P. 2025. Smart agriculture: Predictive modeling of fertilizer requirements using neural networks. Ashish Kumar, Jai Prakash Verma, Rachna Jain (Eds) Emerging smart agricultural practices using artificial intelligence, 189–208. Wiley

Davidson, D. W., Verma, M. S., & Gu, F. X. (2013). Controlled root targeted delivery of fertilizer using an ionically crosslinked carboxymethyl cellulose hydrogel matrix. *Springerplus*, 2(1), 318.

DeRosa, M. C., Monreal, C., Schnitzer, M., Walsh, R., & Sultan, Y. (2010). Nanotechnology in fertilizers. *Nature Nanotechnology*, 5(2), 91–91.

Dincă, L. C., Grenni, P., Onet, C., & Onet, A. (2022). Fertilization and soil microbial community: A review. *Applied Sciences*, 12(3), 1198.

Du, W., Yang, J., Peng, Q., Liang, X., & Mao, H. (2019). Comparison study of zinc nanoparticles and zinc sulphate on wheat growth: From toxicity and zinc biofortification. *Chemosphere*, 227, 109–116. https://doi.org/10.1016/j.chemosphere.2019.03.168

Egas, D., Azarkamand, S., Casals, C., Ponsá, S., Llenas, L., & Colón, J. (2023). Life cycle assessment of bio-based fertilizers production systems: Where are we and where should we be heading? *The International Journal of Life Cycle Assessment*, 28(6), 626–650.

Eghbali Babadi, F., Yunus, R., Masoudi Soltani, S., & Shotipruk, A. (2021). Release mechanisms and kinetic models of gypsum–sulfur–zeolite-coated urea sealed with microcrystalline wax for regulated dissolution. *ACS Omega*, 6(17), 11144–11154.

El Gharrak, A., Essamlali, Y., Amadine, O., Aboulhrouz, S., Hafnaoui, A., Ghalfi, H., & Zahouily, M. (2022). Tunable physicochemical properties of lignin and rapeseed oil-based polyurethane coatings with tailored release property of coated NPK fertilizer. *Progress in Organic Coatings*, 170, 106982.

El Idrissi, A., Dardari, O., Metomo, F. N. N. N., Essamlali, Y., Akil, A., Amadine, O., Aboulhrouz, S., & Zahouily, M. (2023). Effect of sodium alginate-based superabsorbent hydrogel on tomato growth under different water deficit conditions. *International Journal of Biological Macromolecules*, 253, 127229.

Entry, J. A., & Sojka, R. E. (2008). Matrix based fertilizers reduce nitrogen and phosphorus leaching in three soils. *Journal of Environmental Management*, 87(3), 364–372.

Entry, J. A., & Sojka, R. E. (2009). The influence of form and rate on the efficacy of matrix-based fertilizers to reduce nutrient leaching. *Applied and Environmental Soil Science*, 2009(1), 374540.

Essa, R., Afifi, A., & Ashry, S. (2021). Influence of sulfur coated urea and algae fertilization on productivity of some leguminous crops in sandy soils. *Bulletin of the National Research Centre*, 45(1), 60.

Ezema, C. A., Ooi, T., & Matsumoto, K. (2022). Heat-press technique for coating ammonium nitrate granules with biodegradable hydrophobic polymer blend of poly (3-hydroxybutyrate-co-3-hydroxyhexanoate) and poly (ε-caprolactone). *ACS Agricultural Science & Technology*, 2(6), 1179–1186.

Firmanda, A., Fahma, F., Syamsu, K., Sari, Y. W., Suryanegara, L., Wood, K., & Saito, Y. (2023). Factors influencing the biodegradability of agro-biopolymer based slow or controlled release fertilizer. *Journal of Polymers and the Environment*, 31(5), 1706–1724.

Frewer, L. J., Gupta, N., George, S., Fischer, A. R. H., Giles, E. L., & Coles, D. (2014). Consumer attitudes towards nanotechnologies applied to food production. *Trends in Food Science & Technology*, 40(2), 211–225.

Fufaeva, V. M., Taran, Y. A., & Strel'nikova, V. O. (2024). Controlled-release fertilizers—a new development in the fertilizer market (a review). *Theoretical Foundations of Chemical Engineering*, 58(2), 287–294.

Gaiero, J. R., McCall, C. A., Thompson, K. A., Day, N. J., Best, A. S., & Dunfield, K. E. (2013). Inside the root microbiome: Bacterial root endophytes and plant growth promotion. *American Journal of Botany*, 100(9), 1738–1750.

Galati, A., Sofia, S., & Crescimanno, M. (2025). Economics and barriers of precision viticulture technologies: A comprehensive systematic literature review. *Information Processing in Agriculture*.

Garmendia-Lemus, S., Moshkin, E., Hung, Y., Tack, J., & Buysse, J. (2024). European farmers' perceptions and intentions to use bio-based fertilisers: Insights from the theory of planned behaviour and perceived utility. *Journal of Cleaner Production*, 434, 139755.

Ghafar, A. S. A., Hajjaj, S. S. H., Gsangaya, K. R., Sultan, M. T. H., Mail, M. F., & Hua, L. S. (2023). Design and development of a robot for spraying fertilizers and pesticides for agriculture. *Materials Today: Proceedings*, 81, 242–248.

Govil, S., Long, N. V. D., Escriba-Gelonch, M., & Hessel, V. (2024). Controlled-release fertiliser: Recent developments and perspectives. *Industrial Crops and Products*, 219, 119160.

Goyal, V., Rani, D., Ritika, M., & S., Deng, C., & Wang, Y. (2023). Unlocking the potential of nano-enabled precision agriculture for efficient and sustainable farming. *Plants*, 12(21), 3744.

Guo, J., Tian, G., Zhou, Y., Wang, M., Ling, N., Shen, Q., & Guo, S. (2016). Evaluation of the grain yield and nitrogen nutrient status of wheat (Triticum aestivum L.) using thermal imaging. *Field Crops Research*, 196, 463–472.

Guo, X., He, Z., Ren, H., Ren, B., Zhang, J., Liu, P., & Zhao, B. (2025). Long-term effects of coated fertilizers: Enhancing soil properties, optimizing microbial structures, and boosting soil fertility and crop productivity. *Agriculture, Ecosystems & Environment*, 385, 109564.

Haider, S., Bibi, K., Munyaneza, V., Zhang, H., Zhang, W., Ali, A., Ahmad, I. A., Muhammad, M., Xu, F., & Yang, C. (2024). Drought-induced adaptive and ameliorative strategies in plants. *Chemosphere*, 364, 143134.

Harish, V., Tewari, D., Gaur, M., Yadav, A. B., Swaroop, S., Bechelany, M., & Barhoum, A. (2022). Review on nanoparticles and nanostructured materials: Bioimaging, biosensing, drug delivery, tissue engineering, antimicrobial, and agro-food applications. *Nanomaterials*, 12(3), 457.

Hauck, R. D. (1985). Slow-release and bioinhibitor-amended nitrogen fertilizers. *Fertilizer Technology and Use*, 293–322.

Hedley, C. (2015). The role of precision agriculture for improved nutrient management on farms. *Journal of the Science of Food and Agriculture*, 95(1), 12–19.

Igiehon, N. O., & Babalola, O. O. (2018). Below-ground-above-ground plant-microbial interactions: Focusing on soybean, rhizobacteria and mycorrhizal fungi. *The Open Microbiology Journal*, 12(1), 261–279. https://doi.org/10.2174/1874285801812010261

Ikram, M., Minhas, A., Al-Huqail, A. A., Ghoneim, A. M., Mahmood, S., Mahmoud, E., Tahira, M., Mehran, M., Maqsood, M. F. K., & Rauf, A. (2024a). Promoting tomato resilience: Effects of ascorbic acid and sulfur-treated biochar in saline and non-saline cultivation environments. *BMC Plant Biology*, 24(1), 1053.

Ikram, M., Rauf, A., Rao, M. J., Maqsood, M. F. K., Bakhsh, M. Z. M., Ullah, M., Batool, M., Mehran, M., & Tahira, M. (2024b). CRISPR-Cas9 based molecular breeding in crop plants: A review. *Molecular Biology Reports*, 51(1), 227.

Jangid, P., Maurya, S., Kushwah, A., Tiwari, S., & Shaikh, A. (2023). *Fertilizer supply chain management using blockchain*. IJSRCSEIT.

Kah, M., Kookana, R. S., Gogos, A., & Bucheli, T. D. (2018). A critical evaluation of nanopesticides and nanofertilizers against their conventional analogues. *Nature Nanotechnology*, 13(8), 677–684.

Kandel, S. L., Joubert, P. M., & Doty, S. L. (2017). Bacterial endophyte colonization and distribution within plants. *Microorganisms*, 5(4), 77.

Kandulu, J., Thorburn, P., Biggs, J., & Verburg, K. (2017). Estimating economic impacts of controlled release fertilisers in sugarcane systems: An economic risk case study analysis. *Proceedings of the Australian Society of Sugar Cane Technologists*, 39, 251–262.

Karp, A., & Richter, G. M. (2011). Meeting the challenge of food and energy security. *Journal of Experimental Botany*, 62(10), 3263–3271.

Kassem, I., Ablouh, E.-H., El Bouchtaoui, F.-Z., Jaouahar, M., & El Achaby, M. (2024). Polymer coated slow/controlled release granular fertilizers: Fundamentals and research trends. *Progress in Materials Science*, 144, 101269.

Khan, N., & Babar, M. A. (2024). Innovations in precision agriculture and smart farming: Emerging technologies driving agricultural transformation. *Innovation and Emerging Technologies*, 11, 2430004.

Kopittke, P. M., Lombi, E., Wang, P., Schjoerring, J. K., & Husted, S. (2019). Nanomaterials as fertilizers for improving plant mineral nutrition and environmental outcomes. *Environmental Science: Nano*, 6(12), 3513–3524.

Kumar, A. A., & Kumar, S. K. (2019). Comprehensive review on pH and nutrients detection sensitive materials and methods for agriculture applications. *Sensor Letters*, 17(9), 663–670.

Kumari, A., Mishra, P., Chaulya, S. K., Prasad, G. M., Nadeem, M., Kisku, V., Kumar, V., & Chowdhury, A. (2024). Estimation of soil nutrients and fertilizer dosage using ion-selective electrodes for efficient soil management. *Communications in Soil Science and Plant Analysis*, 55(13), 1920–1941.

Kurniawati, A., Stankovics, P., Hilmi, Y. S., Toth, G., Smol, M., & Toth, Z. (2023). Understanding the future of bio-based fertilisers: The EU's policy and implementation. *Sustainable Chemistry for Climate Action*, 3, 100033.

Lakshani, N., Wijerathne, H. S., Sandaruwan, C., Kottegoda, N., & Karunarathne, V. (2023). Release kinetic models and release mechanisms of controlled-release and slow-release fertilizers. *ACS Agricultural Science & Technology*, 3(11), 939–956.

Lawrencia, D., Wong, S. K., Low, D. Y. S., Goh, B. H., Goh, J. K., Ruktanonchai, U. R., Soottitantawat, A., Lee, L. H., & Tang, S. Y. (2021). Controlled release fertilizers: A review on coating materials and mechanism of release. *Plants*, 10(2), 238.

Lebeis, S. L. (2014). The potential for give and take in plant–microbiome relationships. *Frontiers in Plant Science*, 5, 287.

Liao, Y., Xu, D., Cao, Y., & Zhu, Y.-G. (2023). Advancing sustainable agriculture: Enhancing crop nutrition with next-generation nanotech-based fertilizers. *Nano Research*, 16(12), 13205–13225.

Linaza, M. T., Posada, J., Bund, J., Eisert, P., Quartulli, M., Döllner, J., Pagani, A., Olaizola, G., & I., Barriguinha, A., & Moysiadis, T. (2021). Data-driven artificial intelligence applications for sustainable precision agriculture. *Agronomy*, 11(6), 1227.

Lippel, J., Bihlmeier, R., & Stuhlsatz, A. (2025). A computer vision approach to fertilizer detection and classification. *Proceedings of the 20th International Joint Conference on Computer Vision, Imaging and Computer Graphics Theory and Applications*, 563–569.

Liu, R., & Lal, R. (2014). Synthetic apatite nanoparticles as a phosphorus fertilizer for soybean (Glycine max). *Scientific Reports*, 4(1), 5686.

Lubkowski, K. (2016). Environmental impact of fertilizer use and slow release of mineral nutrients as a response to this challenge. *Polish Journal of Chemical Technology*, 18(1).

Maaz, T. M., Dobermann, A., Lyons, S. E., & Thomson, A. M. (2025). Review of research and innovation on novel fertilizers for crop nutrition. *Npj Sustainable Agriculture*, 3(1), 25.

Majeed, Z., Ramli, N. K., Mansor, N., & Man, Z. (2015). A comprehensive review on biodegradable polymers and their blends used in controlled-release fertilizer processes. *Reviews in Chemical Engineering*, 31(1), 69–95.

Manjunatha, R. L., Naik, D., & Usharani, K. V. (2019). Nanotechnology application in agriculture: A review. *Journal of Pharmacognosy and Phytochemistry*, 8(3), 1073–1083.

Mansouri, H., Ait Said, H., Noukrati, H., Oukarroum, A., Ben Youcef, H., & Perreault, F. (2023). Advances in controlled release fertilizers: Cost-effective coating techniques and smart stimuli-responsive hydrogels. *Advanced Sustainable Systems*, 7(9), 2300149.

Mariyam, S., Upadhyay, S. K., Chakraborty, K., Verma, K. K., Duhan, J. S., Muneer, S., Meena, M., Sharma, R. K., Ghodake, G., & Seth, C. S. (2024). Nanotechnology, a frontier in agricultural science, a novel approach in abiotic stress management and convergence with new age medicine-A review. *Science of the Total Environment*, 912, 169097.

Mathuriya, A. S., & Yakhmi, J. V. (2017). Polyhydroxyalkanoates: Biodegradable plastics and their applications. In L. M. T. Martínez, O. V. Kharissova, & B. I. Kharisov (Eds.), *Handbook of ecomaterials* (pp. 1–29). Springer.

ME Trenkel, T. (2021). *Slow-and controlled-release and Stabilized Fertilizers: An option for enhancing nutrient use effiiency in agriculture*. International Fertilizer Industry Association (IFA).

Mehran, M., Ashraf, M., Shahzad, S. M., Shakir, M. S., Azhar, M. T., Ahmad, F., & Alvi, A. (2023). Growth, yield and fiber quality characteristics of Bt and non-Bt cotton cultivars in response to boron nutrition. *Journal of Cotton Research*, 6(1), 1.

Mehran, M., Huang, L., Geng, M., Gan, Y., Cheng, J., Zhu, Q., Ahmad, I. A., Haider, S., & Mustafa, A. (2025). Co-utilization of green manure with straw return enhances the stability of soil organic carbon by regulating iron-mediated stabilization of aggregate-associated organic carbon in paddy soil. *Soil and Tillage Research*, 252, 106624.

Mehran, M., Ikram, M., Ghoneim, A. M., Alotaibi, M. O., Ghafar, S., Rafique, M. T., Ahmad, I. A., Haider, S., Safiullah, D., & Shah, M. N. (n.d.). Plant health, soil structure, and fertility: Developing a sustainable future.

Michael, I. (2025). Integration of GNSS and RTK for high-precision farming automation.

Mohan, S. S., Ajay, A., & Jayan, P. R. (2021). GPS and sensor based technologies in variable rate fertilizer application.

Moon, H.-D., Cho, Y., Jo, E., Kim, H., Kim, B., Jeong, H., Kwon, D., & Cho, J. (2022). Applications of thermal imaging camera to detect the physiological states caused by soil fertilizer, shading growth, and genetic characteristic. *Korean Journal of Remote Sensing*, 38(6–1), 1101–1107.

Moshkin, E., Lemus, S. G., Bamelis, L., & Buysse, J. (2023). Assessment of willingness-to-pay for bio-based fertilisers among farmers and agricultural advisors in the EU. *Journal of Cleaner Production*, 414, 137548.

Mueller, N. D., Gerber, J. S., Johnston, M., Ray, D. K., Ramankutty, N., & Foley, J. A. (2012). Closing yield gaps through nutrient and water management. *Nature*, 490(7419), 254–257.

Munyaneza, V., Zhang, W., Haider, S., Ali, A., Kant, S., Xu, F., Wang, C., Wang, S., Cai, H., & Shi, L. (2025a). Silicon dioxide nanoparticles mitigate aluminum toxicity in Brassica napus via cell wall modification, reduced aluminum accumulation, and enhanced antioxidant defense. *Pedosphere*.

Munyaneza, V., Zhang, W., Haider, S., Ren, L., Ali, A., Kant, S., & Ding, G. (2025b). Mitigating metal toxicity in plants using nanoparticles: Mechanisms and implications for sustainable agriculture. *Journal of Agricultural and Food Chemistry*, 73(28), 17374–17386.

Musanase, C., Vodacek, A., Hanyurwimfura, D., Uwitonze, A., & Kabandana, I. (2023). Data-driven analysis and machine learning-based crop and fertilizer recommendation system for revolutionizing farming practices. *Agriculture*, 13(11), 2141.

Mustafa, A., Athar, F., Khan, I., Chattha, M. U., Nawaz, M., Shah, A. N., Mahmood, A., Batool, M., Aslam, M. T., & Jaremko, M. (2022). Improving crop productivity and nitrogen use efficiency using sulfur and zinc-coated urea: A review. *Frontiers in Plant Science*, 13, 942384.

Nardi, P., Ulderico, N., Di Matteo, G., Trinchera, A., Napoli, R., Farina, R., Subbarao, G. V., & Benedetti, A. (2018). Nitrogen release from slow-release fertilizers in soils with different microbial activities. *Pedosphere*, 28(2), 332–340.

Naz, S., Arshad, A., Shoukat, A., Shaheen, C., Aslam, R., Nawaz, A., Ahmed, S., Saeed, S., Qadir, M. F., & Mehran, M. (2025). Innovative approaches in wastewater management: A comprehensive review. *International Journal of Current Microbiology and Applied Sciences*, 14(3), 140–164.

Othaman, N. N. C., Isa, M. N. M., Hussin, R., Ismail, R. C., Naziri, S. Z. M., Murad, S. A. Z., Harun, A., & Ahmad, M. I. (2021). Development of soil electrical conductivity (EC) sensing system in paddy field. *Journal of Physics: Conference Series*, 1755(1), 12005.

Padhiary, M., Saha, D., Kumar, R., Sethi, L. N., & Kumar, A. (2024). Enhancing precision agriculture: A comprehensive review of machine learning and AI vision applications in all-terrain vehicle for farm automation. *Smart Agricultural Technology*, 8, 100483.

Pain, B., & Menzi, H. (2011). Glossary of terms on livestock manure management. *Darmstadt: Association for Technology and Structures in Agriculture (KTBL).*

Pandeya, S., Gyawali, B. R., & Upadhaya, S. (2025). Factors influencing precision agriculture technology adoption among small-scale farmers in Kentucky and their implications for policy and practice. *Agriculture*, 15(2), 177.

Pawase, P. P., Nalawade, S. M., Walunj, A. A., Bhanage, G. B., Kadam, P. B., Durgude, A. G., & Patil, M. R. (2024). Comprehensive study of on-the-go sensing and variable rate application of liquid nitrogenous fertilizer. *Computers and Electronics in Agriculture*, 216, 108482.

Phong, N. T., Kim, J., Joe, E.-N., Tag, M., Lee, J.-G., Choi, M., Jang, K.-S., Yoon, H. Y., & Jeon, J.-R. (2025). Crop root exudate-responsive fertilizers fabricated by structural conversion of wastewater activated sludges into phenolic polymer-incorporated mineral-like particles. *Chemical Engineering Journal*, 507, 160536.

Plimmer, J. R., & Gammon, D. W. (2003). Insecticides: Overview and introduction. In *Encyclopedia of agrochemicals*. Wiley.

Prasad, R., & Shivay, Y. S. (2021). Enhanced efficiency fertilizers or slow-release and NI/UI blended nitrogen fertilizers. *Indian Journal of Fertilisers*, 17, 316–321.

Prosekov, A. Y., & Ivanova, S. A. (2018). Food security: The challenge of the present. *Geoforum*, 91, 73–77.

Pundhir, S. K. S., Yadav, S. K., Vijay, A., & Kumar, S. (2026). The role of artificial intelligence and machine learning models to optimize fertilizer consumption for sustainable agricultural growth: AI for precision agriculture. In *Precision and intelligence in agriculture: Advanced technologies for sustainable farming* (pp. 273–306). IGI Global Scientific Publishing.

Radicioni, F., Stoppini, A., Brigante, R., Brozzi, A., & Tosi, G. (2020). GNSS network RTK for automatic guidance in agriculture: Testing and performance evaluation. *International Conference on Computational Science and Its Applications*, 19–35.

Rajani, & Meena, R. K. (2025). Formulation of nano-fertilizers using various carriers. In P. Kumar & R. C. Dubey (Eds.), *Nanofertilizers in agriculture: Synthesis, mechanisms, and effect on plants* (pp. 53–74). Springer.

Rakhimol, K. R., Thomas, S., Kalarikkal, N., & Jayachandran, K. (2021). Nanotechnology in controlled-release fertilizers. In F. B. Lewu, T. Volova, S. Thomas, & K. R. Rakhimol (Eds.), *Controlled release fertilizers for sustainable agriculture* (pp. 169–181). Elsevier.

Rani, S., Dahiya, R., Kumar, V., Berwal, P., Sihag, S., & Jatrana, A. (2024). Development of novel zeolite-based controlled-release zinc fertilizers: Synthesis, characterization, and release kinetics. *ACS Agricultural Science & Technology*, 4(6), 664–672.

Ravindran, A., Manivannan, A. C., Kandaiah, R., Kulanthaisamy, M., Indirathankam, S. C., Nachimuthu, G., Panneerselvan, L., & Palanisami, T. (2025). Advancements and challenges in controlled-release fertilisers: An approach to integrate biopolymer-based strategies. *Industrial Crops and Products*, 233, 121349.

Rico, C. M., Majumdar, S., Duarte-Gardea, M., Peralta-Videa, J. R., & Gardea-Torresdey, J. L. (2011). Interaction of nanoparticles with edible plants and their possible implications in the food chain. *Journal of Agricultural and Food Chemistry*, 59(8), 3485–3498.

Rodriguez, C., & Miller, B. (2023). Biotechnology for the production of bio-based fertilizers. *American Journal of Biotechnology*, 4(4), 6–10.

Rosenblueth, M., & Martínez-Romero, E. (2006). Bacterial endophytes and their interactions with hosts. *Molecular Plant-Microbe Interactions*, 19(8), 827–837.

Sahu, D., Priyadarshani, I., & Rath, B. (2012). Cyanobacteria–as potential biofertilizer. *CIBTech Journal of Microbiology*, 1(2–3), 20–26.

Santi, C., Bogusz, D., & Franche, C. (2013). Biological nitrogen fixation in non-legume plants. *Annals of Botany*, 111(5), 743–767.

Schut, A. G. T., Traore, P. C. S., Blaes, X., & de By, R. A. (2018). Assessing yield and fertilizer response in heterogeneous smallholder fields with UAVs and satellites. *Field Crops Research*, 221, 98–107.

Shaghaleh, H., Hamoud, Y. A., Xu, X., Wang, S., & Liu, H. (2022). A pH-responsive/sustained release nitrogen fertilizer hydrogel based on aminated cellulose nanofiber/cationic copolymer for application in irrigated neutral soils. *Journal of Cleaner Production*, 368, 133098.

Sharma, N., & Singh, A. (2019). A review on changes in fertilizers: From coated controlled release fertilizers (CRFs) to nanocomposites of CRFs. *International Journal of Agricultural Science and Research*, 9(2), 53–74.

Shaviv, A. (2001). Advances in controlled-release fertilizers.

Shu, F., Jiang, B., Yuan, Y., Li, M., Wu, W., Jin, Y., & Xiao, H. (2021). Biological activities and emerging roles of lignin and lignin-based products—a review. *Biomacromolecules*, 22(12), 4905–4918.

Singh, M., Goswami, S. P., Sachan, P., Sahu, D. K., Beese, S., & Pandey, S. K. (2024). Nanotech for fertilizers and nutrients-improving nutrient use efficiency with nano-enabled fertilizers. *Journal of Experimental Agriculture International*, 46(5), 220–247.

Sun, G., Wang, M., & Dong, Y. (2023). Enhanced-efficiency nitrogen fertilizers coated with sulfur, inhibitor, and epoxy resin: Preparation and effects on maize growth and nitrogen utilization. *Journal of Soil Science and Plant Nutrition*, 23(3), 4091–4103.

Symanowicz, B., Toczko, R., & Toczko, M. (2022). Enzymatic activity of soil after applying mineral fertilizers and waste lignite to maize grown for silage. *Agriculture*, 12(12), 2146.

Tao, H., Wu, T., Aldeghi, M., Wu, T. C., Aspuru-Guzik, A., & Kumacheva, E. (2021). Nanoparticle synthesis assisted by machine learning. *Nature Reviews Materials*, 6(8), 701–716.

Timilsena, Y. P., Adhikari, R., Casey, P., Muster, T., Gill, H., & Adhikari, B. (2015). Enhanced efficiency fertilisers: A review of formulation and nutrient release patterns. *Journal of the Science of Food and Agriculture*, 95(6), 1131–1142.

Tong, X., He, X., Duan, H., Han, L., & Huang, G. (2018). Evaluation of controlled release urea on the dynamics of nitrate, ammonium, and its nitrogen release in black soils of northeast China. *International Journal of Environmental Research and Public Health*, 15(1), 119.

Toselli, M., Baldi, E., Ferro, F., Rossi, S., & Cillis, D. (2023). Smart farming tool for monitoring nutrients in soil and plants for precise fertilization. *Horticulturae*, 9(9), 1011.

Ukhurebor, K. E., Adetunji, C. O., Olugbemi, O. T., Nwankwo, W., Olayinka, A. S., Umezuruike, C., & Hefft, D. I. (2022). Precision agriculture: Weather forecasting for future farming. In A. Abraham, S. Dash, J. J. P. C. Rodrigues, B. Acharya, & S. K. Pani (Eds.), *AI, edge and IoT-based smart agriculture* (pp. 101–121). Elsevier.

Usman, M., Farooq, M., Wakeel, A., Nawaz, A., Cheema, S. A., ur Rehman, H., Ashraf, I., & Sanaullah, M. (2020). Nanotechnology in agriculture: Current status, challenges and future opportunities. *Science of the Total Environment*, 721, 137778. https://doi.org/10.1016/j.scitotenv.2020.137778

Van Loon, J., Speratti, A. B., Gabarra, L., & Govaerts, B. (2018). Precision for smallholder farmers: A small-scale-tailored variable rate fertilizer application kit. *Agriculture*, 8(4), 48.

Vasile, C., & Baican, M. (2023). Lignins as promising renewable biopolymers and bioactive compounds for high-performance materials. *Polymers*, 15(15), 3177.

Vejan, P., Khadiran, T., Abdullah, R., & Ahmad, N. (2021). Controlled release fertilizer: A review on developments, applications and potential in agriculture. *Journal of Controlled Release*, 339, 321–334.

Velázquez-Antunez, J., Olivares-Perez, J., Olmedo-Juárez, A., Rojas-Hernandez, S., Villa-Mancera, A., & RomeroRosales, T. (2023). Biological activity of the secondary compounds of guazuma ulmifolia leaves to inhibit the hatching of eggs of haemonchus contortus. *Pakistan Veterinary Journal*, 43(1), 55–60.

Wang, B., Wang, Y., Wang, H., Mao, H., & Zhou, L. (2022). Research on accurate perception and control system of fertilization amount for corn fertilization planter. *Frontiers in Plant Science*, 13, 1074945.

Wester-Larsen, L., Müller-Stöver, D. S., Salo, T., & Jensen, L. S. (2022). Potential ammonia volatilization from 39 different novel biobased fertilizers on the European market–a laboratory study using 5 European soils. *Journal of Environmental Management*, 323, 116249.

Xia, X., Zhang, R., Ma, L., Su, J., Yi, T., Zhang, L., & Chen, X. (2025). Optimization of unmanned aerial vehicle operational parameters to maximize fertilizer application efficiency in rice cultivation. *Journal of Cleaner Production*, 514, 145762.

Yaseen, M., Aziz, M. Z., Manzoor, A., Naveed, M., Hamid, Y., Noor, S., & Khalid, M. A. (2017). Promoting growth, yield, and phosphorus-use efficiency of crops in maize–wheat cropping system by using polymer-coated diammonium phosphate. *Communications in Soil Science and Plant Analysis*, 48(6), 646–655.

Ying-hui, G. U., Hong-dan, L. I. U., Wen-cheng, L. I. U., Min, Z. H. U., Tian-bing, L. I., Chun-yan, L. I., Jin-feng, D., Wen-shan, G. U. O., & Xin-kai, Z. H. U. (2021). Effects of different ratios of sulfur coated slow-release fertilizer on yield, quality and nitrogen efficiency of red-skinned strong gluten wheat. *Journal of Southern Agriculture*, 52(9), 2382–2390.

Yuste, J. C., Fernandez-Gonzalez, A. J., Fernandez-Lopez, M., Ogaya, R., Penuelas, J., Sardans, J., & Lloret, F. (2014). Strong functional stability of soil microbial communities under semiarid Mediterranean conditions and subjected to long-term shifts in baseline precipitation. *Soil Biology and Biochemistry*, 69, 223–233.

Zulfiqar, F., Navarro, M., Ashraf, M., Akram, N. A., & Munné-Bosch, S. (2019). Nanofertilizer use for sustainable agriculture: Advantages and limitations. *Plant Science*, 289, 110270.

3

Bioactive-Based Fertilizers: Harnessing Biostimulants, Nutrients, and Signaling Molecules for Enhanced Plant Growth

Muhammad Nauman Khan[1], Tooba[2], Nasir Assad[2], Alevcan Kaplan[3], Yasir Assad[4], Rao Muhammad Faisal Iqbal[5], and Shah Fahad[6]

[1] *Department of Botany, University of Chakwal, Punjab, Pakistan*
[2] *Institute of Chemistry, University of Sargodha, Sargodha, Pakistan*
[3] *Department of Pharmaceutical Botany, Dicle University, Diyarbakır, Turkey*
[4] *Department of Zoology, Hazara University Mansehra, Mansehra, Pakistan*
[5] *Department of Chemistry, The University of Lahore, Lahore, Pakistan*
[6] *Department of Agronomy, Abdul Wali Khan University Mardan, Mardan, Khyber-Pakhtunkhwa, Pakistan*

3.1 Introduction to Bioactive-Based Fertilizers

Sustainable agriculture is driven by both the need to produce more food and to reduce degradation of the environment (Brodt et al., 2011). The traditional fertilizer practices, which thrive on the use of synthetic fertilizers, for example, nitrogen, phosphorous, and potassium, have surely played a part in increasing the yield in the past decade (Ayala & Rao, 2002). These inputs, however, are often linked to low nutrient use efficiency (NUE) levels, land degradation, pollution of water bodies, and alteration of soil microbiota. Such effects have led to the need to come up with more sustainable nutrient management options, which are in congruence with the plant physiology and ecological processes (Shanmugavel et al., 2023). One of these offers is the use of bioactive-based fertilizers as they are progressive, as well as scientifically sound (Kussmann et al., 2023). Bioactive-based fertilizers are characterized by the incorporation of biologically active ingredients including biostimulants, plant-available nutrients, and signaling molecules that act synergistically to promote the best growth of plants, adaptability to stressful conditions, and assimilation of nutrients (Mącik et al., 2020). In contrast to conventional fertilizers that have a more direct nutrient-like effect, bioactive-based formulations have multivalent actions on plant metabolism, soil–plant–microbe interactions, and the overall functionality of agroecosystems (Bano et al., 2023).

Biostimulants are the most important central concepts of bioactive fertilizers, which are a heterogeneous category of naturally derived compounds capable of regulating plant physiological functions regardless of whether they provide any nutrients (Du Jardin, 2015). These are characterized by humic and fulvic acids, protein hydrolysates, seaweed extracts, microbial inoculants, and useful fungi (Rouphael & Colla, 2020). By means of hormonal regulations, modulation of membrane permeability, and the activations of enzymes, biostimulants improve the nutrient uptake capacity, the growth of the roots, and the resistance of the plant to abiotic and biotic stressors (Baltazar et al., 2021; Hasanuzzaman et al., 2021). They are especially important in less than optimal situations when they inhibit stress-related yield decline without the need to use artificial chemical means (Franzoni et al., 2022). While nutrient delivery systems incorporated into bioactive formulations have also been refined and should not be left out (Kyriakoudi et al., 2021). The conventional fertilizers are usually plagued by losses of nutrients; these losses are through leaching, volatilization, or precipitation (Ashitha & Rakhimol, 2021). Comparatively, bioactive-based fertilizers employ superior technology like nanocarriers (Singh et al., 2020), chelated complexes (Areche et al., 2023), and controlled-release mechanisms (Palansooriya et al., 2025) in their applications, ensuring more accurate delivery of nutrients. The use of such systems increases nutrient bioavailability, coordinates release with the demand of the plants, and decreases environmental losses, which ultimately improves the efficiency and sustainability of fertilizer applications.

Bioactive fertilizers are further distinguished by the addition of plant signaling molecules including salicylic acid (SA), jasmonic acid (JA), abscisic acid (ABA), nitric oxide (NO), and others (Arasimowicz & Floryszak-Wieczorek, 2007). Such molecules are vital in the control of the growth of plants, defenses, as well as their adaptations. They can induce cascades of signaling, following their exogenous usage through fertilizers to increase systemically acquired resistance and resistance to stressors in the environment. The specificity of these molecules allows the ability to target modification of plant gene expression and metabolism without genetic modification and thereby gives a biologically specific crop improvement tool. One thing that makes bioactive fertilizer-based products especially powerful is the synergy of effect among the elements of the products (Rouphael & Colla, 2018). When used in combination, biostimulants, nutrients, and signaling molecules have the ability to work across numerous plant physiological levels, including the transport of ions within cells, to regulate gene expression, and to lead to both augmented growth rates and higher nutrient efficiency and resilience to stress (Martínez-Lorente et al., 2024). The combined action of this functionality contributes not only to raising the yields but also to the better structure of the soil, microbial diversity, and the sustainability of the farming activity in the long term. Figure 3.1 illustrates the integrated roles of bioactive compounds in supporting plant growth, nutrient efficiency, and resilience as discussed earlier.

The fertilizers based on bioactive compounds are the intersection of the values of plant science, biotechnology, and ecology (Pereira, 2019). They stand on the frontier of climate-resettled agriculture, where they provide aid to achieve productivity objectives with the sustainability of the surroundings (Chojnacka et al., 2020). With an ever-growing concern shift toward the resource efficiency, ecological health, and adaptability to climate extremes of farming systems, the importance of bioactive-based fertilizers will be core to the nutrient management plans of many countries on the globe (Chaudhary et al., 2022). This chapter gives an in-depth discussion of the bioactive-based fertilizers, starting with the principal functions of biostimulants, the carriers of nutrients, and the signal compounds. Subsections consider the biochemical interplay of these inputs, advances in the formulation technology, field implementations of these inputs across cropping systems,

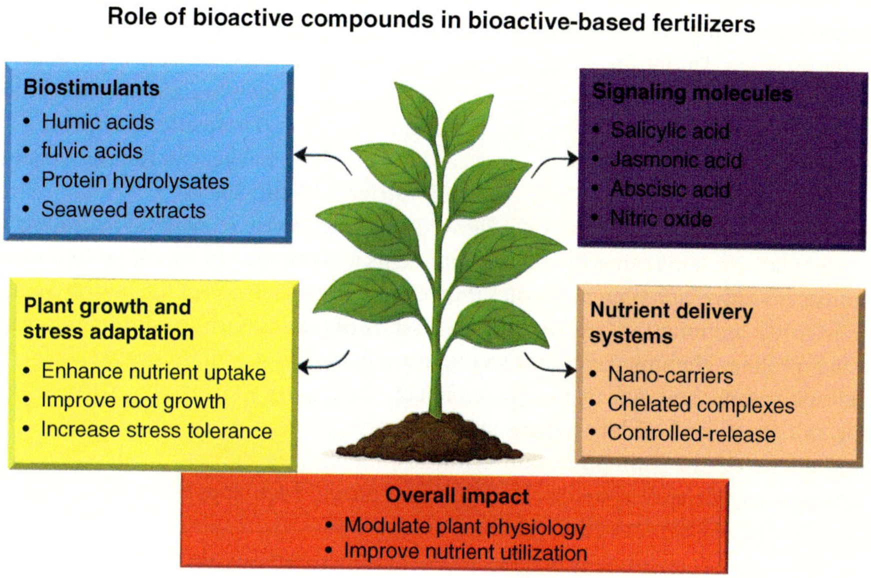

Figure 3.1 Role of bioactive compounds in bioactive-based fertilizers, showing the interaction of biostimulants, signaling molecules, and nutrient delivery systems in enhancing plant growth, nutrient uptake, and stress adaptation.

and issues of scalability and regulation. The discussion ends with an eye on the future of what these fertilizers could do in changing paradigms of plant nutrition and enabling the move toward sustainable and efficient crops.

3.2 Biostimulants: Unlocking the Power of Natural Growth Regulators

Biostimulants have become a major pillar in sustainable, contemporary agricultural production owing to their multifactorial properties of promoting plant growth, development, and exposure to a range of environmental stresses. In contrast to traditional fertilizers, which deliver quantities of vital nutrients, biostimulants are regulators of plant physiology. They work through the activation of climatic processes in plants and the soil, increase the efficiency of nutrient use, trigger natural defense mechanisms, and finally, enhance yield and quality indicators without inflicting any ecological load. They are registered to be used not only as an add-on to fertilization but also as a part of the intention to maximize the plant-level performance under both ideal and less-than-ideal circumstances (Zulfiqar et al., 2024).

The microorganisms and biologically active compounds include a variety of chemical substances, such as humic and fulvic acids, protein hydrolysates, free amino acids, seaweed extracts, beneficial fungi and bacteria, and complex mixtures of natural sources. All the types of biostimulants exert their effects through different but complementary mechanisms, which have the combined effect of enhancing nutrient uptake, hormonal regulation, and physiology of stress (Du Jardin, 2015). Humic and fulvic acids, which are the byproducts of breaking down of organic matters, interact with root plasma membranes, alter ion channels, and enhance the process of chelation of micronutrients. These efforts improve root growth, active growth of lateral roots, more capacity of ion exchange, catalysis of enzymes that boost up internal supply, and absorption of nutrients (Pettit, 2004).

Amino acids and protein hydrolysates derived by the stepwise enzymatic hydrolysis of animal and plant proteins can act both as components and building blocks of protein structures and as bioactive signaling molecules (Sun et al., 2024). They affect the metabolism of nitrogen, carbon assimilation, and hormonal crosstalk of plants. Certain amino acids, such as glycine, proline, glutamic acid, and so on, are reported to stimulate stomatal control, osmotic adaptation, and reactive oxygen radical scavenging (Ali et al., 2019). Under drought, salt stress, and oxidative stress situations, exogenous application of amino acids will help to control the osmoprotectant synthesis, increase capacity to retain water, and trigger antioxidant defense systems via antioxidant enzyme activities (such as superoxide dismutase and catalase) production (Hussain et al., 2019).

Some of the most commercially achievable biostimulants are sea algal extracts, whereby the majority of applications are based on brown algae: *Ascophyllum nodosum*, *Ecklonia maxima*, and *Laminaria digitate*. Those extracts are rather complicated mixtures of phytohormones (auxins, cytokinins, and ABA), laminarins, alginates, oligosaccharides, and polyphenols. They improve root and shoot biomass, postpone senescence, and increase chlorophyll concentration. It is especially interesting how they render systemic acquired resistance (SAR) effects and priming components against the pathogens upon the fluctuation of the pathogenesis-related (PR) proteins and activation of jasmonate and salicylate pathways. Abiotic stress also helps to be supported by seaweed extracts, which helps to reduce the occurrence of lipid peroxidation and also enhance the expression of heat shock protein and aquaporin.

It is also important to include the microbial agents, including plant growth-promoting rhizobacteria (PGPR), arbuscular mycorrhizal fungi (AMF), endophytes, and positive actinomycetes, which are equally imperative to the efficacy of the biostimulant. PGPR strains such as *Bacillus*, *Pseudomonas*, or *Azospirillum* species promote plant growth by releasing phytohormone, that is, indole-3-acetic acid, solubilizing insoluble phosphates by excretion of organic acids, and fixing nitrogen in the atmosphere. These microbes tend to secrete siderophores, which enhance iron uptake, and volatile organic compounds (VOCs), which affect gene expression of the plant host. On the other hand, mycorrhizal fungi develop extensive systems of hyphae that can substantially expand the radius of exploration of roots and essentially enable the absorption of nutrients that are relatively immobile, individuals as phosphorus and zinc, as well as provide drought resistance and protection against pathogen invasion through regulation of root exudates and immune response signaling (Khoso et al., 2024).

Microbial biostimulants have both direct advantages of plant nutrition and immunity and also an essential aspect of the reconstruction of the soil microbiome. They enhance the microbial diversity, establishment of beneficial consortia, and degradation of soil pollutants by producing enzymes. Their biofilms enhance the structure and porosity of the soil, which enhances aeration and water retention. Furthermore, pathogen-signaling molecules like lipo-chitooligosaccharides (LCO), quorum sensing autoinducers, and peptidoglycan fragments have the ability to trigger certain gene expression patterns that favor root branching, activation of nutrient transporters, and anti-pathogen defenses in plants (Castiglione et al., 2021).

Biostimulants often act by affecting hormonal networks in plants and usually in balancing and synergistic modes. Their components may improve or imitate the effect of auxins, gibberellins, ABA, cytokines, ethylene, SA, and jasmonates. With the help of these interactions, biostimulants impact the process of cell division, cell elongation, flowering, fruit ripening, and also stress adaptation. As an example, they can adjust the trade-off between growth and defense, that is, those plants can allocate resources in the best manner having regard to the environment. Various biostimulants are capable of increasing the expression of genes related to flavonoid biosynthesis, lignin synthesis, and terpenoid detoxification, thus strengthening both primary and secondary metabolisms (Omoarelojie et al., 2021). The numerous field experiments and greenhouse trials on the effects of biostimulants conducted over the years have proven that they enhance the quality and quantity of crop yield in many different crops, such as cereals, legumes, fruits, and vegetables. Significant consequences are shown in larger roots, earlier flowering, better fruit set, higher sugar and protein content in grains, and better postharvest preservability. They implied agronomic advantages with low and little environmental interference since most of the biostimulants are biodegradable and capable of fitting into organic and integrated production systems (Mandal et al., 2023).

The context-specific nature of biostimulants is one of their most characteristic features. They are affected by species of plants, stage of growth, weather systems, and soil characteristics. When used in this specific manner, precision farming becomes a possibility whereby biostimulant solutions and combinations are customized to the physiological requirements of the plant and its local ecosystem. To provide the greatest delivery, optimum stability, and bioavailability of active ingredients, innovations in formulation technology, including microencapsulation, mini-carriers, and controlled-release matrices, are being developed. Notably, the addition of biostimulants to bioactive-based fertilizers will give the system an upper hand since they interact with nutrients and signal molecules to help plant health optimally. Biostimulants may alleviate the necessity to use excessive inputs of chemicals when they are integrated with micronutrients and macronutrients, improve the efficiency of the utilization of nutrients, and help maintain soil fertility in the long perspective. Their application to formulations of sustainable fertilizer is a strategic step in climate-resilient farming due to their overall effects of compromising on greenhouse gas emissions, reducing nutrient leaching, and enhancing stress tolerance of crops (du Jardin et al., 2025).

On the whole, biostimulants are not supplementary elements but living biological stimulators that modify the physiological context of cropping systems. Their capacity to adjust the expression of genes, enhance the dynamics of nutrients, reinforce the antioxidant defenses, and rearrange microbial communities renders them inexhaustible in the search for resistant, productive, and sustainable agricultural management. Biostimulants are becoming an increasingly key component of the next generation of smart fertilizer technologies owing to the increasing understanding of their molecular and ecological mode of action.

3.3 Nutrient Delivery Systems: Enhancing Bioavailability and Uptake

The development of bioactive-based fertilizers introduces a game-changing shift in the distribution manner of nutrients in plants—gross delivery to a tailor-made operational delivery system. Within this new framework, formulations of nutrients are no longer merely passive adjuncts but active entities that are designed to respond to biological and environmental cues so that plants are supplied with the correct type of nutrients, at the correct dose, at the correct level, and at the correct time. Conventional fertilizers have suffered efficiency problems including leaching, runoff, volatilization, and soil fixation even though they boosted food production during the green revolution. Such shortcomings contribute not only to financial inefficiency and an inability to achieve good yields but also serious destruction to the environment. By way of contrast, the bioactive nutrient delivery systems are aimed at eliminating these drawbacks by improving their bioavailability and plant uptake pathways using controlled delivery technologies, nanomaterials, chelation, microbial inoculants, and biochemical enhancers. All of these mechanisms are unique in terms of enhancing accessibility and efficiency of nutrient use, reducing ecological footprints, and promoting long-term soil health.

Controlled-released technology is one of the main pillars in the modern nutrient management systems. These systems control the release of nutrients at the right time of the crop cycles and physiological need. Nutrients can be based on biodegradable polymers that are encapsulated or coated with a variety of materials such as organic waste that has been transformed into an organic carbon-rich product such as biochar or zeolite communicated matrices. Such barriers degrade or permit nutrient diffusion in response to the environmental stimuli including temperatures, moisture, or microbial activity. This practically implies that nutrients are discharged over time instead of being a single release, and this prevents quick spikes that can cause volatilization or nitrogen leaching. As an example, the ammonium or nitrate in nitrogenous fertilizers that may be coated with a sulfur layer or polymer layers may continuously liberate it over time and thereby provide stable availability within the root zone as well as reduce the loss of nitrogen to either the groundwater or atmosphere. Such a technique not only does increase NUE but also increases the availability of nutrients during the entire cropping cycle, and this process increases the interval of application of the fertilizers and the labor-associated costs (Rehana et al., 2022).

The use of nanoscale materials, or nano-fertilizers, is another highly effective solution to enhance the uptake of nutrients. Nanoparticles have much higher surface area and are extremely small in size, are able to interact with root cells of the plants better, enhance solubility of otherwise immobile nutrients, and can be controlled more selectively. Nanoparticles of such elements like zinc, iron, phosphorus, and silica have the ability to enter root and foliar tissues via stomata or the root epidermis. Their dissolution in the rhizosphere or on the surfaces of the leaves is controlled, and due to this very fact, nutrients are released in the bioavailable state or form, and losses are also reduced. Besides, nanoformulations are designed to be responsive to certain stimuli like pH changes, light, or even enzyme activity, and this would be the time and place when the nutrients are released when most needed by the plant.

This intelligent response makes the process of delivery more efficient with minimum exposure to the environment and toxicity. Besides, nano-clays and carbon nanomaterials, including graphene oxide, are tested with regard to their potential to adsorb macronutrients, such as potassium and phosphorus, and release them at slow rates, thereby forming nanoscale stores of nutrients (Rameshaiah et al., 2015).

Chelation techniques are critical in solving the problem of micronutrient availability in the soil, usually when soils are alkaline or calcareous, causing molecules of nutrients such as iron, manganese, zinc, and copper to form insoluble complexes. Both synthesized such as *ethylenediaminetetraacetic acid (EDTA)* and natural (such as citric acid or humic substances) chelating agents form complexes with these ions and maintain them in a soluble and plant-available states. Such chelates can persist in the soil solution and are taken up through the membranes of root cells using specific transporters. When absorbed in the body, they are enzymatically broken down, and the active metal ions are set free to be utilized as metabolites. Chelated micronutrients are therefore highly available and do not interact with other soil components that may precipitate or become immobilized. Besides, natural chelates produced by microbial siderophores or exudates of plants provide eco-friendly alternatives with reduced environmental persistence and higher rates of biodegradation, which is consistent with an objective of sustainable agriculture (Wallace, 1963).

Microbial inoculants, especially plant growth-enhancing accompaniments of rhizobacteria and helpful fungi such as arbuscular mycorrhizae, improve nutrient uptake because of multiple mutually supporting ways. These microorganisms occupy the rhizosphere and root surfaces, increasing the functional surface area that is available to absorb nutrients. They stimulate the solubilization of phosphorus resources unavailable due to their insolubility, and they trigger the release of enzymes to release potassium and mobilize micronutrients, which they do by the production of organic acids or siderophores (Hajji-Hedfi et al., 2025). Nitrogen-fixing bacteria like Azotobacter and Rhizobium transform atmospheric nitrogen into forms that are available to plants so that the use of fertilizers decreases sharply due to the substitution of the former with nitrogen-fixing bacteria. The mycorrhizal fungi have complex hyphal systems that reach an immobile type of nutrient beyond the root depletion zone and feed the host plant with it in exchange for carbohydrates. Besides the effect on mobilization of nutrients, these microbes also release phytohormones such as auxins and gibberellins, which increase root development; hence enhancing the ability of the microbes to deplete the environment of nutrients within their range. Their use together with the bioactive-based fertilizers has led to a higher use efficiency of nutrients and the overall improved plant vigor and stress tolerance (Aasfar et al., 2021).

Biochemical growth promoters such as humic and fulvic acids, protein hydrolysates, and some amino acids are involved with signaling molecules and metabolic catalysts in nutrient uptake and assimilation. These chemicals enhance the penetrability of the cell membranes of the roots and nutrient exchanger capacity of the roots and activate the genes of transporters of nutrients to the tissues. As an example, humic substances are able to chelate micronutrients, enhance cation exchange capacity, and enhance root growth (Mackowiak et al., 2001). Peptides and amino acids, which are abundant in protein hydrolysate, have been reported to increase nitrate and ammonium transporters and the

Table 3.1 Overview of nutrient delivery technologies in bioactive-based fertilizers, highlighting their mechanisms and roles in improving nutrient bioavailability, uptake efficiency, and environmental sustainability.

Technology/ component	Mechanism	Main benefits
Controlled-release technology	Nutrients encapsulated/coated with biodegradable polymers, biochar, or zeolite; release triggered by environmental cues (moisture, temperature, microbial activity)	Matches nutrient release to crop demand, reduces leaching/volatilization, prolongs availability, lowers labor costs
Nanofertilizers	Nanoscale particles (Zn, Fe, P, Si) penetrate roots/leaves, release nutrients in controlled, stimulus-responsive manner	Increased solubility, targeted delivery, reduced environmental loss, improved uptake efficiency
Chelation	Synthetic (EDTA, DTPA) or natural (citric acid, humic acids, amino acids) agents keep micronutrients soluble and plant-available	Prevents precipitation/immobilization, enhances uptake in high-pH soils, provides eco-friendly alternatives
Microbial inoculants	PGPR and AMF enhance nutrient solubilization, fixation, mobilization; expand root absorption area	Improves N, P, K, and micronutrient uptake, enhances stress tolerance, promotes root growth
Biochemical growth promoters	Humic/fulvic acids, protein hydrolysates, amino acids enhance membrane permeability, activate nutrient transporter genes	Increases nutrient assimilation, improves root development, enhances microbial activity in rhizosphere

nitrogen assimilatory processes. Some of the amino acids are in addition to metal carriers, which facilitate the transport of trace elements via the phloem and xylem. In addition, there are biochemical supplements that serve to regulate the activity of microbes in the rhizosphere to further enhance the mineralization and availability of nutrients. These enhancers can enhance physiological and biochemical nutrient dynamics thereby ensuring an increase in growth efficiency among the plants even in situations whereby the environment is nutrient-limiting or prone to stress conditions (Heidarzadeh, 2025). Table 3.1 summarizes the main nutrient delivery technologies discussed above, their mechanisms, and key benefits in sustainable agriculture.

Each of these technologies goes toward proving an overall paradigm shift in nutrient handling and delivery in agriculture. They indicate a change from indiscrimination to incorporation of chemical, biological, and environmental intelligence into systems. With such mechanisms, you have a synergistic effect, making the nutrients more available and reducing wastage and providing healthier soils on a long-term scale. This holistic approach at the system level is the framework of the future-ready agricultural products that are aimed at enhancing the productivity and minimizing the environmental loads. The coming of high-tech systems of nutrient delivery in combination with the bioactive-based fertilizers represents the possibility to redefine the field of plant nutrition to a more sustainable and resilient food production chain.

3.4 Signaling Molecules: Facilitating Plant Growth and Stress Resistance

Modern agriculture depends more and more on the extent to which crops can sense and respond to their environment. The key to this flexibility is a highly advanced internal communication network that is controlled through signaling molecules but are potent chemical messengers that coordinate growth, development, and defensive interactions that occur within plants. These are the molecules that allow the plants to sense biotic and abiotic stimuli and translate them into biochemical cascades and refine physiological responses to nutrient acquisition, root and shoot development, flowering, senescence, and adaptability to stress. Incorporated into fertilizers based on bioactive compounds, signaling molecules are, however, not simple supplements but strategic bodyguards of plant response in terms of precision and resilience. The plants can be engineered with their intended usage, which will rewire plant metabolism and gene expression, which will optimize energy allocation and increase the effectiveness of the product in withstanding drought, salinity, and pathogen attacks, among other environmental stresses, without the loss of production or quality.

One of the signaling agent cornerstone pieces used in such fertilizers is known as phytohormones (Ali & Iqbal, 2024). Such naturally occurring compounds (auxins, gibberellins, cytokinins, ABA, ethylene, brassinosteroids, SA, and JA) are master regulators of the plant physiology. Auxins promote cell expansion and orthogonal root growth that increases water and nutrient absorption, particularly that of stress situations. Gibberellins promote stem elongation, seed germination, and reproductive growth, and this enables plants to reach maturity with ease despite the hostile condition. Cytokinins stimulate vegetative bud formation and prevent senescence of leaves, which contain the photosynthetically active leaf area essential to continuing biomass accumulation. The brassinosteroids play the role of promoting cell division, vascular differentiation, and the reaction to heating. In the meantime, ABA can serve as a regulating stress hormone, which can result in stomatal movements, osmotic balancing, and expression of stress-related genes in cases of drought and salinity. Whereas ethylene mediates fruit ripening, abscission, and responding to mechanical stress, JA and SA mediate immune signaling networks in response to insect herbivory and microbial infection in a frequently antagonistic or synergistic manner that varies according to the threat.

Such hormonal cues are supplemented by a distinct group of nonhormonal messengers called secondary messengers, which include calcium ions (Ca^{2+}), reactive oxygen species (ROS), and reactive nitrogen species (RNS), which provide plasticity by amplifying signal and cross-talk among signaling cascades. As an example, calcium influx is the early indicator after environmental stress, activating downstream kinase and transcription factor induction (Ren et al., 2023). Hydrogen peroxide (H_2O_2) and, in general, ROS are lethal waste products and, at the same time, important signals coordinating oxidative stress responses and SAR. On the same note, NO regulates many events such as root growth, stomatal behavior, and activation of defense genes. These messengers also build complex networks with phytohormones, which contributes more specificity and coordination in the responses to stress. This has been achieved by developing stable delivery systems for such labile compounds through encapsulation or slow-release carriers, so applications of these products in soil (and foliar) formulations can be controlled and sustained.

Together with these classical signaling molecules, peptides and oligosaccharides have been found to be very specific plant regulators. Short peptides include phytosulfokines (PSKs), systemins, RALFs (rapid alkalinization factors), and CLE (CLAVATA3/EMBRYO SURROUNDING REGION) peptides, which regulate patterning of tissues, root growth, and immune-system activation (Li et al., 2025). Plant cells have receptor-like kinases (RLKs) in their cell membrane that interact with these peptides to trigger a highly selective process of cell signaling that overdetermines cell fate and adaptation to stress. On the same note, oligosaccharides released by the cell wall, for example, oligogalacturonides (OGs), are damage-associated molecular patterns (DAMPs) for defense priming and production of secondary metabolites. Their use in the form of bioactive fertilizers can increase the resistance to diseases and tolerance to mechanical and pathogenic stresses, and this does not pose a metabolic cost to the plant.

The other area of signaling research is the use of VOCs and microbial signals as tools to alter plant physiology. Examples of signaling molecules of microbial origin are LCOs produced by rhizobia and N-acyl homoserine lactones (AHLs) produced by gram-negative bacteria, and they influence root architecture, nodulation, and plant immunity (Rush et al., 2020). On the same note, microbial volatile organic compounds (mVOCs) like acetoin and 2,3-butanediol may also induce the growth and systemic resistance of adjoining plants via the air-based signaling (Poveda, 2021). These natural cues are currently being developed as microbial consortia and biostimulant-type fertilizers to improve cross-kingdom signaling and improve plant fitness as the plant grows both in the field and in greenhouse environments.

The intracellular response is represented by the impact of these signaling molecules on complicated networks of genes through the regulation of such transcription factors as WRKY, MYB, NAC, and bZIP families. These elements control the expressivity of stress-adaptive genes, transporter proteins, and metabolizing enzymes, which improve the security of plants, water-use efficiency, and nutrient uptake. Further, stress memory may be encoded by epigenetic modifications of some signaling molecules, as the induction of epigenetic histone modifications, such as methylation, and the induction of epigenetic DNA methylation in plants enable them to respond more rapidly and intensely when subjected to the same stress again. Biofertilizer formulations that stimulate such epigenetic priming are under development to make transgenerational stress tolerance, and this innovation is promising for the long-run sustainability of agriculture (Schulman, 2014).

The fine-tuning of the plant response is also mediated by the cross-talking or the interaction between the signaling molecules. As another example, ABA works with ethylene to respond to drought and to balance root growth and activate objects in tandem so as to calibrate immunity in response to certain pathogens (Chen et al., 2021). The newer generation of bioactive fertilizers utilizes multisignal approaches, whereby only a specific amount of signaling agents is sequentially or triggered by the environmental signals to have the same pattern of the natural control in the plant. These kinds of technologies give rise to a dynamic nutrient signal interface that is able to assist in high-yield production as well as the ability to harden to stress.

In a nutshell, the signaling molecules are smart activators in bioactive-based fertilizers, which convert these types of fertilizers into communicative bases of plant guidance and regulation. These new generation inputs can also tune the plant physiology to suit both the developmental and the environmental needs by exploiting phytohormones, peptides, oligosaccharides, ROS/RNS messengers, and microbial signals. Their job is not just to foster development, but they also contribute to producing crops that are more resilient and adaptive to fit the more uncertain farmlands.

3.5 Inorganic and Organic Molecules in Bioactive Fertilizers

The functions of bioactive fertilizers are significantly determined by the coordinated action of inorganic and organic substances (Bhardwaj et al., 2014). These molecular components are the active framework node and the regulatory unit of up-to-date biofertilizers, making them highly effective nutrient-giving and plant metabolism-regulating and giving them the ability to dynamically connect with the rhizospheric microbiome. The combined utilization of these elements is not only used as a nutritional supplement but also enhances the performance of the fertilizers to a biosynthetic level, supporting the growth of the plants in spite of the prevailing environmental conditions.

Inorganic compounds as molecules are mainly responsible for supplying the essential minerals that fundamental physiological and biochemical functions of plants rely on. Macronutrients like nitrogen, phosphorus, and potassium have been found to play the roles of amino acid synthesis, ATP generation, and water balance, respectively. Calcium plays an important role in the maintenance of the membrane stability and signaling, magnesium has a central role in chlorophyll and also enzymatic steps involved in photosynthesis, and sulfur plays vital roles in some amino acids and cofactors. The micronutrients are iron, zinc, copper, manganese, boron, molybdenum, and nickel. Although they are in lesser amounts, but they are also very necessary for redox level, hormone metabolism, viable pollen, and cell wall integrity (Steward, 2012).

To achieve maximum nutrient efficiency, a number of technological inputs have been applied in the formulation of fertilizers. Stable-release technologies, such as the controlled release technology, are aimed at coordinating the availability of nutrients with the requirements of the plants as time goes by. These formulations are usually based on polymer coatings, biodegradable carriers, such as starch-based matrices, or nanomaterials as the means of controlling the release of nutrients via dissolution and diffusion. This avoids nutrient loss through leaching, minimizes volatilization, and eliminates overfertilizing problems, besides making the nutrients available over longer periods in the root zone (Adepu & Ramakrishna, 2021). Nanofertilizers as well as nano-implanted minerals such as nano-ZnO and nano-Fe_2O_3 have even longer-lasting effects due to them having high surface area, controllable release rates, and better root–soil interfaces. These nanoparticles may penetrate the tissues of a plant through the stomata or enter the cells through endocytosis without enduring the processes of usual plant uptake (Nongbet et al., 2022).

Chelation approaches are yet another way through which micronutrient distributions can be enhanced, particularly in high-pH soils or salty soils where the ionic forms may be unavailable due to the precipitation or immobilization process. Chelators take the form of EDTA, EDHA, DTPA, and naturally derived organic acids, which are seen as plant-available soluble forms of the metal ions. The new methods include amino acid chelation, where micronutrients are chelated with glycine, proline, or glutamic acid and can be absorbed as a nutrient and metabolism enhancer. The effect of these chelated complexes is not only to increase the solubility and mobility of the compound in soil but also to make it readily available to the plants in the form of lipophilic interaction with the cuticle (Madhupriyaa et al., 2024).

Organic molecules serve to assist in the process of providing many different functions other than nutrient supply. These are the bioorganic components, that is, the amino acids, organic acids, humic matter, polysaccharides, peptides, lipids, vitamins, and phytohormone precursors. Most of these compounds are signaling, root growth stimulators, metabolic boosters, or sequestering ligands. When amino acids are exogenously supplied, they ameliorate osmoprotective ability and scavenge ROS as well as amplify the nitrogen assimilation pathway and chlorophyll biosynthesis. To take an example, proline acts as one of the osmoprotectants and membrane stabilizers under drought or salt stress (Singh et al., 2015).

Acidic compounds like citric, malic, fumaric, and oxalic acid have two purposes: they lower the pH of the rhizosphere to release phosphate and micronutrients and provide sources of energy on which beneficial microorganisms feed and thrive. The compounds can also affect the profile of root exudation and root hair elongation, increasing the capacity to take up nutrients. Humic and fulvic acids are the products of the decomposition activities of microbes and plants, which have been known to increase cation exchange capacity, increase root growth, and increase expression of nutrient transporter genes. They also have a great number of functional groups (e.g., carboxyl, phenolic) on their molecular structure which interact with mineral ions, stabilize soil colloids, and enhance the soil water-holding capacity (Priecina & Karklina, 2015).

Polysaccharides like alginates, cellulose derivatives, and chitosan are added as mediums of conveying as well as elicitors. Brown algal alginates make up hydrogels that keep nutrients on board and release them gradually to better coordinate with plant utilization of those nutrients. Chitosan, a natural biopolymer produced through the breakdown of chitin, can serve as a fertilizer load and trigger an immune response in the plants such as activation of corresponding genes of defense and providing resistance to pests. It is also able to coordinate metals and help in absorbing them besides encouraging the growth of lignin and other secondary metabolites in plants (Benalaya et al., 2024).

The application of microbial inoculants and peptides is being completed by the formulations containing not only inorganic nutrients but also organic substances. Those biotic elements increase nutrient solubilization, nitrogen fixation, and phytohormone synthesis, providing an active mechanism to upsurge nutrient content and plant prosperity. Microbial hydrolysates can give rise to peptides that mimic signaling molecules, which stimulate plant defenses or growth-promoting cascades in the nanomolar range. PGPR and AMF are also known to be used to form a symbiosis interaction with the roots of the plant to increase the use of phosphorus and micronutrients, as well as the plant being less reliant on chemicals (Sharma et al., 2023).

The manufacturing of hybrid fertilizers, which encapsulate inorganic ions into organic materials, is a process that is still developing at a very high rate. In the case of micronutrients bioimmobilized through biochar, the process not only helps to combat nutrient runoff through leaching but also helps to support microbial colonization and boost carbon storage. On the same note, seasoning complexes of lignosulfonates or amino acid–polysaccharide conjugates enhance the stability, soil binding, and release dynamics of the nutrients

Table 3.2 Comparative roles of inorganic and organic components in bioactive fertilizers, highlighting their nutrient delivery, soil health impacts, plant quality benefits, and synergistic effects for sustainable agriculture.

Parameter	Inorganic components	Organic components (including bioorganic and inoculants)
Nutrient supply	Rapidly provide essential macro- and micronutrients (e.g., N, P, K, Ca, Mg, Fe, Zn)	Supply nutrients slowly; include amino acids, vitamins, and precursors that support metabolism
Release dynamics	Soluble and fast-acting but prone to leaching/volatilization; controlled forms minimize loss	Slow-release, enhance soil structure and nutrient retention over time
Soil health and microbial activity	Limited effect; excessive use can degrade soil structure and pollute water	Boost soil organic carbon (total organic carbon (TOC) / soil organic carbon (SOC)), aggregate stability, microbial biomass, and enzyme activity
Plant quality and stress tolerance	Enhance yield and fruit size rapidly	Improve nutritional quality, antioxidant levels, and stress resilience
Synergistic effects	Alone gives immediate nutrient boost.	When combined with inorganic options, supports soil–plant systems and nutrient synergy

(Easwaran et al., 2024). The interaction between the two classes of molecules, mineral ions and organic bioeffectors, is boosted by the fact that the two interact with the endogenous regulatory networks of the plant. As an example, nutrient acquisition oftentimes signals sets of transcription factors and transporters that often determine the expression of transporters, and organic molecules have a strong tendency to prestimulate such mechanisms to be used to an optimal extent. In addition to that, the combined use of organic molecules with mineral nutrients usually demonstrates the additive effect on the root density, chlorophyll index, and yield quality, and even the synergetic (Kleber et al., 2021). Table 3.2 here summarizes the preceding discussion.

In short, the combination of organic and inorganic molecules in bioactive fertilizers is a comprehensive approach to enhance nutrient transport, improve soil life, and alter plant physiological dynamics. This molecular incorporation not only fulfills the short-term nutritional requirements of crops but also induces a set of biochemical and ecological mechanisms that result in sustainable and resilient agricultural systems. Further advancement in scientific knowledge will lead to customized molecular compositions, depending on the crop, soil, and environment stressing factors, opening the door to more efficient biofertilizers with little impact on the environment.

3.6 Revolutionizing Nutrient Delivery with Nanoscale Precision

Nanotechnology is an interdisciplinary field that manipulates matter at dimensions between approximately 1 and 100 nm, where unique phenomena enable novel applications (Ghaffar et al., 2023; Hussain et al., 2025; Khan et al., 2023). At this scale, materials exhibit remarkable physical, chemical, and biological properties compared to their bulk counterparts—such as increased surface area, reactivity, and quantum effects (Assad et al., 2024; Jabbar et al., 2023). These properties are now being harnessed across various sectors, including electronics, energy, medicine, and increasingly, agriculture (Assad et al., 2025a, b; Afzal et al., 2024). The controlled design and engineering of nanostructures provide innovative tools for addressing long-standing challenges in food production, environmental conservation, and plant health (Hashmi et al., 2024, Wajid et al., 2024).

Agriculture faces complex challenges—rising food demand, declining soil health, nutrient inefficiencies, and climate-induced stress. Traditional farming practices, heavily reliant on synthetic fertilizers and pesticides, often lead to nutrient losses, soil degradation, and environmental contamination. In this context, nanotechnology offers a promising solution. The application of nanomaterials in agriculture—termed nano-agriculture—includes nano-based fertilizers, pesticides, sensors, and soil conditioners. These nano-enabled inputs are engineered to enhance precision, efficiency, and sustainability in crop production by interacting with plant systems at the molecular level.

Among the most impactful innovations in nano-agriculture are nanofertilizers—fertilizer formulations that contain nutrients either encapsulated within nanoparticles, coated onto nanomaterials, or synthesized as nanostructured materials. Unlike conventional fertilizers, which often suffer from poor NUE due to leaching, volatilization, or soil fixation, nanofertilizers offer controlled release, targeted delivery, and minimal losses. Their enhanced solubility, high surface area-to-volume ratio, and ability to penetrate plant tissues efficiently make them potent tools for improving nutrient uptake and promoting plant growth.

Nano-fertilizers function through several key mechanisms. First, their nanoscale size enables better interaction with plant roots and leaf stomata, facilitating more efficient uptake. Second, their surface characteristics allow them to adhere to plant surfaces or soil particles, prolonging their availability. Third, they can be engineered for controlled release, responding to stimuli such as moisture, pH, enzyme activity, or root exudates—ensuring nutrients are released only when needed. This reduces the frequency of application, minimizes environmental runoff, and significantly enhances NUE. Additionally, nanofertilizers can act as carriers for multiple nutrients, combining macro- and micronutrients in a single formulation for balanced plant nutrition.

Nanofertilizers are generally classified into three categories: (i) nanoscale nutrient materials, such as nano-ZnO, nano-Fe_2O_3, and nano-hydroxyapatite, where nutrients themselves exist in nanoparticle form; (ii) nutrient-loaded nanocarriers, where traditional nutrients are encapsulated within nanomaterials like chitosan, silica, or carbon nanotubes; and (iii) nanostructured formulations, which are composite materials engineered at the nanoscale to improve delivery and uptake. These formulations can be applied via soil, foliar, or seed treatments, depending on crop requirements and environmental conditions.

Extensive research has been conducted on the nanoformulation of essential nutrients. For instance, nano-zinc oxide (ZnO) is known for improving root development, enzymatic activity, and photosynthesis. Nano-iron (Fe_2O_3) enhances chlorophyll synthesis and alleviates iron deficiency in calcareous soils. Nano-hydroxyapatite ($Ca_{10}(PO_4)_6(OH)_2$) serves as a slow-release phosphorus source, minimizing phosphate fixation in soil. Nano-silica (SiO_2) strengthens cell walls and improves resistance to drought and pests. These nanoparticles not only fulfill nutrient requirements but also stimulate plant immune responses, enhance stress tolerance, and promote overall vigor.

One of the most important features of nanofertilizers is the controlled-release mechanism achieved through nano-encapsulation. Nutrients can be enclosed in biodegradable polymers such as chitosan or alginate, which gradually degrade in the soil, releasing nutrients over time. This slow release aligns with crop growth stages, preventing nutrient overloading and minimizing wastage. Additionally, stimuli-responsive nanocarriers release nutrients in response to external triggers such as pH shifts, root exudates, or microbial activity, offering site-specific and time-dependent delivery. Such intelligent systems are particularly beneficial in precision agriculture, where micro-dosing can significantly improve yield and resource efficiency.

Nano-fertilizers interact with plants in unique ways. Upon application, nanoparticles can enter plant systems through root epidermis, root hairs, or foliar stomata. Inside the plant, they are translocated through xylem and phloem to target tissues. Studies using electron microscopy have shown nanoparticles accumulating in chloroplasts, mitochondria, and vacuoles, indicating their active participation in metabolic processes. While this enhanced interaction is advantageous, it also necessitates careful evaluation of dose, exposure time, and particle properties to avoid phytotoxicity.

Nano-fertilizers offer significant environmental benefits. Their reduced application rates and higher efficiency result in lower nutrient runoff into water bodies, mitigating eutrophication and contamination. The use of biodegradable nanocarriers and reduced chemical load supports soil microbial health and ecological balance. Economically, although nano-fertilizers may have higher initial costs, their long-term benefits—such as reduced input frequency, enhanced yield, and minimized environmental remediation costs—make them attractive for both smallholders and large-scale farmers. Moreover, integration with precision farming technologies enables site-specific application, reducing labor and resource expenditure.

Despite their potential, nano-fertilizers are not without challenges. The long-term fate of nanoparticles in soil and plant systems is still under investigation. Concerns about bioaccumulation, nanoparticle toxicity, and unintended interactions with beneficial soil microbes must be addressed through rigorous risk assessment and standardized testing protocols. Additionally, regulatory frameworks for nano-fertilizers remain underdeveloped in many countries, complicating market entry and adoption. Public perception and lack of awareness about the safety and benefits of nanotechnology in agriculture further hinder widespread implementation.

3.7 Synergistic Effects: Combining Biostimulants, Nutrients, and Signaling Molecules

Plant nutrition science and technology are moving toward the understanding that a decrease in functionality of a crop is not only related to the provision of microelements or macroelements in isolation. Instead, it can be described as a multidimensional interaction between several biochemical and biological factors that collaborate synergistically to promote plant growth, improve resistance to stresses, and maximize the use of available resources. These include (among others) the synergistic combination of bioactive components, biostimulants, nutrients, and signaling molecules into integrated bioactive-based fertilizers, which is a paradigm shift, changing the input-driven approach to fertilization (as is currently the case through the use of bioactive-based fertilizers) to systems that mimic the intended and more agile actions of the plant–soil interactions as seen in the natural environment.

Examples of biostimulants include humic substances, protein hydrolysates, seaweed extracts, microbial inoculants, and amino acids; they are reported to have regulatory effects on plant physiology instead of being a direct source of nutrients (Nardi et al., 2016). They affect their mechanisms by altering the expression of genes, balancing hormones, and stimulating root growth and enzymes that join in the primary and secondary metabolism. When these have been applied along with some vital nutrients, a positive feedback loop is triggered. As an example, humic acids are able to stimulate plasma membrane ATPases, which increases the inflow of nutrients due to the presence of an electrochemical gradient. Protein hydrolysates and amino acids can also be used as metabolic pathway precursors and complexing agents that enhance micronutrient mobility and absorption similarly.

The nutrient side lies in the fact that the accessibility of macro- and micronutrients is at the focus of metabolic processes in plants. Macronutrients such as nitrogen and phosphorus are needed in substantial quantities in order to produce nucleic acids and proteins associated with the transfer of energy. Iron, zinc, and manganese are examples of micronutrients that are cofactors to the enzyme systems important in photosynthesis, respiration, and the synthesis of hormones. Nevertheless, absorbency of nutrients can be inefficient, owing to loss of nutrients through leaching, volatilization, and soil fixation. This disadvantage can be overcome with co-application with biostimulants and signaling molecules by increasing NUE, through changing the profile of root exudates, setting up transporter activity as well as enlarging root–soil interface dynamics.

The release technologies are also crucial elements in this synergistic approach, for example, the use of controlled-release technologies. These are polymer-coated fertilizers, nanocarriers, and biochar-loaded nutrients, which make it possible to release the nutrients steadily and over a long time at the pace that the plant can use them. The combination of these technologies with biostimulants and signaling molecules makes the availability of nutrients continuous and predictable. As an illustration, nanostructured carriers can target micronutrients like Zn^{2+} or Fe^{2+} to root zones accurately and also concomitantly carry phytohormones, auxins, or SA that precondition the root system to take up more (Rani et al., 2024). Such spatial–temporal coordination has the effect of minimizing nutrient waste, limiting the environmental costs incurred, and ensuring that the needs of the plants physiologically in terms of resource supply to the plants are kept up throughout developmental stages.

This interaction becomes even more complicated by having signaling molecules integrated into the picture. Molecules such as SA, JA, NO, ABA, strigolactones, and ROS affect internal and external stimuli organization of plants and manage their reaction regarding them. An example is SA, which takes part in triggering the SAR in plants to overcome the attack by pathogens. Jasmonates coordinate the mechanisms of response to insect herbivory and abiotic stresses, as well as how roots elongate and how they behave in terms of nutrient foraging. ABA controls the closing of the stomata and osmotic balance during drought, and NO controls root architecture and serves as a molecular signal in nutrient signaling pathways. These signaling agents, when delivered together with nutrients and biostimulants, do so to create a plant system that would be ready to respond rapidly, efficiently, and adaptively to altered environmental and nutritional conditions.

Microbial inoculants are also a great part of this synergy. Plant growth-promoting PGPR, such as *Bacillus* species, *Pseudomonas* species, and *Azospirillum* species, are nitrogen-fixing rhizobacteria, and they solubilize phosphate, synthesize phytohormones, and induce systemic resistance. AMF are symbiotic symbionts of the roots, which increase the absorption surface area by a factor of extraradical hyphae and make immobile nutrients such as phosphorus and zinc available and accessible. By implementing these microbial communities in parallel with biostimulants and signaling molecules, the result is a rhizosphere environment with high biochemical crosstalk, an increase of nutrient flux, perception of stress, and diversity of the microbes.

The synergetic effect is also fortified by chelation plans. The micronutrients combine with organic ligands of EDTA (ethylenediaminetetraacetic acid), DTPA (Diethylenetriaminepentaacetic acid), and natural chelators such as amino acids and citric acid that form soluble complexes, preventing the precipitation of micronutrients or fixation in alkaline soils. These chelates also create translocation of membranes and vascular tissues when used together with biostimulants to enable the nutrients to arrive at their respective intracellular destinations. Enhancement of plant metabolic pathways at the cellular level occurs through bionutrients–biochemical enhancers. Bionutrient–biochemical enhancers are naturally occurring producers of plant metabolic pathways, namely enzyme activators, secondary metabolites, and vitamins. These agents facilitate some catalytic cycles that are required in nutrient absorption, the redox status, and energy transfer. As an illustration, vitamin B complexes and flavonoids regulate antioxidative enzyme action, whereas some polyphenols have the capacity to scavenge high levels of ROS produced during exposure to abiotic stress. These enhancers keep the cells healthy with the provision of mineral nutrients and biostimulants to ensure cellular homeostasis and the absence of metabolic dysfunction.

Plants that are treated with these synergistic formulations will, at the physiological level, have better characteristics; they will have deeper and more branching root systems, greater leaf area index, greater photosynthetic rate, better stomatal regulation, and more effective osmotic adjustment. There is a larger increase in yield-related parameters, for example, fruit number, size, and quality, for example, sugar content, antioxidant levels, and shelf life. Moreover, it is seen that crops are also more resistant to both biotic and abiotic stresses such as drought, salinity, heat, and pest burden.

The connotations of these synergistic interactions do not end at the plant level but rather enter the health of the soil and support of the ecosystem. The increase in the microbial activity will make organic matter easier to decompose and create humus, taking the soil structure, the degree of aeration, and water retention as major advantages. Minimized reliance on manmade inputs reduces the possibility of mixing soils and water with chemicals, which means that these bioactive formulations can fit the concept of sustainable and regenerative agriculture. At the molecular level, transcriptomics and proteomics studies have shown that these synergistic inserts trigger a broad repertoire of genes related to the promotion of growth, response to defense, hormone signaling, and transport of nutrients (Manzoni et al., 2018). Among the major upregulated pathways, there are MAPK signaling cascades, calcium ion physiology, ROS homeostatic processes, and gene-encoding transporter proteins like NRT1 (nitrate transporter), ZIP (zinc transporter), and HMA (heavy metal ATPase).

In short, synergistic interaction of biostimulants, nutrients, and signaling molecules in bioactive-based fertilizers is an integrative and progressive approach to sustainability in crop production. These formulations go beyond the shortcomings of traditional fertilizers by using a well-orchestrated process of molecular, biochemical, and physiological processes. Not only do they boost and strengthen plant growth, but they also help to restore soil ecosystems, an aspect that will lead to long-term agricultural sustainability and food security as the global climate is increasingly becoming difficult to work with.

3.8 Applications of Bioactive Fertilizers in Modern Agriculture

A paradigm shift is taking place in modern agriculture, where yield-focused application of inputs through modern-day agriculture is shifting toward resource-efficient and sustainable crop management practices. It is within this dynamic environment that bioactive fertilizers have been introduced as a multipurpose technology, which is more than just a nutrition source, as it incorporates biostimulants, signaling molecules, micronutrients, and beneficial bacteria to regulate the plant metabolism, soil health, and environmental robustness. Such fertilizers are increasingly being used in many different farming systems including high-input monocultures, regenerative agriculture, precision agriculture, and controlled-environment agriculture. The use of bioactive fertilizers in field crop production is one of the most common uses of bioactive fertilizers because large-scale cereal crops which include wheat, rice, and maize require high nutrient input. Conventional fertilizers tend to cause nutrient depletion and deficits. Bioactive formulations, and more so those enriched with humic acids, amino acids, and microbial consortia, enhance NUE through better root outgrowth, the activation of transporter proteins, and greater solubility of nutrients in the rhizosphere. It leads to better tillering, grain fill, and total biomass production without overreliance on synthetic chemicals.

Bioactive fertilizers have two purposes in the production of vegetable and fruit crops: to stimulate vegetative growth and to improve the biochemical quality of the product. As an example, foliar sprays with SA or seaweed extracts in tomatoes and bell peppers result in early flowering, improved fruit set, and lycopene and vitamin C levels (Massimi et al., 2023). The use of fertilizers with a high content of amino acids enhances fruit firmness and aroma, hence marketability, and prolongs postharvest fruit life in strawberries. Also, zinc, iron, and boron chelate micronutrient-enhanced formulations enhance synchronization of flowering and evenness of fruits in crops such as mangoes, grapes, and citrus. Another area, which is rapidly developing, is soilless cultivation systems (hydroponics and aquaponics), where bioactive fertilizers are also very important. Such systems do not have a natural rhizosphere, and their management depends greatly on exact nutrient management. Bioactive fertilizers ensure a continuous supply of the readily available nutrients and growth regulators, which increase the kinetics of nutrient uptake and balanced vegetative-reproductive growth. Nano-formulated micronutrients in hydroponics, for example, nano-Fe, nano-Cu, and nano-Mn, enhance the synthesis of chlorophyll, avoid deficiencies, and improve photosynthetic efficiency of leafy greens, for example, lettuce, kale, and spinach.

Largely benefiting are also the so-called protected cultivation systems like greenhouses and vertical farms, where bioactive fertilizers are used. In this case, manufactured by combining controlled-release nutrient systems with biostimulant complexes, it makes it possible to control plant physiological processes with meticulousness. Protein hydrolysates and trace elements are used in liquid formulation, which are fertigated using fertigation systems, leading to greater nutrient assimilation and less wastage of the input. Cucumbers and herbs produced using vertical farms have also used JA and NO donors, which are known to aid in regulating the stomatal conductance of plants to prevent loss of water, hence making the crops more productive. Bioactive fertilizers are more specific to leguminous crops natively able to fix atmospheric nitrogen but with the aim of strengthening the symbiotic relationship. Nodulation occurs early and effectively when rhizobial inoculants are combined with flavonoid signaling molecules, and this enhances nitrogen fixation potential. The process of biological nitrogen fixation is enhanced when supplemented with molybdenum and cobalt, which are nitrogenase enzyme cofactors. These fertilizers are specifically good at enhancing protein level and seed weight of crops like soybean, chickpea, and lentil.

Potatoes, carrots, and onions, referring to tuber and root crops, are among the crops that also have significant advantages from using bioactive fertilizers (Kazimierczak et al., 2021). They are crops with great nutrient requirements, and they are susceptible to nutrient imbalances. Phosphate-solubilizing microorganisms (phosphate-solubilizing bacteria) and potassium-mobilizing fungi contained in biofertilizers have been found to increase the bulking and uniformity of the tubers. The formulations enriched with seaweed extract assist in the control of cell division and expansion that leads to improved root growth and sugar content. Bioactive fertilizers play a role in the pasture and forage system, where lack of soil microbial and nutrient depletion are major issues to manage, especially when an increase in biomass production is required. Use of microbial consortia enhances the dynamics of the rhizosphere and also stimulates the growth of grasses like alfalfa, Bermuda grass, and clover. In addition, the gibberellic acid and fulvic acid have been added to the formulation that has enhanced quicker shoot extension, augmented dry matter production, and enhanced fodder grade within the livestock system.

Other systems such as rice paddies that are nutrient-intensive and environmentally sensitive systems also have better results when using bioactive fertilizers. Inoculants of microbes like Azospirillum and Anabaena enhance nitrogen economy and root growth, and silicon-formulated fertilizers enhance the cell wall of the plant cells, decreasing pest and disease pressure. Such injections also decrease the amount of methane emissions since they encourage aerobic microbial communities in submerged soils. In the manageable soils, root growth and nutrient uptake are impaired, which is the case in saline and degraded soils; bioactive fertilizers including halotolerant bacteria, and osmoprotectants, such as proline, glycine betaine, and polyamines, will support plants in the maintenance of ionic balance and membrane stability in the saline and degraded soils. These supplements make crops like barley, sorghum, and quinoa be resistant to extreme salt stress by increasing Na^+ exclusion and K^+ retention rates at the root-soil level.

Canola, sunflower, and groundnut are oilseed crops; their partitioning of nutrients and biosynthesis of lipids need to be highly efficient. Boron and calcium chelates and plant hormones such as brassinosteroids enhance pollination, pod setting, and accumulation of oil. The activities of important enzymes like the acetyl-CoA carboxylase that is important in the production of fatty acids are also regulated by the use of bioactive fertilizers. In addition, bioactive fertilizers can be used in practice during seed priming and nursery phases, and it is necessary to ensure early root development, stress tolerance, and seedling vigor. It is possible to break dormancy, activate antioxidant defense systems, and condition seedlings to thrive following transplantation by priming agents that contain signaling peptides, micronutrients, and microbial elicitors. This has proved to enhance survival and decrease lag time in the establishment of the field.

In bioactive fertilizers, in the case of intercropping and polyculture, the distribution of nutrients and communication between microorganisms of different species are harmonized. The balance between nitrogen and phosphorus in such formulations of maize/beans and cotton/groundnut systems enhances the activities of microbes in the rhizosphere led by root exudates of maize and beans/cotton and groundnut to create an interactive rhizosphere. Bioactive fertilizers are used in agroforestry and afforestation activities to make the plants grow faster and become more resilient. Mycorrhizal-enriched fertilizers can increase root colonization in woody plants, which improves drought resistance and nutrient mining in soils deficient in nutrients. The applications are essential in restoring the degraded lands and enhancing biodiversity corridors. Bioactive fertilizers are also used in climate-resilient agriculture to regulate the response of plants to changes in temperatures, periods of drought, and even exposure to heavy metals. NO donors, silicon-based type formulations, and ROS-scavenging compounds (polyphenols) can have a positive effect on promoting plant immune reactions, increasing antioxidant enzyme expression, and decreasing oxidative stress indicators. In short, the flexibility of bioactive fertilizers can be measured by the extent to which they are applicable in various cropping systems, climatic conditions, and agricultural technologies. This multifunctional feature, which can be observed in nutrient, stress, and plant health provision as well as environmental conservation, makes them must-have elements in contemporary agriculture. Table 3.3, summary of bioactive fertilizer applications across different cropping systems, is compiled from the preceding discussion. As further research and innovation proceed, their use is likely to be extended further, helping ensure food security and the sustainability of agroecological systems on a global scale.

Table 3.3 Applications of bioactive fertilizers across cropping systems and their key benefits.

Agricultural system/ crop type	Key bioactive components used	Main benefits/outcomes	Agricultural system/crop type
Field crops (wheat, rice, maize)	Humic acids, amino acids, microbial consortia	Improved nutrient use efficiency, better root growth, enhanced grain fill and biomass, reduced chemical dependence	Field crops (wheat, rice, maize)
Vegetables and fruits (tomato, bell pepper, mango, grapes, citrus, strawberries)	Salicylic acid, seaweed extracts, amino acids, chelated Zn/Fe/B	Early flowering, improved fruit set, enhanced nutritional quality (lycopene, vitamin C), better firmness, aroma, and postharvest life	Vegetables and fruits (tomato, bell pepper, mango, grapes, citrus, strawberries)

Table 3.3 (Continued)

Agricultural system/ crop type	Key bioactive components used	Main benefits/outcomes	Agricultural system/crop type
Soilless systems (hydroponics, aquaponics)	Nano-Fe, nano-Cu, nano-Mn, growth regulators	Continuous nutrient supply, improved chlorophyll synthesis, deficiency prevention, balanced growth	Soilless systems (hydroponics, aquaponics)
Protected cultivation (greenhouses, vertical farms)	Controlled-release nutrients, protein hydrolysates, jasmonic acid, nitric oxide donors	Precise physiological control, higher nutrient assimilation, better water-use efficiency	Protected cultivation (greenhouses, vertical farms)
Leguminous crops (soybean, chickpea, lentil)	Rhizobial inoculants, flavonoids, molybdenum, cobalt	Enhanced nodulation, nitrogen fixation, higher protein content and seed weight	Leguminous crops (soybean, chickpea, lentil)
Tuber and root crops (potato, carrot, onion)	Phosphate-solubilizing bacteria, potassium-mobilizing fungi, seaweed extract	Better bulking and uniformity, improved root growth, higher sugar content	Tuber and root crops (potato, carrot, onion)
Pasture and forage (alfalfa, Bermuda grass, clover)	Microbial consortia, gibberellic acid, fulvic acid	Faster shoot growth, higher dry matter yield, improved fodder quality	Pasture and forage (alfalfa, Bermuda grass, clover)
Rice paddies	Azospirillum, Anabaena, silicon formulations	Improved nitrogen economy, reduced pest/ disease pressure, lower methane emissions	Rice paddies
Saline/degraded soils (barley, sorghum, quinoa)	Halotolerant bacteria, proline, glycine betaine, polyamines	Improved ionic balance, salt stress tolerance, Na^+ exclusion, K^+ retention	Saline/degraded soils (barley, sorghum, quinoa)
Oilseed crops (canola, sunflower, groundnut)	Boron/Ca chelates, brassinosteroids	Enhanced pollination, pod setting, oil accumulation, regulation of fatty acid synthesis	Oilseed crops (canola, sunflower, groundnut)
Seed priming and nurseries	Signaling peptides, micronutrients, microbial elicitors	Early root development, stress tolerance, higher seedling vigor	Seed priming & nurseries
Intercropping and polyculture	Balanced N:P formulations, microbial enhancers	Improved microbial synergy, balanced nutrient cycling	Intercropping & polyculture
Agroforestry and afforestation	Mycorrhizal-enriched fertilizers	Increased root colonization, drought resistance, nutrient mining	Agroforestry & afforestation
Climate-resilient farming	Nitric oxide donors, silicon-based fertilizers, ROS scavengers	Enhanced stress tolerance, antioxidant enzyme activity, reduced oxidative damage	Climate-resilient farming

3.9 Challenges in Developing Bioactive Fertilizers

Even though bioactive fertilizers pose an epistemic revolution toward sustainable agricultural practices, they are challenged by numerous challenges including scientific, technical, financial, and regulatory aspects in the development and commercialization of bioactive fertilizers. The issue is that the technical challenge here involves the formulation of mixing many functional components, including biostimulants, minerals, signaling compounds, and beneficial microbes, into an effective, single, and consistent mix. Most of these components tend to have different physicochemical characteristics, sensitivities, and modes of action, leading to antagonistic interactions or degradation during formulation, storage, or use in the field. As an example, microbial inoculants must have the right conditions to remain viable; biochemical agents such as peptides and phytohormones can be broken down when subjected to light, heat, or changes in pH (Kumari et al., 2022). Formulating carrier systems or encapsulation matrices capable of protecting and also delivering these agents together without compromising their independent activity is indeed a big formulation challenge.

Besides the complexity of formulation, another obstacle that may severely impair the formulation is the uneven performance of bioactive fertilizers in various environmental systems. Although the products may demonstrate positive outcomes in terms of their performance in greenhouse tests or in an experimental field, the efficacy of these products may plunge to zero upon their application in an actual agricultural environment. Soil heterogeneity, indigenous microbial population, climatic conditions, irrigation management, and crop genotypes all contribute to the variability in the bioavailability, persistence, and activity of the bioactive constituent. An example is that microbial inoculants might not establish well or be functional in the presence of a pre-existing microbiome that outcompetes the inoculants or in the presence of abiotic stress such as drought or soil salinity that prevents activities of the inoculants. This environment-based dependency complicates generalizing or predicting results, thus limiting farmer confidence and the adoption.

Additionally, the fact that there is little mechanistic knowledge on how some biostimulants and signaling molecules work at the molecular level is also a huge bottleneck. Although their functions in the improvement of nutrient uptake, root growth, or stress tolerance are supported by empirical data, the plant schematics of signal transduction, gene expression regulations, and intermicrobial interactions still remain understudied in many cases. In the absence of the clear picture of the mechanisms, the design of the precise and crop-specific formulations turns out to be more speculative than scientific. Also, the lack of predictive models or decision-support systems hampers the possibility of suggesting what combinations or at what timing to apply to specific agroecological zones, causing either inefficient performance in the field or nonoptimal performance. Provision of bioactive fertilizers involves using special infrastructure on the manufacturing side to provide stability and ensure activity of delicate nutritional and hormonal compounds during operations, packaging, and storage. Use of sterile environments, regulated fermentation systems, refrigerated transportation, and more advanced encapsulation or coating technology imposes a major complexity and cost factor on the production. It is especially true in the case of products that contain live microbial cultures, for example, nitrogen-fixing bacteria or phosphate-solubilizing fungi, as retaining spore viability and metabolic activity after prolonged storage is a technically demanding process. The shelf life of many of these products is often short, and they are therefore susceptible to the degradation of their strength, therefore limiting the ability to use them in remote or resource-scarce agricultural regions.

Another great challenge is the regulatory environment concerning bioactive fertilizers (Ochieng, 2015). Numerous nations do not have either thoughtful equal definitions or regulation mechanisms for products that do not fit within familiar fertilizer or pesticide definitions. This uncertainty puts a strain on the registration, compliance, and labeling of products, and many times it prevents entry into the market and is expensive to start. Also, when no standard testing procedures or effectiveness standards are put in place, no product comparison is possible nor rapid quality control present. This lack of regulation has also opened the marketplace to cheap or ineffective products, and this weakens farmer confidence and creates a bad image of the industry in general. The other important obstacle is awareness and low uptake by farmers especially in the areas where they have treated conventional chemical fertilizers as the guiding principle of the last several decades. Some of the growers do not know about the mechanism of action of bioactive fertilizers, the difference between bioactive fertilizers and traditional ones, and the long-term impact they may have on the soil, crop resistance, and environmental sustainability. In addition, they do not necessarily elicit prompt visual effects that synthetic fertilizers generate, and this can make the farmers doubtful since they depend on quick profits to carry on with their lives. This barrier needs to be defeated through rampant field demonstration, focused extension services, and information campaigns in order to inculcate confidence in the users and educate them to practice informed adoption.

A major role is also played by economic factors. Bioactive fertilizers typically have an expensive burden of cost up front but far outweigh the expense of conventional fertilizers in terms of complexity of formulation, storage, and distribution. This cost difference may be a deterrent to adoption in price-sensitive markets, especially the developing countries where the perceived risk of unreliability in performance is also factored in. To most of the smallholder farmers, the initial cost of bioactive fertilizers might not be viable without clear intended, measurable advantages, or subsidized by an incentive scheme or a government program. The economies of scale which may minimize costs are a tricky task and can only be achieved after bulk production or optimization of the supply chain is completed. Although environmental and biosafety concerns are usually less intensive in comparison with those of synthetic agrochemicals, they can still be applied to the use of bioactive fertilizers. The infiltration of the non-native or modified microbial strains in the soil ecosystem might cause destabilization of the native microbial community or have unwanted ecological effects. On the same note, there are still unknown long-term impacts of the repeated application of the signaling molecules or the hormone analogs on the soil fauna, the beneficial arthropods, or the plant–microbe interactions. Conditioned to the guarantee of ecological compatibility and mitigation of unintentional effects, rigorous risk assessment, biosafety examination, and post-release surveillance programs should be carried out in place.

There are also barriers to innovation due to intellectual property issues and restrictions in the field of open-access research. Most bioactive fertilizers are proprietary mixtures of microorganisms, biochemical routes, or delivery agents that are patent-protected. This limits knowledge freedom of sharing, and it hinders cooperation among the campus, startups, and agritech organizations. In addition, in most regions of the world, there is yet to be access to high-throughput screening platforms, omic-based technologies, and precision phenotyping tools to identify and validate new bioactive agents, which remains another snag to speeding up research and product development. Finally, there is a practical challenge in implementing bioactive fertilizers into the current system of agronomic practices, such as precision farming and operations of mechanized farming uses. Some bioactive products have use limitations that include a requirement to be applied at certain times of the year, at certain soil conditions, or in certain ways, and this does not match the existing practice or existing infrastructure. The new formulations may require compatibility optimization or redesigning with other equipment, such as existing irrigation systems, fertilizer spreaders, or seed treatment machines. In the absence of smooth integration, the use of bioactive fertilizer disruption will be seen as labor-intensive or disruptive by the growers.

In short, the possibilities of transformative action with bioactive fertilizers and sustainable and resistant agriculture growth are obvious, complicated by a combination of formulation science, environmental conditions, technological constraints, regulatory ambiguity, economic barriers, acceptance, and adoption of bioactive fertilizers by farmers. Resolving those problems demands strong collaboration between researchers, policymakers, industry stakeholders, and agricultural communities. By means of interdisciplinary myriad, niche innovation, and eased policy structures, these obstacles can be surmounted and the potential of bioactive fertilizers realized to the full in contemporary agricultural systems.

3.10 The Future of Bioactive-Based Fertilizers

The growing need for global food security, environmental sustainability, and climate resilience is driving a shift from traditional agrochemicals to more sustainable, biological solutions. Bioactive fertilizers, which combine biostimulants, targeted nutrients, and signaling molecules, offer innovative approaches to crop nutrition, productivity, and resilience. These fertilizers have the potential to revolutionize agriculture by improving plant–soil–microbe interactions and enhancing plant health at a molecular level.

The development of bioactive fertilizers will be closely linked to advancements in omics technologies, systems biology, and AI. High-throughput tools like metagenomics and transcriptomics will help identify functional genes and microbial diversity, enabling the creation of customized fertilizers. AI and machine learning can further optimize nutrient formulations based on real-time data from sensors and weather patterns, making farming more precise and resource-efficient (Nyagadza et al., 2025).

Nanotechnology plays a significant role in improving the effectiveness of bioactive fertilizers. Smart nanocarriers can control the release of nutrients and other bioactive substances, improving NUE and protecting unstable compounds. Future developments may include multilayered nanocomposites that provide time-controlled growth stimulation throughout the crop's lifecycle.

Synthetic biology offers another path for enhancing fertilizer performance by customizing microbial consortia to perform multiple functions, such as nitrogen fixation or biostimulant production. These genetically modified microbes could also serve as biosensors to monitor environmental conditions and stress levels.

Bioactive fertilizers can also support regenerative farming practices by encouraging soil health, improving carbon storage, and enhancing soil structure. They can be combined with conservation tillage, crop rotation, and cover cropping to reduce erosion and improve soil fertility.

For commercial scalability, innovations in production, such as fermentation bioreactors and encapsulation technologies, will help reduce costs and improve the stability of bioactive fertilizers. However, regulatory and policy frameworks need to adapt to these multifunctional products, ensuring sustainability and safety.

To ensure the successful adoption of bioactive fertilizers, education and support for farmers are essential. Extension networks should embrace digital platforms and participatory research to disseminate knowledge and gather feedback from users.

Overall, bioactive fertilizers hold immense potential to support food security, environmental sustainability, and climate resilience. By leveraging advanced technologies and scientific innovations, these fertilizers will play a critical role in transforming agriculture into a more sustainable and adaptive system.

3.11 Conclusion

Bioactive-based fertilizers mark a significant evolution in agricultural nutrient management, aligning with the global imperative for food security, environmental stewardship, and climate resilience. Unlike conventional fertilizers that primarily act as nutrient suppliers, these advanced formulations integrate biostimulants, targeted nutrients, and signaling molecules to interact synergistically with plant physiological processes, soil microbiota, and environmental cues. This multicomponent approach allows for precise modulation of plant growth, metabolism, and stress responses, ensuring not only higher yields but also improved crop quality, resilience to biotic and abiotic stressors, and long-term soil health. The chapter has highlighted the central role of biostimulants—humic substances, seaweed extracts, amino acids, microbial inoculants, and beneficial fungi—as natural growth regulators capable of enhancing root architecture, nutrient assimilation, and stress tolerance. When paired with optimized nutrient delivery systems, such as nano-formulations, chelated micronutrients, and controlled-release carriers, the efficiency of nutrient uptake increases substantially, reducing losses through leaching, volatilization, or fixation. The incorporation of signaling molecules like SA, JA, NO, and phytohormones further elevates the capacity of crops to respond adaptively to environmental fluctuations, prime defense pathways, and optimize energy allocation for growth and reproduction. Synergistic interactions between these components amplify their individual benefits, enabling a system-level impact on plant health and soil functionality. The addition of microbial consortia, such as PGPR and AMF, enhances nutrient mobilization, disease resistance, and rhizosphere diversity, reinforcing the sustainability of farming systems. This integrative approach supports regenerative agriculture by improving organic matter content, promoting carbon sequestration, and maintaining ecological balance. Bioactive fertilizers are effective across diverse systems—from field crops to hydroponics—boosting nutrient efficiency, crop quality, and postharvest life. Their adoption faces challenges like formulation stability, variable field performance, regulatory gaps, costs, and low awareness. Advancements in formulation, precision tools, and technologies such as omics, AI, nanotech, and synthetic biology, supported by strong policies and quality standards, can unlock their full potential. Bioactive-based fertilizers can become central to climate-smart, regenerative agriculture by uniting plant physiology, soil ecology, and technology. They boost productivity sustainably, replacing input-heavy models with resource-efficient nutrient management that enhances profitability, protects ecosystems, and ensures resilient farming for the future.

References

Aasfar, A., Bargaz, A., Yaakoubi, K. et al. (2021). Nitrogen fixing azotobacter species as potential soil biological enhancers for crop nutrition and yield stability. *Frontiers in Microbiology* 12: 628379.

Adepu, S. and Ramakrishna, S. (2021). Controlled drug delivery systems: Current status and future directions. *Molecules* 26: 5905.

Afzal, M., Ullah, S., Assad, N. et al. (2024). Photocatalytic degradation of organic dyes and biological potentials of biogenic zinc oxide nanoparticles synthesized using the polar extract of cyperus scariosus R. Br.(Cyperaceae). *Green Processing and Synthesis* 13: 20240038.

Ali, B. and Iqbal, J. (2024). *New insights into phytohormones*. Bod–Books On Demand.

Ali, Q., Haider, M.Z., Shahid, S. et al. (2019). Role of amino acids in improving abiotic stress tolerance to plants. In: *Plant tolerance to environmental stress*, 1est edition, 1–488 (ed. M. Hasannuzzaman, M. Fujita, H. Oku, and M.T. Islam), 1–30. Crc Press.

Arasimowicz, M. and Floryszak-Wieczorek, J. (2007). Nitric oxide as a bioactive signalling molecule in plant stress responses. *Plant Science* 172: 876–887.

Areche, F.O., Aguilar, S.V., More López, J. et al. (2023). Recent and historical developments in chelated fertilizers as plant nutritional sources, their usage efficiency, and application methods. *Brazilian Journal of Biology* 83: E271055.

Ashitha, A. and Rakhimol, K. (2021). Fate of the conventional fertilizers in environment. In: *Controlled release fertilizers for sustainable agriculture* (ed. F.B. Lewu, T. Volova, and K.R. Rakhimol), 25–39. Elsevier.

Assad, N., Abbas, A., Ur Rehman, M.F., and Naeem-Ul-Hassan, M. (2024). Photo-catalytic and biological applications of phyto-functionalized zinc oxide nanoparticles synthesized using a polar extract of equisetum diffusum D. *RSC Advances* 14: 22344–22358.

Assad, N., Laila, M.B., Hassan, M.N.U. et al. (2025a). Eco-friendly synthesis of gold nanoparticles using equisetum diffusum D. Don. with broad-spectrum antibacterial, anticancer, antidiabetic, and antioxidant potentials. *Scientific Reports* 15: 19246.

Assad, N., Naeem-Ul-Hassan, M., Ajaz Hussain, M. et al. (2025b). Diffused sunlight assisted green synthesis of silver nanoparticles using cotoneaster nummularia polar extract for antimicrobial and wound healing applications. *Natural Product Research* 39: 2203–2217.

Ayala, S. and Rao, E.P. (2002). Perspectives of soil fertility management with a focus on fertilizer use for crop productivity. *Current Science* 82: 797–807.

Baltazar, M., Correia, S., Guinan, K.J. et al. (2021). Recent advances in the molecular effects of biostimulants in plants: An overview. *Biomolecules* 11: 1096.

Bano, A., Qadri, T.A., and Khan, N. (2023). Bioactive metabolites of plants and microbes and their role in agricultural sustainability and mitigation of plant stress. *South African Journal of Botany* 159: 98–109.

Benalaya, I., Alves, G., Lopes, J., and Silva, L.R. (2024). A review of natural polysaccharides: Sources, characteristics, properties, food, and pharmaceutical applications. *International Journal of Molecular Sciences* 25: 1322.

Bhardwaj, D., Ansari, M.W., Sahoo, R.K., and Tuteja, N. (2014). Biofertilizers function as key player in sustainable agriculture by improving soil fertility, plant tolerance and crop productivity. *Microbial Cell Factories* 13: 66.

Brodt, S., Six, J., Feenstra, G. et al. (2011). Sustainable agriculture. In: *Nat. Educ. Knowl*, 3.

Castiglione, A.M., Mannino, G., Contartese, V. et al. (2021). Microbial biostimulants as response to modern agriculture needs: Composition, role and application of these innovative products. *Plants* 10: 1533.

Chaudhary, P., Singh, S., Chaudhary, A. et al. (2022). Overview of biofertilizers in crop production and stress management for sustainable agriculture. *Frontiers in Plant Science* 13: 930340.

Chen, H., Bullock, D.A. Jr., Alonso, J.M., and Stepanova, A.N. (2021). To fight or to grow: The balancing role of ethylene in plant abiotic stress responses. *Plants* 11: 33.

Chojnacka, K., Moustakas, K., and Witek-Krowiak, A. (2020). Bio-based fertilizers: A practical approach towards circular economy. *Bioresource Technology* 295: 122223.

Du Jardin, P. (2015). Plant biostimulants: Definition, concept, main categories and regulation. *Scientia Horticulturae* 196: 3–14.

Du Jardin, P., Brown, P.H., Dejong, T.M. et al. (2025). *Unlocking the black box of plant biostimulants*. Elsevier.

Easwaran, C., Christopher, S.R., Moorthy, G. et al. (2024). Nano hybrid fertilizers: A review on the state of the art in sustainable agriculture. *Science of the Total Environment* 929: 172533.

Franzoni, G., Cocetta, G., Prinsi, B. et al. (2022). Biostimulants on crops: Their impact under abiotic stress conditions. *Horticulturae* 8: 189.

Ghaffar, S., Abbas, A., Naeem-Ul-Hassan, M. et al. (2023). Improved photocatalytic and antioxidant activity of olive fruit extract-mediated zno nanoparticles. *Antioxidants* 12: 1201.

Hajji-Hedfi, L., Dali, S., Rhouma, A. et al. (2025). Microbial inoculants for plant resilience performance: Roles, prospects and challenges. *Journal of Plant Diseases and Protection* 132: 88.

Hasanuzzaman, M., Parvin, K., Bardhan, K. et al. (2021). Biostimulants for the regulation of reactive oxygen species metabolism in plants under abiotic stress. *Cells* 10: 2537.

Hashmi, S.S., Ibrahim, M., Adnan, M. et al. (2024). Green synthesis of silver nanoparticles from olea europaea L. extracted polysaccharides, characterization, and its assessment as an antimicrobial agent against multiple pathogenic microbes. *Open Chemistry* 22: 20240016.

Heidarzadeh, A. (2025). Role of amino acids in plant growth, development, and stress responses: A comprehensive review. *Discover Plants* 2: 1–31.

Hussain, S., Rao, M.J., Anjum, M.A. et al. (2019). Oxidative stress and antioxidant defense in plants under drought conditions. In: *Plant Abiotic Stress Tolerance* (ed. M. Hasanuzzaman, K. Hakeem, K. Nahar, and H. Alharby). Cham: Springer.

Hussain, S.B., Zahra, S., Hussain, S. et al. (2025). Characterization and comparative antibacterial activities of gold nanoparticles synthesized by tagetes patula L.(Asteraceae) flower extract. *Microscopy Research and Technique* 88: 1869–1880.

Jabbar, A., Abbas, A., Assad, N. et al. (2023). A highly selective Hg 2+ colorimetric sensor and antimicrobial agent based on green synthesized silver nanoparticles using equisetum diffusum extract. *RSC Advances* 13: 28666–28675.

Kazimierczak, R., Średnicka-Tober, D., Barański, M. et al. (2021). The effect of different fertilization regimes on yield, selected nutrients, and bioactive compounds profiles of onion. *Agronomy* 11: 883.

Khan, Z.U.R., Assad, N., Naeem-Ul-Hassan, M. et al. (2023). Aconitum lycoctonum L.(Ranunculaceae) mediated biogenic synthesis of silver nanoparticles as potential antioxidant, anti-inflammatory, antimicrobial and antidiabetic agents. *BMC Chemistry* 17: 128.

Khoso, M.A., Wagan, S., Alam, I. et al. (2024). Impact of plant growth-promoting rhizobacteria (Pgpr) on plant nutrition and root characteristics: Current perspective. *Plant Stress* 11: 100341.

Kleber, M., Bourg, I.C., Coward, E.K. et al. (2021). Dynamic interactions at the mineral–organic matter interface. *Nature Reviews Earth and Environment* 2: 402–421.

Kumari, M., Swarupa, P., Kesari, K.K., and Kumar, A. (2022). Microbial inoculants as plant biostimulants: A review on risk status. *Life* 13: 12.

Kussmann, M., Abe Cunha, D.H., and Berciano, S. (2023). Bioactive Compounds For Human And Planetary Health. *Frontiers in Nutrition* 10: 1193848.

Kyriakoudi, A., Spanidi, E., Mourtzinos, I., and Gardikis, K. (2021). Innovative delivery systems loaded with plant bioactive ingredients: Formulation approaches and applications. *Plants* 10: 1238.

Li, X., Zhu, L., Wang, H. et al. (2025). Peptide hormone-mediated regulation of plant development and environmental adaptability. *Advanced Science* 12: E06590.

Mącik, M., Gryta, A., and Frąc, M. (2020). Biofertilizers in agriculture: An overview on concepts, strategies and effects on soil microorganisms. *Advances in Agronomy* 162: 31–87.

Mackowiak, C., Grossl, P., and Bugbee, B. (2001). Beneficial effects of humic acid on micronutrient availability to wheat. *Soil Science Society of America Journal* 65: 1744–1750.

Madhupriyaa, D., Baskar, M., Sherene Jenita Rajammal, T. et al. (2024). Efficacy of chelated micronutrients in plant nutrition. *Communications in Soil Science and Plant Analysis* 55: 3609–3637.

Mandal, S., Anand, U., López-Bucio, J. et al. (2023). Biostimulants and environmental stress mitigation in crops: A novel and emerging approach for agricultural sustainability under climate change. *Environmental Research* 233: 116357.

Manzoni, C., Kia, D.A., Vandrovcova, J. et al. (2018). Genome, transcriptome and proteome: The rise of omics data and their integration in biomedical sciences. *Briefings in Bioinformatics* 19: 286–302.

Martínez-Lorente, S.E., Martí-Guillén, J.M., Pedreño, M.Á. et al. (2024). Higher plant-derived biostimulants: Mechanisms of action and their role in mitigating plant abiotic stress. *Antioxidants* 13: 318.

Massimi, M., Radócz, L., and Csótó, A. (2023). Impact of organic acids and biological treatments in foliar nutrition on tomato and pepper plants. *Horticulturae* 9: 413.

Nardi, S., Pizzeghello, D., Schiavon, M., and Ertani, A. (2016). Plant biostimulants: Physiological responses induced by protein hydrolyzed-based products and humic substances in plant metabolism. *Scientia Agricola* 73: 18–23.

Nongbet, A., Mishra, A.K., Mohanta, Y.K. et al. (2022). Nanofertilizers: A smart and sustainable attribute to modern agriculture. *Plants* 11: 2587.

Nyagadza, B., Bashar, A., Ligaraba, N. et al. (2025). Artificial intelligence (Ai), machine learning (Ml) and big data analytics' impact on frugal digital marketing firms. In: *Disruptive Frugal Digital Innovation In Africa* (ed. B. Nyagadza, A. Bashar, N. Ligaraba, et al.). Emerald Publishing Limited.

Ochieng, R. (2015). Towards a regulatory framework for increased and sustainable use of bio-fertilizers in Kenya.

Omoarelojie, L.O., Kulkarni, M.G., Finnie, J.F., and Van Staden, J. (2021). Modes of action of biostimulants in plants. In: Gupta, S. and Staden, J. V. (edns) Biostimulants for crops from seed germination to plant development. Elsevier. 445–459. https://doi.org/10.1016/B978-0-12-823048-0.00015-0

Palansooriya, K.N., Dissanayake, P.D., El-Naggar, A. et al. (2025). Biochar-based controlled-release fertilizers for enhancing plant growth and environmental sustainability: A review. *Biology and Fertility of Soils* 61: 701–715.

Pereira, G.C. (2019). Application of biotechnology in producing plant bio-active compounds. In: Akhtar, M., Swamy, M. (eds) Natural Bio-active Compounds. Springer, Singapore. https://doi.org/10.1007/978-981-13-7438-8_3

Pettit, R.E. (2004). Organic matter, humus, humate, humic acid, fulvic acid and humin: Their importance in soil fertility and plant health. *CTI Research* 10: 1–7.

Poveda, J. (2021). Beneficial effects of microbial volatile organic compounds (Mvocs) in plants. *Applied Soil Ecology* 168: 104118.

Priecina, L. and Karklina, D. (2015). Composition of major organic acids in vegetables and spices. *Cbu International Conference Proceedings* 3: 447–454.

Rameshaiah, G., Pallavi, J., and Shabnam, S. (2015). Nano fertilizers and nano sensors–an attempt for developing smart agriculture. *International Journal of Engineering Research and General Science* 3: 314–320.

Rani, D.V., Singh, A., and Jakhwal, R. (2024). Nanobiostimulants for improving plant roots regulations and nutrient uptake. In: Singh, V., Bhat, R.A., Dar, G.H. (eds) Nanobiostimulants. Emerging strategies for agricultural sustainability. Springer, Cham. https://doi.org/10.1007/978-3-031-68138-7_20

Rehana, M., Gladis, R., and Joseph, B. (2022). Controlled release of nutrients for soil productivity—a review. *Current Journal of Applied Science and Technology* 41: 34–46.

Ren, H., Zhang, Y., Zhong, M. et al. (2023). Calcium signaling-mediated transcriptional reprogramming during abiotic stress response in plants. *Theoretical and Applied Genetics* 136: 210.

Rouphael, Y. and Colla, G. (2018). Synergistic biostimulatory action: Designing the next generation of plant biostimulants for sustainable agriculture. *Frontiers in Plant Science* 9: 1655.

Rouphael, Y. and Colla, G. (2020). *Biostimulants in agriculture*. Frontiers Media Sa.

Rush, T.A., Puech-Pagès, V., Bascaules, A. et al. (2020). Lipo-chitooligosaccharides as regulatory signals of fungal growth and development. *Nature Communications* 11: 3897.

Schulman, H. (2014). Intracellular signaling. In: *From molecules to networks. An Introduction to Cellular and Molecular Neuroscience*, Thirde (ed. J.H. Byrne, R. Heidelberger, and M.N. Waxham). Elsevier.

Shanmugavel, D., Rusyn, I., Solorza-Feria, O., and Kamaraj, S.-K. (2023). Sustainable smart fertilizers in agriculture systems: A review on fundamentals to in-field applications. *Science of the Total Environment* 904: 166729.

Sharma, P., Bano, A., Singh, S.P., and Tong, Y.W. (2023). Microbial inoculants: Recent progress in formulations and methods of application. In: *Microbial inoculants, Recent Progress and Applications Developments in Applied Microbiology and Biotechnology* (ed. V.K. Sharma, A. Kumar, and V.K. Singh), 1–28. Elsevier.

Singh, M., Kumar, J., Singh, S. et al. (2015). Roles of osmoprotectants in improving salinity and drought tolerance in plants: A review. *Reviews in Environmental Science and Bio/Technology* 14: 407–426.

Singh, M.R., Singh, D., Kanwar, J., and Chauhan, N.S. (2020). *Advances and avenues in the development of novel carriers for bioactives and biological agents.* Academic Press.

Steward, F.C. (2012). *Inorganic nutrition of plants.* Elsevier.

Sun, W., Shahrajabian, M.H., Kuang, Y., and Wang, N. (2024). Amino acids biostimulants and protein hydrolysates in agricultural sciences. *Plants* 13: 210.

Wajid, M., Uzair, M., Muhammad, G. et al. (2024). Sulfamethoxazole-derived schiff bases: Synthesis, characterization, biological activities, molecular docking, dft, and adme studies. *Journal of Molecular Structure* 1312: 138640.

Wallace, A. (1963). *Role of chelating agents on the availability of nutrients to plants.* Wiley Online Library.

Zulfiqar, F., Moosa, A., Ali, H.M. et al. (2024). Biostimulants: A sufficiently effective tool for sustainable agriculture in the era of climate change? *Plant Physiology and Biochemistry* 211: 108699.

4

The Nano-fertilizer Advantage: Tiny Particles, Big Impact on Agriculture

Nasir Assad[1], Muhammad Naeem-ul-Hassan[1], Muhammad Nauman Khan[2], Sharia Fatima[1], Sher Ullah[1], and Syeda Zainab Batool[1]

[1] *Institute of Chemistry, University of Sargodha, Sargodha, Pakistan*
[2] *Department of Botany, University of Chakwal, Punjab, Pakistan*

4.1 Introduction to Nanotechnology in Agriculture

Agriculture is undergoing a transformative shift, with nanomaterials (NMs) emerging not as replacements for existing practices but as innovative complements that enhance precision, efficiency, and environmental sustainability in agricultural production. NMs, strongly characterized by high surface area-to-volume ratio, solubility, and reactivity, can be used to solve old agrarian problems (Assad et al., 2025a). NMs are updating contemporary agriculture by facilitating a more effective delivery of nutrients, increasing the resilience of crops to environmental stress, and maximizing input utilization (Bhandari et al., 2023).

Nanoparticles (NPs) such as iron oxide (Fe_3O_4) are small enough to penetrate plant tissues and promote seed germination more effectively than conventional supplements. Similarly, silica (SiO_2) NPs in the soil act as nutrient reservoirs, reducing nutrient loss through leaching (Ghaffar et al., 2023; Jabbar et al., 2023). In addition to fertilizers, nanopesticides offer selective, environmentally friendly pest management with a low chemical load and contamination of the natural environment. Most of them use pH-dependent nanocarriers that cannot respond under neutral environments but activate within the alkaline stomachs of pests, so they can maximize their effects and minimize their effects on beneficial insects and pollinators (Jayasoorya & Kumar, 2024).

NMs, seed-priming specifically, have demonstrated great success in the enhancement of stress tolerance, with zinc oxide (ZnO) in particular boosting plant resilience to drought, salinity, and heat. Among all these innovations, the nanosensors can be the most revolutionary. Such devices are capable of monitoring the changes in the soil state in real-time conditions; that is, nutrient concentrations, moist levels, pH, and pathogen presence (El-Sayed et al., 2025). Usually made of high-tech materials, including graphene or silicon, nanosensors enable farmers to make evidence-based decisions, dramatically increase the efficiency of their inputs, and react instantly to environmental hazards. The shift toward precision agriculture is not only result-oriented but also decreases the use of particular resources and the direct effect on the environment.

Nanotechnology in agriculture has long-term implications that have yet to be fully investigated. Uncontrolled or excessive use of NPs has the potential of interfering with the photosynthesis system, causing oxidative stress to plants, and depositing in the soil and water systems. Besides, regulatory systems on agricultural NMs are in their early phases, and as such, safety, environmental protection, and equal availability are in question. A tangible danger exists that such sophisticated technologies will continue to be in the possession of large-scale, well-funded farms, a further technological and economic gap between the rich and powerful farmers and their smallholder counterparts in marginalized nations.

It is in this context that our book chapter dwells on one of the most innovative ways of utilizing agricultural nanotechnology; nano-fertilizers (Pikula et al., 2024). It looks at how the formulation came about, their mode of action, and their benefits relative to traditional fertilizers. Particular focus is put on the physicochemical properties of NMs, including particle size, shape, composition, and surface properties, which make it feasible to regulate nutrient release, have high bioavailability, and promote effective soil–plant interaction (Quintarelli et al., 2024). Figure 4.1 demonstrates these properties, where they establish how NPs behave and perform in agricultural systems. Field case studies about the growth

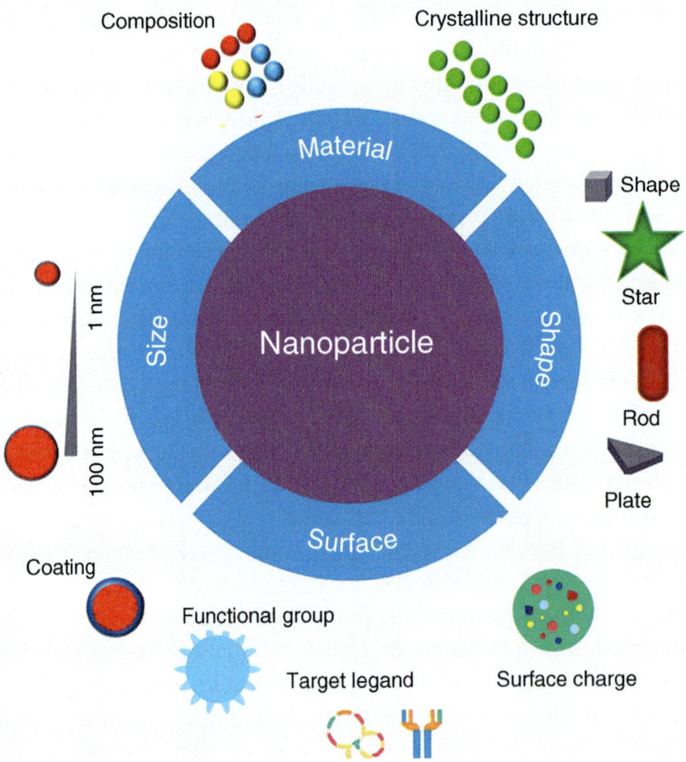

Figure 4.1 Fundamental attributes of NPs, including size, shape, material type, and surface properties, which influence their interactions with soil, plants, and the surrounding environment in agricultural applications.

of crop production, scantiness to stress, and soil wellness are also included in this chapter (Balusamy et al., 2023).

Besides the scientific matters, this chapter covers wider implications, such as regulatory voids, safety issues, and socioeconomic implementation difficulties. We discuss the idea of nano-fertilizers being used in a climate-resilient, environmentally sustainable, and cost-effective way of growing crops (Abbas et al., 2024, Nadporozhskaya et al., 2022). Nano-fertilizers are not described as a panacea, but they are part of an important toolset in the modernized 21st-century farming shed. We are also at the crossroad of nanoscience and agricultural innovation, and the path taken by nano-fertilizers is just taking off. Through intensive research, cautious implementation, and fair policymaking, nanotechnology has the potential of achieving an economically more productive crop, safeguarding the use of modern science tools that reach the farms, both big and small, in the developed and developing worlds alike (Arif et al., 2024).

4.2 Understanding Nanoparticles: Structure and Function

Nanotechnology deals with imagining and making materials at the atomic and molecular levels, which normally falls between 1 and 100 nm. Since materials in this size regime are referred to as NPs, the materials show unique physical, chemical, and biological behavior relative to bulk materials (Assad et al., 2024, 2025d). The above distinctive features largely occur since they have a large surface-area-to-volume ratio, quantum confinement effects, and tunable surface chemistry. The activity related to such nanoscale phenomena contributes to higher solubility, reactivity, and functionality of nanoparticles, which makes them effective tools in agriculture (Khan et al., 2023; Saurabh et al., 2024). This small dimension makes it easier to interact with cells, enabling them to enter the plant tissues via roots or the leaves to precisely transport nutrients or active substances in a high-efficiency method. Exploitation of such properties makes NPs an effective means of maximizing nutritional use of both macro- and micronutrients, resulting in fewer losses of fertilizer and damaged environments.

It is imperative to realize their structural categories to better understand how they can contribute to agriculture. Depending on dimensionality, NPs are broadly classified into four categories, namely zero-dimensional (0D) (such as quantum dots), one-dimensional (1D) (such as nanorods), two-dimensional (2D) (such as nanosheets), and three-dimensional (3D) (such as nanocomposites). Different classes possess different properties to suit different uses in detecting and imaging and catalysis and smart delivering systems (Arora et al., 2024, Assad et al., 2025d). The means of synthesis is also crucial in deciding their applicability and eco-friendliness. Typical strategies incorporate physical top-down and chemical or biological bottom-up techniques (connecting with milling and laser ablation). Among them, biological (green) synthesis has become a trending fact in agricultural practices because it is environmentally friendly and sustainable in nature (Hashmi et al., 2024, Hussain et al., 2025). The fact that it uses plant extracts, bacteria, or fungi is secondary in regard; this applies to plant and microbially made chemicals as well. The structural classifications, synthesis approaches, and key agricultural applications of NPs are summarized in Table 4.1.

Table 4.1 Structural classifications, synthesis methods, and key properties of nanoparticles relevant to agricultural applications, along with safety considerations.

Parameter	Description	Agricultural context/example
Dimensional classification	Nanoparticles are categorized into zero-dimensional (0D), one-dimensional (1D), two-dimensional (2D), and three-dimensional (3D) types based on structure.	—0D (e.g., quantum dots): useful for plant imaging and sensing.—1D (e.g., nanorods, carbon nanotubes): ideal for nutrient/pollutant sensing.—2D (e.g., graphene nanosheets): applied in smart soil sensors, stress monitoring. - 3D (e.g., nanocomposites, clusters): integrated into smart fertilizers for controlled nutrient delivery.
Synthesis approaches	Two principal methods exist: - top-down (physical/ mechanical): breaking larger structures (e.g., laser ablation, milling). - bottom-up (chemical or biological): building from atoms/ molecules via processes like sol–gel, electrochemical, or green synthesis (microbe/plant mediated).	Green synthesis is especially valued in agriculture, offering eco-friendly, low-toxicity nanoparticle production suitable for sustainable farming.
Physicochemical traits	Due to their nanometer-scale size (1–100 nm), nanoparticles have high surface area, surface reactivity, distinctive optical, and catalytic properties.	These characteristics enhance nutrient solubility and uptake, facilitate controlled delivery, and enable use in sensors or coatings for precision agriculture.
Characterization techniques	Standard methods include TEM, SEM, XRD for morphology and structure, and spectroscopic tools (UV–Vis, FTIR, Raman) for chemical and surface analyses.	Accurate characterization helps tailor nanoparticles to crop-specific needs and ensures safety and effectiveness in the field.
Surface functionalization	Surface charge, hydrophilicity/ hydrophobicity, and functional groups can be modified—e.g., with polymers or ligands—to enhance stability, targeting, and biocompatibility.	Functionalized surfaces aid in creating controlled-release fertilizers, nano-pesticides, or growth-regulating agents with precision and minimized off-target effects.
Risk and safety considerations	While offering utility, nanoparticles pose ecological risks—phytotoxicity, uptake to nontarget sites, long-term soil, or biodiversity effects.	Rigorous toxicological and environmental assessments are essential before field deployment to avoid negative impacts.

NP characterization is paramount in the determination of the use of the NPs in agriculture. They can be visualized using techniques including transmission electron microscopy (TEM), scanning electron microscopy (SEM), and X-ray diffraction (XRD) to see their size, morphology, and the crystalline structure. UV–Vis, FTIR, and Raman spectroscopic instruments can elucidate the information about the chemical composition and surface dynamics of these NPs, which is crucial in tailoring NPs to satisfy a given crop or environmental requirement (Elshayb et al., 2024).

Surface chemistry of NP becomes significant, especially since it determines the interaction of NP with soil particles, plant roots, and the microbial community. Products may be charged, hydrophobic, or functionalized with bio-relevant functional groups by using polymers or organic ligand attachment to increase stability, biocompatibility, and targeting. Surface functionalization in this process is essential in developing effective controlled-release fertilizers, nano-pesticides, and plant growth regulators in high precision and effectiveness (Assad et al., 2025c; Sahoo et al., 2024).

Although the field of possible applications of NPs in various branches is large enough, there is a high necessity in considering possible risks when deploying this technology in the agricultural sphere. They are highly reactive and small, which can result in unintended effects with the soil ecosystem. As an example, long-term impacts of NP buildup in soil may change biodiversity in microbes or their nutrient cycling patterns (An et al., 2022). Consequently, toxicological and ecological risk assessments based on extensive studies are obligatory before their wide use in fields to guarantee that their utilization is efficient and not harmful to the environment.

The adherence to the responsible use of the potential is the first approach in comprehending the main structural and functional characteristics of NPs. Through the understanding of the versatility of these nanomaterials in scale, this chapter will, therefore, dwell on how these qualities have been extended in the realization of superior nano-enhanced fertilizers. They are formulated to fulfill the goals of environmental stewardship, food security, and climate-smart agriculture by converting the process of nutrient application to crops (Dilnawaz et al., 2024; Gul et al., 2025).

4.3 Nano-enhanced Fertilizers: Improving Nutrient Delivery

Nano-enhanced fertilizers: They are a technological breakthrough in farming as the agricultural sector faces the most significant pressure of climate change, declining soil fertility, and an increased demand to supply more food with limited resources (Assad et al., 2025b; Pudhuvai et al., 2024). The conventional fertilizers are central to the green revolution, albeit they are subject to very high inefficiencies. They are easily released as nutrients cause a high proportion of losses by leaching, volatilization, and runoff representing not only a waste of money on farmers but also a serious environmental hazard such as greenhouse gases and eutrophication of water bodies. This wastefulness can lead farmers to get into a situation of oversupply with ever-decreasing returns at a high ecological price. The hope of an alternative solution is presented by nano-enhanced fertilizers, which can be seen as a combination of material science and the sustainability-oriented agriculture crop-reducing potential (Thangavelu et al., 2024). Such advanced formulations take advantage of the superior attributes of NM's—including ZnO, SiO_2, and hydroxyapatite—to be used toward encapsulation or binding of nutrients. They have a naturally large surface area and ease of tuning chemistry, as well as the ability to interact at the cellular level, resulting in a significant increase in nutrient use efficiency (NUE) in that a higher percentage of applied nutrients are absorbed by the plant.

In addition to enhancing the simple delivery of nutrients first, nano-fertilizers possess some sophisticated practical capabilities that help in the enhancement of healthier and

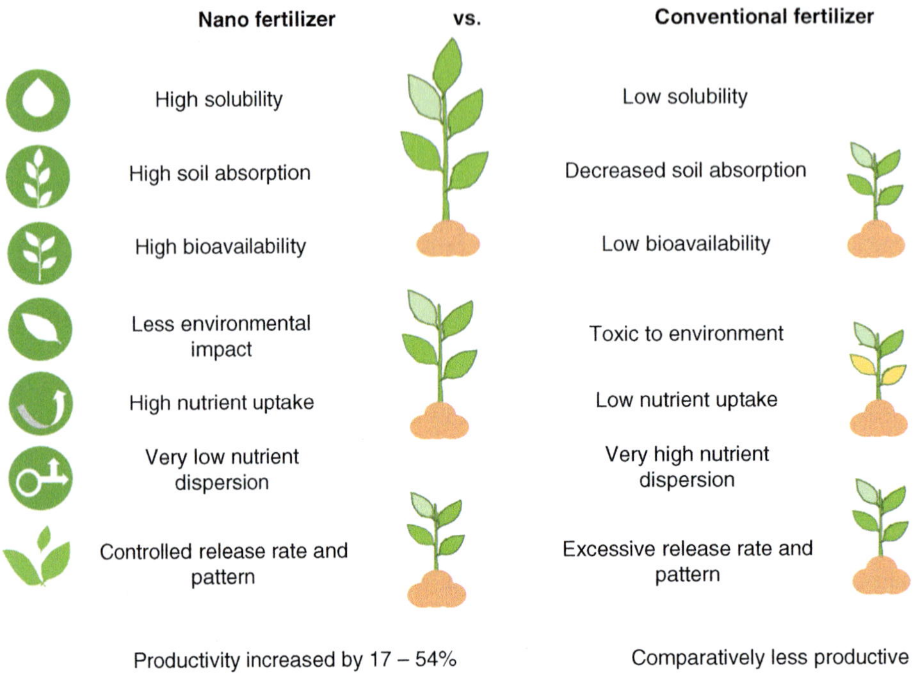

Figure 4.2 Comparative analysis of key properties, efficiency, and environmental impacts of nano-fertilizers versus conventional fertilizers.

productive agricultural ecosystem, as shown in Figure 4.2. One of the main strengths is that they have the constant and controlled release that aligns nutrient availability with the most vital periods of the growth of a plant (Das, 2021).

Benefits of nano-fertilizers do not just end with nutrition of the plants but impact other aspects of the soil–plant system positively. Some of these NPs have the potential to promote positive plant growth-promoting rhizobacteria (PGPRs), producing positive micro-environments or acting as microbial inoculant vehicles, a process that would then increase nutrient recycling and long-term soil vitality. This effect can be observed even at the stage of initial growth when the so-called nano-priming of seeds has been demonstrated to enhance germination rate and root strength. NP entry allows to increase the extent of activity of major hydrolytic enzymes like 6-3 amylase and protease in seed, which triggers more effective release of the stored energy to support the benchmark growing embryo, resulting in stronger and more stress-tolerant seedlings (Ingle et al., 2022; Simonin & Richaume, 2015).

One of the most promising developments is the application of natural-polymer-based biodegradable nanocarriers, for example, chitosan. These carriers have a positive surface charge (cationic nature) at acidic to neutral pH, which further helps electrostatically bind the carriers with negatively charged plant root surfaces, as well as colloids in soil environments, resulting in localized delivery of nutrients into the plant and minimizing losses. Moreover, the chitosan NP also has a twofold role, as it can support a controlled delivery of nutrients in combination with biostimulation capable of producing plant responses to release systemic resistance to plant pathogens through signaling pathway activation

(Dilnawaz et al., 2024). The decomposition of these natural polymers is harmless, leaving the visions of a circular economy and sustainable agriculture that under this type of farming, the goal is not to increase farm productivity by compromising the health of the planet (Ansabayeva et al., 2025; Mwaanga, 2018).

A comparative overview of nano-fertilizers and conventional fertilizers is presented in Table 4.2. Amid local and worldwide attempts to produce enough food to satisfy the demand without using up all the natural resources the world has to offer, the inclusion of nanotechnology, especially in the definition of intelligent fertilizer systems, appears to be an essential way forward. The latter part of this chapter goes into further detail with these advancements, including case studies, experimental findings, and future views regarding the support of a more resilient, sustainable, and more efficient food production system using nano-enabled nutrient delivery system.

Table 4.2 Key parameters and associated benefits of nano-fertilizers in enhancing crop yield, quality, and sustainability.

Parameter	Key benefit	Example/study
Nutrient solubility and bioavailability	Nanoscale size and high surface area increase solubility, reactivity, and uptake efficiency, reducing nutrient loss via leaching and volatilization.	Enhanced uptake efficiency in cereals and horticultural crops.
Controlled nutrient release	Timed, targeted nutrient delivery ensures constant availability during critical growth phases, minimizing wastage and environmental contamination.	Gradual nutrient release matching crop demand.
Yield performance	Increased grain and biomass yields even with reduced conventional fertilizer doses.	Higher maize and rice yields with reduced nitrogen application.
Nutritional quality	Enhanced protein, carbohydrate content, and micronutrient accumulation; improved flavor and shelf life.	Improved fruit quality and vegetable nutrient content.
Stress tolerance	Improved drought, salinity, and heat tolerance via physiological and biochemical modulation.	Better survival and productivity under drought and saline conditions.
Seed germination and vigor	Nano-priming accelerates germination, increases root–shoot length, and improves seedling establishment.	Faster and more uniform seedling emergence.
Soil health and microbiology	Supports beneficial microbes, enhances nutrient cycling, and sustains soil fertility over time.	Increased microbial diversity and improved soil structure.
Environmental footprint	Lower GHG emissions, reduced nutrient runoff, and minimized chemical input requirements.	Reduced nitrogen losses and improved water quality.
Economic efficiency	Lower fertilizer input volumes and energy cost of production; reduced overall cultivation costs.	Reduced fertilizer costs with equal or higher yields.

4.4 Nanotechnology for Environmental Sustainability

Nanotechnology has become a powerful tool in the quest to achieve environmental sustainability by introducing new-fangled ways of resolving looming environmental problems. Water/soil contamination, waste management, and air pollution control are all areas where the manipulation of materials at the nano-level has opened up new opportunities to accomplish the same. The advancements present efficient, cost-effective, and environment-friendly alternatives to the traditional methods, which are in line with the global practices in meeting the objective of achieving sustainable development goals (Karri et al., 2022). Water purification is one of the most vital areas in which nanotechnology comes in handy in environmental sustainability. Various developments such as uncontrolled population, industrial effluents, and agricultural runoffs worsen the access to clean water existing as a huge human concern in the world. NPs, like graphene oxide, carbon nanotubes and titanium dioxide, have been employed in generating membranes and filters that perform better. The examples of graphene-founded membranes have exceptional permeability and mechanical and selective properties, allowing them to clear heavy metals, pathogens, organic waste, and even salts in the course of desalination (Al-Maliki et al., 2022). The membranes appear to be more efficient than the conventional reverse osmosis plants because they not only need less energy but also have higher throughput and hence can be applied on a large scale, both in developed and developing nations.

Besides purifying water, nanotechnology is also central in terms of remedying the soil. Industrialization, mining, and widespread use of agrochemicals cause soil contamination, which accumulates heavy metals and persistent organic pollutants, which brings lot of danger to agricultural production and the health of people (Mukhopadhyay et al., 2022). To address the soil contamination problem, nanoremediation, where engineered NMs like nanoscale zero-valent iron (nZVI) can be deployed, is gaining popularity. These NPs have high surface area and reactivity and therefore are able to degrade the harmful substances, such as polychlorinated biphenyls (PCBs), pesticides, and petroleum hydrocarbons, to less harmful substances. nZVI particles have been found most effective in immobilizing lead, cadmium, arsenic, and other metals under redox reactions and sorption processes so that the metals do not leach into the groundwater (Majewska & Hanaka, 2025).

Nanotechnology has the potential of bringing new contributions in waste management as well. The global growth in solid waste generation, especially due to agricultural and food industry wastage, raises the idea of novel valorization. Nanocatalysts have now become abundantly common, through which it is possible to convert organic waste into luxurious products like biofuels, bio-degradable plastics, and biochar. This can be achieved by the fact that nanocatalysts have higher catalytic activities and surface properties. A conversion of this type is not only potentially helpful in eliminating the increase in the landfills but also capable of contributing to the development of a circular economy. In addition, it is also possible to monitor such parameters of waste processing processes as the degree of gas microbial activity and decomposition rates in real time so that they could manage and bring waste processing processes to perfection (Sridhar et al., 2024). Air pollution is another big menace to both the environment and human beings, and the solution to this problem is being addressed by purification systems that operate with the assistance of nanotechnologies. NMs are materials that are available in size as NPs (e.g., magnesium oxide (MgO), ZnO, cerium oxide (CeO_2)) that have been used as a photocatalyst and as an adsorbent to trap and neutralize airborne pollutants. To convert carbon monoxide, nitrogen oxide, and volatile organic compounds (VOCs) into nonhazardous contents, nanofiber catalysts in MgO are highly efficient to work out as oxidization reactions. Besides, the materials can be incorporated into the building materials, paints, and textiles to provide passive control of pollution in the city. With such inventions like self-cleaning paint or air-purifying curtain, this is a realistic illustration of how nanotechnology can be applied to reduce air pollution within the city and improve the health of the inner-city communities.

The benefits of nano know-how in environmental technology are often outweighed by the possibility of an ecological and toxicological effect of NMs once incorporated into environmental technology. Lacking proper regulation, NPs that are so small and therefore very reactive and have bio-accumulation, cytotoxicity, and/or genotoxicity potentials and hence are released in the unregulated environmental condition. As an example, silver and titanium dioxide NPs seem to work well in antimicrobial and photocatalytic activities; they were reported to be harmful to aquatic organisms and microorganisms in the soil in conditions of uncontrolled exposure. Concerted risk times, lifecycle analyses, and eco-toxicological testing are thus essential prior to large-scale implementation of such materials. To facilitate safe and responsible use, there is a need to develop sound regulatory frameworks, which include the production of these materials, their use, and the disposal of engineered NMs. They should establish norms on how to characterize NMs, label them, set exposure limits, and conduct monitoring in the environment. In addition, sensitization by the society and involvement of interested parties will be essential to enhance acceptance and responsible science innovation about the nature of environmental solutions using nanotechnology. In the future, the development of nanotechnology in environmental sustainability remains hopeful, with new trends now being geared toward green synthesis, biodegradable NMs, and multifunctional composite materials. The paradigm of research is being expected to transform into the production of eco-friendly nanostructures via plant- or microbe-based pathways and incurs minimal environmental mark (ecological footprint) because of common synthesis techniques. Also, the intersection of nanotechnology with artificial intelligence (AI), biotechnology, and renewable energy will only serve to increase its contribution to sustainable development.

Summing it up, nanotechnology has a lot of potential in redefining applications of environmental sustainability in the aspects of water, soil, waste, and air. Several tools are advanced and achieve greater efficiency in the detection, prevention, and remediation of pollution using its application compared to conventional methods. Nevertheless, its potential can be used to its full extent only after a balanced approach is exercised toward it—an approach involving both scientific innovative strides and environmental ethical and regulated policies (Takhar & Singh, 2025). With the ongoing development of the field, an interdisciplinary cooperation between the scientists, policymakers, and industry members should ensure the full potential of nanotechnology in relation to a cleaner and healthier planet.

4.5 The Benefits of Nano-fertilizers in Crop Yield and Quality

The rising global population combined with the urgent need for environmental conservation has propelled agricultural innovation to the forefront of scientific inquiry. In this context, nano-fertilizers—advanced formulations that encapsulate or coat nutrients within nanoscale carriers—have emerged as a breakthrough solution. These fertilizers represent a convergence of materials science and agronomy, designed to enhance nutrient delivery while diminishing environmental harm. Recent studies underscore their significant potential in improving nutrient uptake, maximizing crop productivity, and elevating crop quality, ultimately redefining the paradigm of sustainable agriculture.

The core mechanism of the efficacy of nano-fertilizers is in their nanoscale size and an extremely high surface area-to-volume ratio. The nanostructure enhances the solubility, reactivity, and bioavailability of nutrients to the plant roots, and thus they can be absorbed more readily than common fertilizers. Nano-nutrient compositions allow precise and regulated delivery, matching the nutrient supply with the needs of plant growth during a certain period. This reduces wastage due to both leaching and volatilization, which are typical disadvantages of regular bulk fertilizer, and the potential environmental impact of contaminating waterways and soil ecologies with extra fertilizer. In addition, the control of nutrient release also leads to sustained nutrition through sensitive growth periods, hence maximizing growth.

There is growing evidence that nano-fertilizers have a marked effect in increasing crop yield. As an example, foliar uptake of nano-urea and supplemental recommended nitrogen–phosphorus–potassium (NPK) soil fertilizer boosted maize grain and stover yields by 1000 to over 3000 $1\ ha^{-1}$—an approximately 66% over unfertilized controls—at the same time, it generated comparable yields compared to full application of traditional fertilizers. More importantly, a half dose of nitrogen (75% NPK + nano-urea) was the same yield as full NPK treatments, thus improving nitrogen use efficiency and reducing chemicals without loss of productivity.

In addition to increases in yield, nano-fertilizers enhance crop quality, such as nutrition and stress tolerance. As an example, calcium-based nano-fertilizers promote the quality of fruits and vegetables as well as enhance disease resistance due to increased stability of the cell walls and hormone signaling. Additional research emphasizes how the application of nano-fertilizers can improve protein and carbohydrate production as well as seed germination and increase photosynthesis, all of which are necessary structural elements when it comes to food quality and energy efficiency in plants.

Further, nano-fertilizers improve physiological properties that favor crop productivity. SiO_2 NPs serve as nutrient reservoirs in the soil to store macronutrients, including nitrogen, phosphorus, and potassium, and reduce leaching loss. Fe_3O_4 NP enhances seed germination and root growth compared to conventional iron supplements since the bioavailability of iron builds up, and the roots grow deep due to the bioavailability of iron. Priming of seeds with nano-formulations also proves beneficial in crop establishment. Seeds treated with Fe_3O_4 NPs have higher rates of emergence and good growth of shoots and roots and deposition of iron in the grains, which is economically and agronomically useful in early growth stages. This kind of priming makes seeds resistant to environmental stresses and has an early advantage in the uptake of nutrients.

Nano-fertilizers also affect the health of the soil and microbial ecology. Their slow-release properties minimize loss of nutrients, enhance nutrient reserves within soil in the long term, and host microorganisms that have significant roles in plant nutrition and nutrient cycling. The overall soil fertility due to the improved microbial activity leads to long-term agricultural productivity and climate resiliency.

The important advantage of nano-fertilizers is environmental sustainability. The technologies decrease the volume of fertilizers applied to the land and minimize wastage of fertilizers, thereby reducing greenhouse gas emissions and energy consumption. In the maize trial, nano-urea treatment with 75% of nitrogen produced a considerably lower level of greenhouse gas (GHG) emissions compared with full NPK fertilization, and negligible additional emissions were caused by the nano-urea. Moreover, the synthesis of nano-fertilizer commonly uses significantly smaller amounts of energy compared to conventional fertilizer manufacturing, which enhances net energy efficiency even further.

Examining broader advantages, nano-fertilizers address pressing challenges such as precision nutrient management, environmental degradation, and resource scarcity. By precisely delivering nutrients in sync with crop needs, they minimize environmental runoff and lower input costs for farmers. Their multifunctionality—supplying calcium, zinc, phosphorus, or combined nutrients—offers targeted solutions for diverse cropping systems and agroecologies. For poly-nutrient needs, nano-formulations can enrich tailored delivery schedules aligned with phenological crop stages. Despite their promise, the adoption of nano-fertilizers faces challenges. Concerns remain regarding phytotoxicity and long-term soil or water accumulation of NPs. Some studies note that excessive concentrations can disrupt photosynthesis, generate oxidative stress, or disturb soil microbial balance, warranting cautious evaluation and regulation. Comprehensive risk assessments, packaging standards, and deployment protocols are necessary to balance benefits with ecological safety.

Nevertheless, agricultural agencies and governments are recognizing nano-fertilizer potential. In India, for instance, the National Farmers' cooperative IFFCO introduced nano liquid urea, reporting nearly 50% reduction in standard urea usage, improved yield, and reduced environmental footprint. Similarly, the Andhra Pradesh agriculture department is promoting nano-urea and nano diammonium phosphate (DAP) along with bio-fertilizers to enhance soil health, reduce micronutrient deficiencies, and improve productivity sustainably. These policy initiatives reflect growing institutional support for nano-enabled agriculture. Integrating nano-fertilizers with conventional practices—such as blending nano- with micronutrients or combining with soil conditioning agents—could further refine nutrient efficiency and economic viability. Ongoing research aims to optimize application methods (foliar spray, seed coating, and soil integration) and dosing schedules to ensure uniform efficacy across soil types and crop needs. Furthermore, developing regulation frameworks that balance innovation with safety will be vital for public confidence and equitable access, particularly among smallholder farmers.

Overall, nano-fertilizers offer transformative benefits for crop yield and quality through enhanced nutrient uptake, controlled nutrient release, physiological enhancements, yield enhancement, soil health improvement, and environmental sustainability. Notable successes like improved maize yields, nutrient efficiency, and reduced emissions validate their potential. Yet, responsible deployment—underpinned by comprehensive evaluations, regulation, and equitable distribution—remains crucial. Ultimately, nano-fertilizers stand poised to become an indispensable component of sustainable, high-quality agriculture in the 21st century. The major parameters and benefits of nano-fertilizers for crop yield, quality, and sustainability are summarized in Table 4.2.

4.6 The Future of Nanotechnology in Fertilizer Development

The nanotechnology future in fertilizer development won the limelight of enacting the agriculture facet by improving NUE and decreasing environmental influence as well as environmentally sustainable agricultural methods (Gao et al., 2023). Developments in NMs and nanotechnology provide new possibilities to counter the disadvantages of the traditional approach to fertilizing the soil, the future of more productive and environmentally sustainable agricultural production. Nano-fertilizers are NPs (whose sizes fall in the nanoscale regime and hence possess large surface areas) with unique properties through which controlled and accurate delivery of nutrients to plants is facilitated (Kanjana, 2017). This accuracy would guarantee the supply of the nutrients to plants at its most appropriate time and therefore avoid any wastage of the nutrients and also avoid the chances of nutrient leaching into the environment. It is an example of how nano-urea formulations improve nitrogen utilization in crops that can be seen in their enhanced growth and improved yields with reduced leakages of nitrogen. Moreover, nanotechnology contributes to the slow and steady release of nutrients as they can be encapsulated in NMs. Not only does this slow-release mechanism help in the steady delivery of nutrients to plants, but it also decreases the number of times applications of fertilizers are made, thus saving farmers' labor costs and material costs (Bannon et al., 2022). Also, the fact that you require less frequent use of the fertilizer further reduces the environmental impact of using fertilizers. The application of precision agriculture with nanotechnology is one of the greatest inventions in farming processes (Nadporozhskaya et al., 2022). With nano-sensors and tracking devices, farmers have a chance to get real-time control of the soil quality, plant health, and environmental parameters (Pandey et al., 2025). Using this data-based method, decisions can be made accordingly to know the best method of applying fertilizer and overall farm management (Pooja & Anandan, n.d.).

Also, instead of manufactured NMs, it is possible to use biogenic NMs, which are naturally derived. These nano-fertilizers become biogenic and therefore have greater biodegradability and are less toxic compared to the other common fertilizers, hence, they are safer for the environment and nontarget organisms (Ingle et al., 2022). The use of such green NMs is in line with the increased focus on using environmentally friendly agricultural techniques, as well as sustainable practices in protecting the environment. Even though nanotechnology has great potential in fertilizer research, there are a number of barriers. NPs' toxicity and environmental effects are still running under investigation. The interaction of NPs with plant systems and their behavior in soils is an understanding, which is essential to provide safe utilization of NPs. Also, scalability and economic viability of nano-fertilizer manufacture require attention in order to provide farmers around the globe with access to these technologies.

To sum it up, the future of nanotechnology as an aspect of fertilizer development is set to revolutionize the agriculture sector toward efficient, sustainable, and eco-friendly practices. Further research and development plus concerted efforts among the scientists, policymakers, and stakeholders in the industry are necessary to eliminate the current problems and achieve the promises of nanotechnology in helping improve food security at global levels and achieve sustainable agriculture (Khan et al., 2021).

4.7 Conclusion: Harnessing the Nano Advantage for Sustainable Agriculture

Nanotechnology, particularly through nano-enhanced fertilizers, is reshaping the future of sustainable agriculture by addressing the long-standing inefficiencies of conventional fertilizers. With superior solubility, controlled and sustained nutrient release, higher bioavailability, and markedly improved NUE, nano-fertilizers not only increase crop yield and quality but also reduce nutrient losses through leaching, volatilization, and runoff, thereby minimizing environmental pollution. Their ability to interact at the cellular level allows for precise nutrient delivery, supporting plant growth at critical developmental stages while enhancing soil health and fostering beneficial microbial activity. Moreover, biodegradable nanocarriers, such as chitosan-based systems, offer eco-friendly solutions aligned with the principles of climate-smart agriculture. Despite these benefits, potential risks remain, including NP accumulation in soil and water systems, disruption of microbial diversity, and possible toxicity to nontarget organisms, all of which necessitate thorough toxicological assessments, ecological risk evaluations, and well-defined regulatory frameworks. Cost-effectiveness and scalability also pose challenges, particularly for smallholder farmers in developing regions, highlighting the need for affordable production methods and region-specific field trials. Ultimately, nano-enhanced fertilizers should be integrated into holistic nutrient management systems rather than replacing conventional approaches entirely. When developed responsibly, governed by science-based policies, and tailored to local agronomic and environmental conditions, they have the potential to significantly contribute to climate-resilient, resource-efficient, and environmentally sound agriculture—supporting the urgent global goal of producing more food with fewer resources while safeguarding ecosystems for future generations.

References

Abbas, R., Luo, J., Qi, X. et al. (2024). Silver nanoparticles: Synthesis, structure, properties and applications. *Nanomaterials* 14: 1425.

Al-Maliki, R.M., Alsalhy, Q.F., Al-Jubouri, S. et al. (2022). Classification of nanomaterials and the effect of graphene oxide (GO) and recently developed nanoparticles on the ultrafiltration membrane and their applications: A review. *Membranes* 12: 1043.

An, C., Sun, C., Li, N. et al. (2022). Nanomaterials and nanotechnology for the delivery of agrochemicals: Strategies towards sustainable agriculture. *Journal of Nanbiotechnology* 20: 11. https://doi.org/10.1186/s12951-021-01214-7.

Ansabayeva, A., Makhambetov, M., Rebouh, N.Y. et al. (2025). Plant growth-promoting microbes for resilient farming systems: Mitigating environmental stressors and boosting crops productivity—a review. *Horticulturae* 11: 260.

Arif, M., Raza, H., and Akhter, T. (2024). Classification, synthesis, characterization, and applications of metal nanoparticle-containing hybrid microgels: A comprehensive review. *RSC Advances* 14: 24604–24630.

Arora, P.K., Tripathi, S., Omar, R.A. et al. (2024). Next-generation fertilizers: The impact of bionanofertilizers on sustainable agriculture. *Microbial Cell Factories* 23: 254. https://doi.org/10.1186/s12934-024-02528-5.

Assad, N., Abbas, A., ur Rehman, M.F., and Naeem-ul-Hassan, M. (2024). Photo-catalytic and biological applications of phyto-functionalized zinc oxide nanoparticles synthesized using a polar extract of Equisetum diffusum D. *RSC Advances* 14: 22344–22358.

Assad, N., Ahmad, H., Abbas, A. et al. (2025a). Green synthesis of selenium nanoparticles using equisetum diffusum: Characterization, antibacterial potential, effects on plant growth. *IEEE Transactions on NanoBioscience* 24 (4): 443–453.

Assad, N., Ansa, Laila, M.B. et al. (2025b). Nanotechnology in oilseed crops: A new frontier for abiotic stress adaptation. In: *Oilseed crops under abiotic stress: Mitigation strategies and future perspectives* (ed. A. .A. .H. .A. Latef), 507–536. Springer.

Assad, N., Laila, M.B., Hassan, M.N.U. et al. (2025c). Eco-friendly synthesis of gold nanoparticles using Equisetum diffusum D. Don. with broad-spectrum antibacterial, anticancer, antidiabetic, and antioxidant potentials. *Scientific Reports* 15: 19246.

Assad, N., Naeem-ul-Hassan, M., Ajaz Hussain, M. et al. (2025d). Diffused sunlight assisted green synthesis of silver nanoparticles using *Cotoneaster nummularia* polar extract for antimicrobial and wound healing applications. *Natural Product Research* 39: 2203–2217. https://doi.org/10.1080/14786419.2023.2295936.

Balusamy, S.R., Joshi, A.S., Perumalsamy, H. et al. (2023). Advancing sustainable agriculture: A critical review of smart and eco-friendly nanomaterial applications. *Journal of Nanbiotechnology* 21: 372. https://doi.org/10.1186/s12951-023-02135-3.

Bannon, C., Rapp, I., and Bertrand, E.M. (2022). Community interaction co-limitation: Nutrient limitation in a marine microbial community context. *Frontiers in Microbiology* 13: 846890.

Bhandari, G., Dhasmana, A., Chaudhary, P. et al. (2023). A perspective review on green nanotechnology in agro-ecosystems: Opportunities for sustainable agricultural practices & environmental remediation. *Agriculture* 13: 668.

Das, B. (2021). Effect of foliar applied nano fertilizers of N, P, K & Zn on growth and yield of wheat (*Triticum aestivum* L.) (PhD Thesis). Punjab Agricultural University Ludhiana.

Dilnawaz, F., Acharya, S., and Kanungo, A. (2024). A clinical perspective of chitosan nanoparticles for infectious disease management. *Polymer Bulletin* 81: 1071–1095. https://doi.org/10.1007/s00289-023-04755-z.

El-Sayed, M., Ayoub, H.A., Helal, M. et al. (2025). Nanotechnology-enabled soil management for sustainable agriculture: Interactions, challenges, and prospective. *Environmental Science: Nano* 12: 2128.

Elshayb, O.M., Ghazy, H.A., Wissa, M.T. et al. (2024). Chitosan-based NPK nanostructure for reducing synthetic NPK fertilizers and improving rice productivity and nutritional indices. *Frontiers in Sustainable Food Systems* 8: 1464021.

Gao, Y., Dong, C., Chen, S. et al. (2023). Effect of nano carbon and nano calcium carbonate application on soil nutrient dynamics in winter wheat (*Triticum aestivum* L.). *Communications in Soil Science and Plant Analysis* 54: 2800–2812. https://doi.org/10.1080/00103624.2023.2241502.

Ghaffar, S., Abbas, A., Naeem-ul-Hassan, M. et al. (2023). Improved photocatalytic and antioxidant activity of olive fruit extract-mediated ZnO nanoparticles. *Antioxidants* 12: 1201.

Gula, H., Khan, H. A., Nasreen, Z., Assad, N., Turab, S. A., Hanif, M. (2025). Exploring the antibacterial, anticoagulant, and hemolytic potential of green-synthesized $Fe_2 O_3$ nanoparticles by Cucurbita pepo pulp. *IEEE Transactions on NanoBioscience*. vol. 24, no. 4, pp. 411–420

Hashmi, S.S., Ibrahim, M., Adnan, M. et al. (2024). Green synthesis of silver nanoparticles from *Olea europaea* L. extracted polysaccharides, characterization, and its assessment as an antimicrobial agent against multiple pathogenic microbes. *Open Chemistry* 22: 20240016. https://doi.org/10.1515/chem-2024-0016.

Hussain, S.B., Zahra, S., Hussain, S. et al. (2025). Characterization and comparative antibacterial activities of gold nanoparticles synthesized by *Tagetes patula* L. (Asteraceae) flower extract. *Microscopy Research and Technique* 88: 1869–1880. https://doi.org/10.1002/jemt.24829.

Ingle, P.U., Shende, S.S., Shingote, P.R. et al. (2022). Chitosan nanoparticles (ChNPs): A versatile growth promoter in modern agricultural production. *Heliyon* 8: e11893.

Jabbar, A., Abbas, A., Assad, N. et al. (2023). A highly selective Hg^{2+} colorimetric sensor and antimicrobial agent based on green synthesized silver nanoparticles using Equisetum diffusum extract. *RSC Advances* 13: 28666–28675.

Jayasoorya, R. and Kumar, P. (2024). Utilization of biodegradable carrier-based nano herbicide formulations for sustainable weed management in agriculture. *Frontiers in Agronomy* 6: 1497041.

Kanjana, D. (2017). Advancement of nanotechnology applications on plant nutrients management and soil improvement. In: *Nanotechnology* (ed. R. Prasad, V. Kumar, and M. Kumar), 209–234. Singapore: Springer Singapore https://doi.org/10.1007/978-981-10-4678-0_12.

Karri, R.R., Koduru, J.R., Mubarak, N.M., and Bandala, E.R. (2022). *Sustainable nanotechnology for environmental remediation*. Elsevier.

Khan, N., Ray, R.L., Sargani, G.R. et al. (2021). Current progress and future prospects of agriculture technology: Gateway to sustainable agriculture. *Sustainability* 13: 4883.

Khan, Z.U.R., Assad, N., Naeem-ul-Hassan, M. et al. (2023). Aconitum lycoctonum L. (Ranunculaceae) mediated biogenic synthesis of silver nanoparticles as potential antioxidant, anti-inflammatory, antimicrobial and antidiabetic agents. *BMC Chemistry* 17: 128. https://doi.org/10.1186/s13065-023-01047-5.

Majewska, M. and Hanaka, A. (2025). Biochar in the bioremediation of metal-contaminated soils. *Agronomy* 15: 273.

Mukhopadhyay, R., Sarkar, B., Khan, E. et al. (2022). Nanomaterials for sustainable remediation of chemical contaminants in water and soil. *Critical Reviews in Environmental Science and Technology* 52: 2611–2660. https://doi.org/10.1080/10643389.2021.1886891.

Mwaanga, P. (2018). Risks, uncertainties, and ethics of nanotechnology in agriculture. *New Visions in Plant Science* 22: 225.

Nadporozhskaya, M., Kovsh, N., Paolesse, R., and Lvova, L. (2022). Recent advances in chemical sensors for soil analysis: A review. *Chemosensors* 10: 35.

Pandey, U., Vashisth, A., Mishra, A., Kumari, M. (2025). AI and IoT-driven soil health restoration: A machine learning approach for sustainable agriculture.

Pikula, K., Johari, S.A., Santos-Oliveira, R., and Golokhvast, K. (2024). Joint toxicity and interaction of carbon-based nanomaterials with co-existing pollutants in aquatic environments: A review. *International Journal of Molecular Sciences* 25: 11798.

Pooja, K., & Anandan, P. (n.d.). Precision agronomy: Leveraging technology for enhanced crop management.

Pudhuvai, B., Koul, B., Das, R., and Shah, M.P. (2024). Nano-fertilizers (NFs) for resurgence in nutrient use efficiency (NUE): A sustainable agricultural strategy. *Current Pollution Reports* 11: 1. https://doi.org/10.1007/s40726-024-00331-9.

Quintarelli, V., Ben Hassine, M., Radicetti, E. et al. (2024). Advances in nanotechnology for sustainable agriculture: A review of climate change mitigation. *Sustainability* 16: 9280.

Sahoo, B.R., Dash, A.K., Mohapatra, K.K. et al. (2024). Strategic management of nano-fertilizers for sustainable rice yield, grain quality, and soil health. *Frontiers in Environmental Science* 12: 1420505.

Saurabh, K., Prakash, V., Dubey, A.K. et al. (2024). Enhancing sustainability in agriculture with nanofertilizers. *Discover Applied Sciences* 6: 559. https://doi.org/10.1007/s42452-024-06267-5.

Simonin, M. and Richaume, A. (2015). Impact of engineered nanoparticles on the activity, abundance, and diversity of soil microbial communities: A review. *Environmental Science and Pollution Research* 22: 13710–13723. https://doi.org/10.1007/s11356-015-4171-x.

Sridhar, A., Sunil, C.K., Sarkar, R., and Sarojini, S. (2024). Greener assembly of Nano catalysts and sustainable applications of magnetically retrievable and plasmonic nano catalysts. *Topics in Catalysis* 67: 265–279. https://doi.org/10.1007/s11244-023-01885-6.

Takhar, V. and Singh, S. (2025). Nanomaterials ROS: A comprehensive review for environmental applications. *Environmental Science: Nano* 12: 2516–2550.

Thangavelu, R.M., da Silva, W.L., Zuverza-Mena, N. et al. (2024). Nano-sized metal oxide fertilizers for sustainable agriculture: Balancing benefits, risks, and risk management strategies. *Nanoscale* 16: 19998–20026.

5

Advancing Fertilizer Technology: Enhancing Nutrient Use Efficiency (NUE) for Sustainable Crop Production

Sameel Hassan[1], Waseem Ahmed Khattak[2], Muhammad Anas[3], Javed Iqbal[4], and Banzeer Ahsan Abbasi[5]

[1] *Department of Agriculture & Forest Science (DAFNE), University of Tuscia, Viterbo, Italy*
[2] *Department of Plant Sciences, Faculty of Biological Sciences, Quaid-i-Azam University, Islmabad, Pakistan*
[3] *Department of Soil Science & Plant Nutrition, Selçuk University, Konya, Türkiye*
[4] *Department of Botany, Bacha Khan University, Charsadda, Pakistan*
[5] *Department of Botany, Rawalpindi Women University, Rawalpindi, Pakistan*

5.1 Introduction

The future of global agriculture is at a crossroads, needing to feed an increasing population while working within the restrictions of finite natural resources and more challenging climate conditions (Anas et al., 2025). The inefficient utilization of traditional fertilizers has been a major problem. Despite favoring crop yield for centuries and enriching soil health, the conventional usage of fertilizers has brought extensive ecological and economic harm (Nakachew et al., 2024). Nitrogen (N), phosphorus (P), and potassium (K), or the NPK fertilizers, are typically applied, either in excess or at the wrong times for crop nutrient demand. Consequently, nutrient use efficiency (NUE) is unacceptably low for global cropping systems, with nitrogen use efficiency <50% in much of the world. In addition to the high cost of production, this inefficiency has also led to serious environmental issues such as the emission of nitrous oxide (N_2O), algal bloom, and lack of biodiversity in agroecosystems (Anas & Quraishi, 2025; Penuelas et al., 2023).

The solution that is the next trend is next-generation fertilizers, which are about more than just macro-nutrient supply, aiming to provide precision, responsiveness, and sustainability. These fertilizers are designed to mirror the physiological phenology of nutrient absorption by the crop and limit nutrient dissipation to the environment. Controlled-release fertilizers (CRFs), such as those coated with a special material or that contain inhibitors, delay the release of nutrients so they are available during a plant's key growth stages (Munir et al., 2024; Sachan et al., 2023). Likewise, nano-fertilizers, by virtue of their high surface area and reactivity, make nutrients more bioavailable and can be applied in lower doses without compromising productivity. Such novel fertilizers are a turning point in the concept of plant nutrition in the fact that they take into account the temporal and spatial synchronization of the nutrient for a sustainable agricultural practice (Panpatte et al., 2016; Raza et al., 2025).

Besides chemical innovations, biological and organic matter amendments are more and more recognized as a promising approach to enhance NUE and soil functioning. Beneficial microorganisms (*Rhizobium*, *Azospirillum*, and phosphate-solubilizing bacterium) containing biofertilizers enhance the nutrient availability, N_2 fixation, and root architecture. Organic applications, including compost, vermicompost, and plant-based biochar, lead to increased soil organic matter and microbial activity, and subsequently to increased nutrient retention and a resilient response to nutrient leaching. The complementary use of these bioagents with modern chemical fertilizers is a deeper and fuller solution that caters to productivity and sustainability goals (Abbasi & Yousra, 2012).

Additionally, precision agriculture technologies have advanced nutrient management by permitting farmers to apply fertilizers according to information collected on the fly from the crop and soil. GPS-based equipment, remote sensing imagery, and in situ local soil sensors can be used to apply nutrients as necessary in a specific location, thus reducing wastage. Currently, there is an increasing use of decision support systems that combine weather predictions, crop growth models, and nutrient balancing calculations to assist with the scheduling of fertilizer. When combined with advanced fertilizers, precision agriculture provides a very efficient nutrient delivery system, allowing crops to excel with minimal inputs (SAIRAM et al., 2023).

Despite their potential application, several limitations make it difficult for next-generation fertilizers to be applied widely. High production cost, farmers' ignorance, ineffective extension system, and low accessibility in the rural markets are the major constraints faced. Furthermore, there are still many countries where regulation regarding newer fertilizer products, such as nano-based and microbial fertilizer formulations, is not well established (Anas et al., 2024). It will take concerted action of researchers, policymakers, private sector, and agricultural communities to fill these knowledge gaps. Investment in education, demonstrative testing, and the redesign of subsidies can accelerate the introduction of more sustainable fertilizer use. In this dynamic context, the development of fertilizer technology is not only a scientific task but also the key to resolving the food problems of the world and its environmental protection (Sachan et al., 2023).

5.2 Nutrient Use Efficiency: Concepts and Global Relevance

NUE is the efficiency with which crops take up, transform, and use the applied nutrients for biomass accumulation and yield formation. It is a complex phenomenon that includes agronomical, physiological, and environmental factors. Agronomic NUE is commonly defined as the increase in crop yield per unit of nutrient applied, whereas physiological NUE measures the efficiency with which utilized nutrients are converted into economic yield. High NUE means that the crops can efficiently utilize the nutrients applied to attain higher yields with lower input costs and reduced environmental losses (Ebbisa, 2022). Urgency for improving NUE worldwide is evident in the unsustainable patterns of fertilization. For example, whereas in parts of South Asia and Sub-Saharan Africa, the fertilizer application rates have almost doubled in the recent two decades, the corresponding yield gain has not been that significant. This increasing discrepancy between the amount of

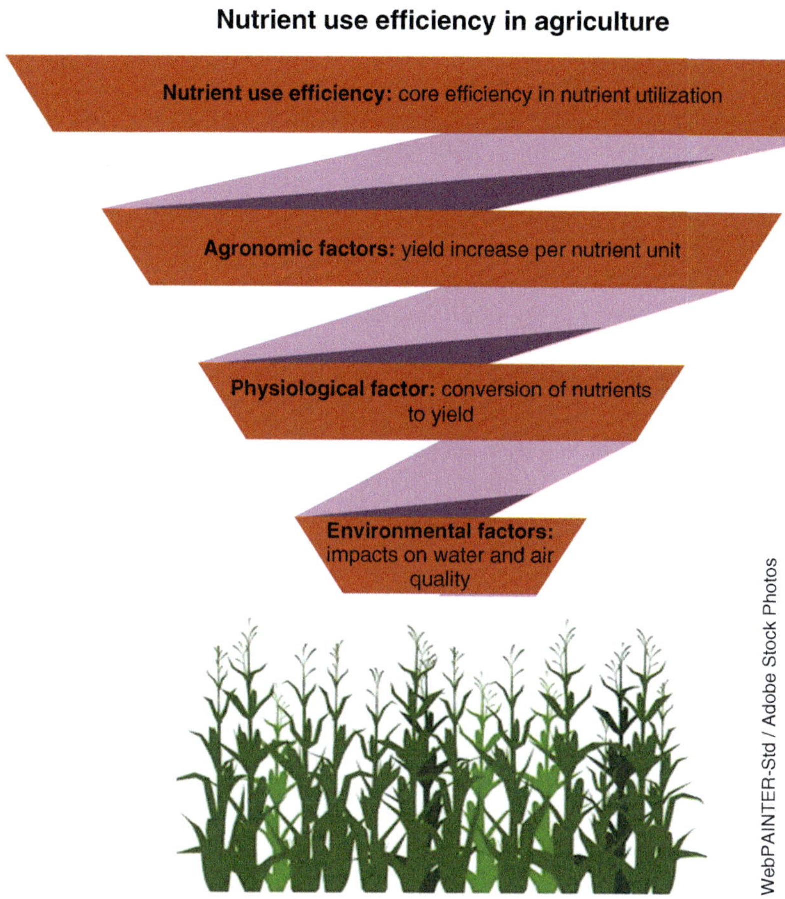

Figure 5.1 Nutrient utilization pathways in agriculture, highlighting physiological conversion to yield and environmental factor interactions.

input used and productivity indicates inefficiencies in nutrient management and the use of blanket fertilizer recommendations instead of nutrient applications. Furthermore, low NUE leads to nutrient loss, which has an adverse impact on water and air quality. Nitrogen loss to groundwater leaching damages human health, with ammonia volatilization and N_2O emission increasing atmospheric pollution and the greenhouse effect (Figure 5.1) (Yadav et al., 2023).

Enhancing NUE is not only a technical necessity but also an economic and environmental necessity. On the only metric that matters—feeding people, it has been a roaring success. In countries where smallholder farmers are the backbone of food systems, fertilizer is typically the single largest input cost. Increasing NUE is producing more from less—getting the most from your investment while preserving natural resources. From a policy point of view, high NUE is in line with several sustainable development goals (SDGs), those linked to zero hunger (SDG 2), clean water (SDG 6), climate action (SDG 13), and life on land (SDG 15). Therefore, approaches to enhance NUE will have to be incorporated into national agricultural development plans and promoted through incentives for sustainable input use (Shah & Wu, 2019). The term NUE also has to be related to a given cropping system, type of soil, and climatic conditions. Soil pH, organic matter, microbial diversity, and water are among the numerous factors that directly affect how efficiently plants can uptake nutrients. Crop-specific distribution of root architecture, growth rate, and nutrient demand adds even more complexity to the equation. Consequently, to enhance NUE, holistic measures, which involve technological advancements in fertilizer materials and agronomic management/practices such as crop rotation, conservation tillage, and intercropping, are necessary (Oyetunji et al., 2022). In the end, the improvement of NUE is the basis of sustainable agriculture. That way, it is a middle road between productivity ambitions and environmental stewardship, the chance to produce more food with fewer things coming from outside. As the world population approaches 10 billion by 2050, it will be crucial to improve NUE to meet food security needs and reduce environmental burdens. In the following sections of this chapter, we discuss how the use of new fertilizer practices and precision management tools is advancing the goal (Islam, 2025). Details are provided in Table 5.1.

5.3 Limitations of Conventional Fertilizers

Traditional fertilizers—consisting of artificial nitrogen, phosphorus, and potassium—have been instrumental in increasing agricultural productivity worldwide. However, due to the excessive and indiscriminate use, a number of agronomic, economic, and socioeconomic problems hampered the cultivation of these pesticides. Another disadvantage is that they have a low efficiency of nutrient recovery. In most agricultural ecosystems, <50% of applied nitrogen, 30% of phosphorus, and 40% of potassium are utilized efficiently by crops. The rest is lost through leaching, surface runoff, volatilization, or chemical fixation in soil, causing not only economic waste but also environmental problems (Ahmed et al., 2019). Nitrogen-containing fertilizers like urea and ammonium nitrate are highly prone to loss through volatilization and denitrification, especially during hot conditions, and at the wrong timing, or under waterlogged conditions. Not only is the nitrogen that is lost no longer available for crops but the losses are also causing greenhouse gas (GHG) emissions, in particular N_2O, one of the leading contributors to climate change (Chien et al., 2009). Also, phosphorus fertilizers are frequently blocked when reacting with the calcium, iron, or aluminum, or the like, in the soil and cannot be absorbed by plants, and the cost of the application thereof must be repeatedly incurred. Such repeated inputs will only further contribute to the accumulation in soils of nutrients not being used that could instead eventually be mobilized and moved into freshwater systems, which would in turn induce eutrophication and a degradation of freshwater biodiversity (Ashitha & Rakhimol, 2021).

In terms of sustainability, the excessive use of conventional fertilizer also resulted in the depletion of soil health. Uninterrupted synthetic fertilizer application without the addition of organic matter reduces soil microbial activities, affects the recycling of nutrients, and induces soil acidification and compaction. The damage disrupts root development and the

Table 5.1 Key dimensions of nutrient use efficiency (NUE), highlighting their functional definitions, relevance in agriculture, and role in enhancing both productivity and environmental stewardship.

NUE aspect	Key description	Relevance to agriculture	Citation
Agronomic NUE	Yield increase per unit of nutrient applied	Indicates the effectiveness of fertilizer use	Salim and Raza (2020)
Physiological NUE	Biomass or yield per unit of nutrient absorbed	Reflects the plant's internal nutrient utilization	Ebbisa (2022)
Nitrogen use efficiency	Efficient uptake and assimilation of nitrogen	Reduces leaching, volatilization, and GHG emissions	Yadav et al. (2017)
Phosphorus use efficiency	Solubilization and uptake of P in soil	Prevents fixation, improves root development	Shenoy and Kalagudi (2005)
Potassium use efficiency	Uptake and transport efficiency of potassium	Enhances water regulation, enzyme activation	Wang and Wu (2015)
Soil-based factors	pH, organic matter, texture, microbial diversity	Influence nutrient retention and availability	Pandao et al. (2024)
Crop-specific demand	Root structure, growth stage, and genetic traits	Determines nutrient timing and form requirements	Gregory (1994)
Environmental influence	Temperature, rainfall, and soil moisture	Affects nutrient mobility and plant access	Cavagnaro (2016)
Economic efficiency	Cost of fertilizer per unit yield gain	Important for profit margins, especially in LMICs	Teklu and Hailemariam (2009)
Ecological efficiency	Reduced nutrient loss to the environment	Essential for sustainable intensification	Pretty (2018)
NUE and food security	Producing more with fewer inputs	Supports productivity with environmental safety	Fan et al. (2012)
NUE and SDGs	Links to SDG 2, 6, 13, 15	Aligns with climate, hunger, and water goals	Gu et al. (2023)

ability of soils to store water, rendering crops more vulnerable to drought and disease. Second, uniform application of traditional fertilizers does not take into consideration the spatial variability of the soil fertility, often resulting in under- or overapplication in varying parts of the field (Singh et al., 2024). In terms of socioeconomic effects, the overreliance on the use of inorganic fertilizers has imposed a heavy cost burden on the farmers, especially in developing countries, as fertilizer expenses make up a huge part of the total cost of cultivation. Price shocks in international fertilizer markets, related to energy prices and supply chain disturbances, also exacerbate the cost and accessibility of the inputs. Moreover, the environmental externalities (e.g., contaminated surface and ground waters, degraded air quality, and eroding ecosystem services) associated with excessive use of fertilizers impose an unaccounted cost on society that is not integrated into typical agricultural practices (Rasul & Thapa, 2004). Considering these constraints, it seems obvious that traditional fertilizers will not meet the sustainability standards of the agriculture of the future. Overcoming their shortcomings will require a transition to integrated nutrient management techniques that incorporate improved fertilization technologies, precision application, and sustainable soil management. The following section details the new developments in fertilizer science that offer a solution to these challenges and facilitate increased NUE in different agro-climatic regions (Gruhn et al., 2000).

5.4 Innovations in Fertilizer Technology

The drawbacks of traditional fertilizers have led to extensive developments of nutrient delivery systems to enhance synchronization between the release of nutrients and the plant uptake. This transformation from bulk nutrient application to precision nutrient delivery has the potential to improve both NUE and environmental losses, and it is key to the sustainability of agriculture for the future. This section covers the most important classes of advanced fertilizers that are shaping the future of nutrient delivery to crops (Sohail, 2024).

5.4.1 Slow- and Controlled-Release Fertilizers

Slow-release fertilizer or CRFs are designed to release nutrients slowly so that they coincide with the crop nutrient uptake rate. These fertilizers are usually surrounded by a surface of sulfur, polymer, or resin to form diffusion barriers. The coatings respond to soil temperature, moisture, and microbial activity to control release rates, so that nutrients are made available to the plant over a digestible period of time without overwhelming it. CRFs can effectively reduce NO_3^- and other nutrient loss through leaching and volatilization and efficiently improve NUE and fertilization frequency (ME Trenkel, 2021). Their efficiency is well shown in intensive systems like rice paddies or on maize and wheat. CRFs reduce the labor costs and the potential environmental hazards of nutrient runoff. Technology also extends the scope for integrated nutrient management through combining CRFs with organic substrates or microbial inoculants. Although more expensive than regular urea, the potential for enhanced yield, less frequent application, and more environmentally friendly practices, in terms of reduced N leaching loss, makes them an attractive tool for precision farming, especially when combined with site-specific application practices (Figure 5.2) (Sohail, 2024).

Figure 5.2 Controlled release versus regular urea fertilizers: contrasting nutrient release patterns, environmental impacts, application frequency, and cost structures for sustainable agriculture.

5.4.2 Nano-fertilizers

Nano-fertilizers are a recent strategy for providing nutrients, which employ nanoparticles to improve the solubility, bioavailability, and immobilization of nutrients. On account of their small size at the nanometer scale (1–100 nm), such materials can penetrate plant tissues well, and they are absorbed at the cellular level. For instance, the nanoparticles of some nutrients (zinc, iron, and nitrogen) can exhibit higher reactivity and uptake efficiencies, even at low dosages. For instance, nano-zinc improves photosynthesis and enzyme activity, and nano-iron increases chlorophyll content and root growth (Iqbal et al., 2019). The traits of nano-fertilizers enable us to use them at less volume, to avoid a decrease in the nutrients when using heavy-rain fawns, and to reach a more stable response of the plants. They can be applied alone or in combination with fertilizer and are ideal for use in both soil and foliar applications in a variety of cropping systems. They also play nicely with sensor-based precision agriculture systems. Nevertheless, their environmental behavior, their soil long-term impacts, and regulation deserve further research and standardization. Notwithstanding these problems, nano-fertilizers have shown high potential to change the crop nutrition scenario, especially in resource-poor and stress-prone habitats (Singh, 2025).

5.4.3 Bioformulated and Coated Fertilizers

Bioformulated fertilizers incorporate living microorganisms such as nitrogen-fixing bacteria (*Azotobacter*, *Rhizobium*), phosphate-solubilizing bacteria (*Bacillus*, *Pseudomonas*), and potassium-mobilizing species. These microbes enhance nutrient availability by catalyzing biological processes in the rhizosphere. For instance, phosphate-solubilizing microbes release organic acids that convert insoluble phosphates into forms absorbable by roots. This biological nutrient mobilization complements mineral fertilizers, improving uptake efficiency and reducing dependency on synthetic sources (Nagpal et al., 2021). Coated fertilizers enriched with humic substances, fulvic acids, or organic extracts stimulate root activity and support beneficial microbial populations. These coatings improve soil structure, enhance microbial colonization, and buffer nutrient release. When used in combination with controlled-release or nano-fertilizers, bioformulated products offer synergistic effects on yield and soil health. Though adoption is limited by shelf-life concerns and temperature sensitivity, continued innovations in formulation and delivery are expanding their application in sustainable farming systems (Khan et al., 2023). Details are provided in Table 5.2.

5.5 Role of Precision Agriculture in Enhancing NUE

Precision farming is a revolution in modern agriculture that uses digital techniques and data analysis to ensure the efficient use of agricultural inputs, such as fertilizers. At its center is to make sure that the right nutrients are applied at the right time and at the right place, with this all geared to the precise requirements of each individual crop and field zone. This specific management contributes to increased NUE by avoiding overapplication, decreasing imbalances in the application to the field as well as the environmental losses,

Table 5.2 Overview of various next-generation fertilizer types, their key advantages, and recommended crop-specific applications aimed at improving nutrient use efficiency (NUE) and sustainability in modern agriculture.

Fertilizer type	Key advantage	Application method/crop	Citation
Controlled-release fertilizer	Timed release reduces nutrient loss	Soil application—rice, wheat	Qiong et al. (2021)
Nano-fertilizer (zinc)	Better uptake at lower doses	Foliar spray—cereals, legumes	Nandy et al. (2025)
Nano-fertilizer (iron)	Boosts chlorophyll and root growth	Foliar spray—vegetables, wheat	Faridvand et al. (2021)
Urease inhibitor-coated urea	Minimizes ammonia volatilization	Soil—maize, oilseeds	Khariri (2016)
Nitrification inhibitor urea	Limits nitrate leaching	Soil—rice, corn	Cui et al. (2022)
Sulfur-coated urea	Gradual nitrogen availability	Basal application—wheat, maize	Mustafa et al. (2022)
Polymer-coated NPK	Customized release for crop stages	Fertigation—high-value crops	Chandra et al. (2019)
Bioformulated N carrier	Enhances biological nitrogen fixation	Seed coating—legumes, cereals	Khan et al. (2023)
Humic acid-coated fertilizer	Improves nutrient uptake and soil health	Soil dressing—pulses, vegetables	GAWDE (2024)
Chelated micronutrient blend	Avoids nutrient fixation	Foliar/drip—fruits, vegetables	Madhupriyaa et al. (2024)
Smart fertilizer	Responds to environmental signals	Precision application—varied crops	Karthik and Maheswari (2021)
Compost-based nano-fertilizer	Combines organics with nano-efficiency	Soil incorporation—vegetables	Dawar et al. (n.d.)

and improving crop response (Gorai et al., 2021). Site-specific nutrient management (SSNM) is one of the basic features of precision agriculture. This practice is based on spatial variability information, gathered from soil sampling, yield maps, and satellite or drone imagery, to guide fertilizer application (Pratyusha et al., 2023). Rather than recommendations for a blanket nutrient rate, SSNM permits tailor-made nutrient rates by field section, which consider differences in soil fertility, texture, moisture content, and history of P and K management. Variable rate technology (VRT) systems enable farmers to apply fertilizers at variable rates in real time by defining the dosage with GPS-based spreaders that facilitate reducing input needs and increasing economic returns (Patel et al., 2023).

Remote sensing systems, such as multispectral and hyperspectral imaging, augment nutrient monitoring by identifying the first symptoms of nutrient deficiencies by the way they affect crop reflectance patterns. From these inputs, fertilization maps are developed using vegetation indices like normalized difference vegetation index (NDVI). Combined with weather forecasting and crop model software, such tools offer predictive analytics for improved nutrient scheduling, which is especially useful in rainfed or climate-stressed areas (Ajith et al., 2025). Soil nutrient-monitoring systems using the Internet of Things (IoT) and soil nutrient sensors are also an integral part of precision farming systems, gaining popularity in modern agriculture implementation. Such sensors can measure and track soil pH, soil moisture, and nutrient levels in situ so that fertilizer treatments can be fine-tuned while the crop is growing. For instance, nitrate sensors installed in the root zone can induce the fertigation apparatus to apply nitrogen only when soil level inceptors drop below a designated level; that is, when the soil levels reach this designated level, nitrogen application is wasteful and no longer synchronized with the plant uptake (Senapaty et al., 2023). Furthermore, precision agriculture supports documentation and traceability for documenting input use efficiency over time and adopting more informed decision-making for subsequent seasons by farmers. Decision Support System (DSS), featuring advanced technologies like AI and ML, analyzes vast amounts of data to suggest nutrient application options that are economical as well as environment-friendly. Needless to say, these systems are indispensable in large installations where it would be impossible to monitor the system manually (Tantalaki et al., 2019). Although its development and adoption have been increasing, particularly in developed countries, it is not applicable to smallholder systems due to the high cost, lack of digital infrastructure, and the absence of technical knowledge. Addressing this gap will involve investment in capacity building, government policy initiatives, and the production of low-cost, user-friendly precision instruments. However, the next generation of fertilizers combined with precision agriculture has the potential to revolutionize nutrient management and support the development of a more efficient, productive, and sustainable world food system (Da Silveira et al., 2023).

5.6 Microbial Interventions and Organic Nutrient Sources

Biological inputs have been paid more attention as a sustainable substitute for a high proportion of high-tech chemical fertilizers to improve NUE. Microbial inoculants and organic amendments do not function only to deliver useful nutrients to plants but they also establish favorable conditions for plants, enhance biological activity in soil, and increase cycling of nutrients (Saha & Bauddh, 2020). These natural inputs are particularly important in low-input or depleted systems in which synthetic fertilizers are not enough or are ecologically expensive. Their incorporation into the conventional nutrient management schemes provides a sustainable and environment-friendly approach toward enhancing crop productivity and soil health (Pahalvi et al., 2021).

Biofertilizers, including either living N_2-fixing (*Rhizobium*, *Azotobacter*, and *Azospirillum*) or P-solubilizing (*Bacillus*, *Pseudomonas*), and K-mobilizing (*Frateuria aurantia*) microorganisms, contribute to nutrient mobilization in the rhizosphere. These microorganisms form symbiotic associations with plant roots, which lead to improvement in nutrient availability because the bacteria and fungi transform non-plant-available nutrients into plant-available forms via biological processes. Phosphate-solubilizing bacteria secrete organic acids that dissolve insoluble phosphates, and nitrogen-fixing microbes fix atmospheric nitrogen into ammonia, which the plant takes up. Their extant proliferation not only provides crop nourishment but also decreases reliance on mineral fertilizer (Kumar et al., 2018).

Arbuscular mycorrhizal fungi (AMF) establish symbiotic relationships with plant roots and substantially increase the absorbing surface by the hyphae networks. Such fungi promote the absorption of less mobile nutrients like phosphorus, zinc, and copper. Additionally, AMF enhances root architecture, water use efficiency, and biotic and abiotic stress resistance. Their symbiosis is frequently associated with an increase in NUE and the nutrient-uptake efficiency, particularly with nutrient-poor soils. Synergistic effects of AMF and slow-release or nano-fertilizer on plant growth and yield have been obtained in different cropping systems (Tang et al., 2022).

Organic sources of nutrients like compost, farmyard manure, vermicompost, and green manure are commonly used for both nutrient supply and improving soil physicochemical properties. These are organic materials that contribute to higher levels of soil organic matter, which increases cation-exchange capacity (CEC), nutrient retention, and microbial activity. Nutrients are slowly released into the soil for a constant feeding of your plants, like long-term fertilizers. Furthermore, biochar (biochar is made from plant biomass via pyrolysis) has been reported to increase NUE due to improved soil aeration, moisture retention, and nutrient sorption (Rahman et al., 2020).

The combination of natural organic inputs and microbial inoculants with second-generation fertilizers results in a comprehensive nutrient management system. This integrated nutrient management (INM) system is based on the combination of the two systems of biological and integrated nutrient management, a holistic approach to more balanced and long-term nutrition. It also contributes to the recovery of soil fertility of intensively cultivated soils that were fertilized for years with chemicals, leading to a loss of soil microbial diversity and organic carbon content. These systems are more robust to fluctuations in environmental variables, which allows them to be productive with reduced input levels (Wu & Ma, 2015). But to be successful, biological interventions must consider soil ecology and crop-specific compatibility of the microbials and recyclers within biofertilizer products. Microbial activity and efficacy may be affected by environmental factors, including temperature, moisture, and pH. Accordingly, research and extension should develop location-oriented bioformulations, validate their field worthiness, and train farmers in application for effective use. And as ecological farming comes of age, microbial and organic nutrient solutions are projected to emerge as one of the key pillars of future-fit farming, alongside sophisticated fertilizer technologies as well as broader global imperatives like food security and protecting the environment (Sarangthem et al., 2023).

5.7 Environmental and Economic Implications

The transition to advanced fertilizers and enhanced NUE has substantial environmental and economic consequences. From an environmental perspective, the introduction of sophisticated fertilizers can significantly reduce the undesirable externalities that are caused by the application of conventional fertilizers. Excessive application of N and P fertilizers has been one of the main causes for NPS pollution by contaminating groundwater, accelerating eutrophication of freshwater bodies, and also elevating the GHG emission in the environment. Synchronized nutrient release with plant demand using slow-release/CRF, nano-fertilizers, and bio-formulations reduces such losses of materials to reduce their environmental footprint (Chien et al., 2009). By lowering nutrient runoff and leaching to bodies of water, we directly contribute to the rehabilitation of water systems, where excess nutrients can lead to harmful algal growth and low oxygen levels, impacting biodiversity and fisheries. In the same way, nitrogen volatilization reduction also reduces the emission of ammonia and nitrous oxide gases, which pollute the environment and act as GHGs. Increased NUE also results in less finite phosphate resource mining and less fossil fuel consumption in fertilizer manufacturing, to be in line with the world's demand for decreasing the environmental impacts of agricultural inputs (Aulakh & Malhi, 2005). From an economic perspective, enhanced NUE means improved input-use efficiency, which means that farmers can apply less fertilizer without significant yield reductions. This efficiency advantage results in lower production costs and wider profit margins, a factor of particular importance for small-hold farmers with little to no excess capital. The increased reliance on next-generation fertilizers in such regions can provide more stable and cost-effective practices to manage nutrients when fertilizer products are commonly affected by unstable prices due to supply chain interruptions or energy market dynamics. What's more, reducing the number of applications lowers labor and transport costs, especially on larger farms (Hossain et al., 2021).

More broadly still, durable nutrient management is pertinent to longer-term agricultural resilience. Productive, healthy soils that have balanced nutrients will give a better nursery over time, which minimizes the need for expensive soil rehabilitation. Governments may also enjoy lower cleanup costs of fertilizer contamination to the environment. As knowledge spreads, there is more potential for public–private partnerships and green investments directed at transferring sustainable fertilizer production, distribution, and education (Solanki et al., 2023). Yet the commercial outlay of next-generation fertilizers tends to be higher than traditional products, hampering adoption by resource-poor farmers. Moreover, without sufficient support from extension services and no compelling financial incentive in a world clouded by uncertain international commodity markets, the risk of moving to new technologies perceived to be unfamiliar might outweigh their benefit. Therefore, policy tools including focused subsidies, credit mechanisms, tax incentives, and certification approaches are the need of the hour to facilitate the advanced and economically feasible fertilizers for all scales of farming (Ahlawat et al., 2024). The environmental and economic value of improved NUE in the context of contemporary fertilizers is clear, but getting the potential completely out of these benefits needs policy commitment, fair access, and sharing of knowledge. As we work to address the twin challenges of climate change and global food security, the world's adoption of these sustainable practices provides a tremendous opportunity to match human ingenuity with human enterprise and ingenuity to develop a more sustainable future (Mgendi, 2024).

5.8 Policy, Regulatory, and Adoption Challenges

Next-generation fertilizers have great scientific and technical potential, but they face multiple interconnected hurdles related to policy, regulations, and implementation that limit their acceptance. The regulations on fertilizers in many countries are still old and are not updated considering the fact of nano-fertilizers, biofertilizers, and polymer-coated fertilizers. This regulatory lag generates an ambiguous situation for manufacturers or importers and hampers the acceptance and commercialization of new products, reducing the options available to farmers (Lam et al., 2022). The lack of harmonized guidelines for the testing of safety, efficacy, and environmental impacts of new fertilizer technologies is a major challenge. As opposed to chemical fertilizers, which can be used in current risk assessment procedures, the new generation of products could entail biological and/or nano-scale interactions, necessitating new risk assessment tools. Especially, the long-term destiny and behavior of NPs in the soil ecosystem remain less clear, causing worries for the ecological aspects. Hence, science-based, transparent regulatory frameworks are essential in accelerating the development and responsible use of fertilizers that balance innovation with environmental and human protection (Figure 5.3) (Smith et al., 2021).

Farm-level acceptance is additionally hindered by a lack of knowledge and an insufficient understanding of the advantages of good quality and proper utilization of advanced fertilizers. Many of the smallholder farmers who make up most farmers in developing countries have no access to trusted extension services or training programs, and demonstration sites are also often not available to demonstrate the practical benefits of these products. Therefore, even if such fertilizers exist, their use is either not

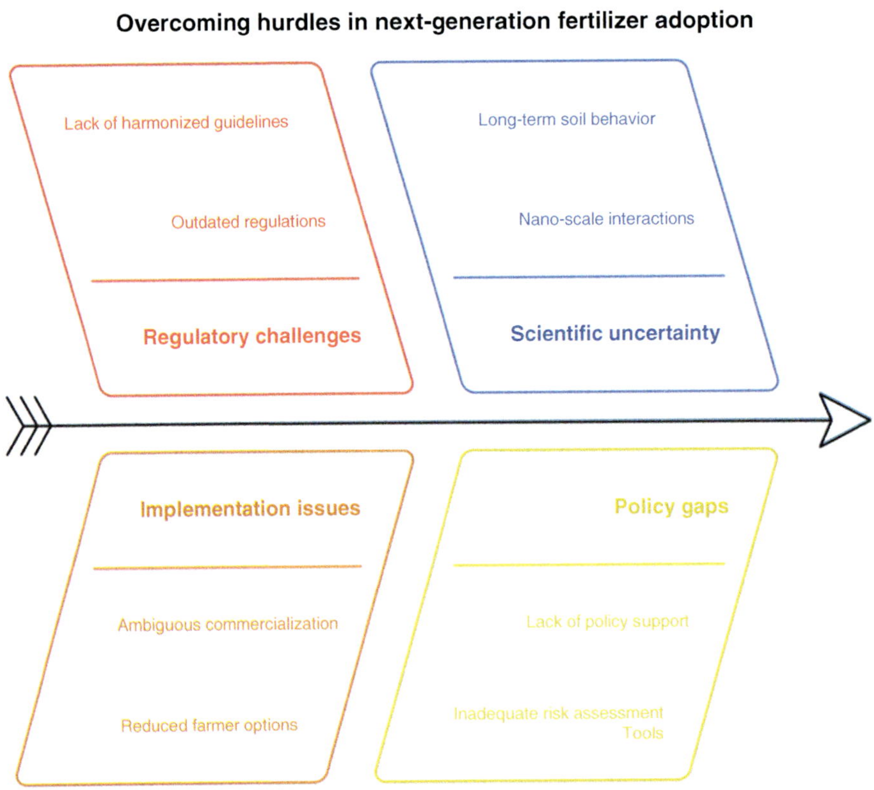

Figure 5.3 Key adoption challenges for advanced fertilizers spanning regulatory, scientific, and commercial domains, highlighting needs for updated policies and better risk assessment tools.

maximized or is employed in the wrong way, thwarting their possible efficacy (Rudnick et al., 2023). There are also economic obstacles. Many newer growers cost more to make because they are made by cumbersome processes and have unique materials. Without the proper subsidies or sharing of the costs, the farmers might not have the incentive to invest in themselves. Furthermore, market fragmentation and poor rural infrastructure and supply chains restrict the timely delivery and availability of these products, aggravating reliance on traditional but low-efficiency chemical fertilizers (Ren et al., 2022).

In order to address these barriers, there is a need for a coherent policy approach. Governments need to invest in context-specific R&D to bring about innovation in fertilizer and continue to support and improve public–private partnerships to increase their production and distribution. There is a need to establish clear regulations regarding registration, labeling, and quality control of the new generation of fertilizers, and international cooperation would be beneficial in ensuring that standards are similar across countries. In addition, incorporating nutrient stewardship concepts into national fertilizer policies, like the 4Rs (right source, right rate, right time, and right place), can provide direction for large-scale sustainable nutrient management (Sukprasert & Phadungkit, 2024). Strong institutional support is also necessary to encourage adoption. This involves training extension officers, field trials to produce site-specific data, and farmer awareness campaigns showcasing the economic and environmental gains of increased NUE in the longer term. Digital technologies and mobile platforms also have a potential role in educating and advising farmers, particularly those in remote or underserved areas. Policymakers and stakeholders can unleash the potential of fertilizer innovations and advance the shift toward a more sustainable and nutrient-efficient agricultural future by tackling these systemic barriers (Tiemann & Douxchamps, 2023).

5.9 Future Perspectives and Research Directions

Emerging fertilization technologies are expected to be dominated by nanotechnology, biotechnology, and intelligent digital technology in the future. Intelligent fertilizers able to react from moment to moment to what is happening in soil and plants (e.g., moisture, pH, and nutrient status) would improve the coordinated release of nutrients, thus reducing loss of nutrients and bettering their recovery. These intelligent formulations, coupled with sensor networks and precision irrigation, can allow automatic and optimal nutrient management at the field level. Multi-nutrient nano-complexes and designed synthetic microbial communities are being developed to cater to the crop-specific requirements and to redress the environmental constraints. One direction of current research lies in utilizing beneficial microbes (endophytes) to endophytically aid in nutrient transport and stress tolerance. Furthermore, the production and use of bio-based crop fertilizers derived from agricultural waste streams may follow a circular economy approach to minimize reliance on finite mineral resources and provide benefits to soil organic matter and microbial activity. Artificial intelligence and machine learning are expected to be key in nutrient management, with systems analyzing large datasets from sensors, satellites, and models. These predictive tools will inform fertilization based on crop phenology, weather, and soil status. Large-scale field tests and environmental risk assessments will be crucial, including those of nano-fertilizers for the safety, scaling, and ecosystem compatibility of new products.

5.10 Conclusion

It is a critical requirement to improve fertilizer technology for increased productivity to sustain the ecological system at the lowest cost. Conventional fertilizers have brought us as far as they are able to go, and next-generation applications provide more efficient, precise, and sustainable alternatives. In combination with digital technologies, product innovations such as nano-fertilizers, microbial inoculants, and controlled-release systems have the potential to transform nutrient management.

These technologies keep input costs down and the environment cleaner while enhancing the quality of crops and soil. Yet, adoption barriers, lack of knowledge/awareness, no regulatory support, and high introduction costs to penetrate the market have to be tackled by integrated policy support, education, and infrastructure investments. It is essential that public–private partnerships and robust extension networks are in place to promote upscaling.

At the end of the day, enhancing the NUE is not only a technological advancement, but it is an agroeconomic need for sustaining a resilient agriculture for the future. With this focus, next-generation fertilizers could play a role in advancing global food systems in the context of population growth, resource constraints, and climate change.

References

Abbasi, M., & Yousra, M. (2012). Synergistic effects of biofertilizer with organic and chemical N sources in improving soil nutrient status and increasing growth and yield of wheat grown under greenhouse conditions. *Plant Biosystems-An International Journal Dealing with all Aspects of Plant Biology*, 146(sup1), 181–189.

Ahlawat, U., Naruka, A., Changdeo, W. B., Rehsawla, R., Sansanwal, R., Mishra, R., & Singh, V. (2024). A review of cutting-edge Biotechnological solutions for next-generation farming. *J Exp Agric Int*, 46(7), 1187–1204.

Ahmed, M., Ahmad, S., Qadir, G., Hayat, R., Shaheen, F. A., & Raza, M. A. (2019). Innovative processes and technologies for nutrient recovery from wastes: A comprehensive review. *Sustainability*, 11(18), 4938.

Ajith, S., Vijayakumar, S., & Elakkiya, N. (2025). Yield prediction, pest and disease diagnosis, soil fertility mapping, precision irrigation scheduling, and food quality assessment using machine learning and deep learning algorithms. *Discover Food*, 5(1), 1–23.

Anas, M. and Quraishi, U.M. (2025). Green-fabricated mno_2 nanoparticles function as dual nanofertilizers and chromium remediators, enhancing antioxidant pathways, ionomic networks, and physiological resilience in wheat. *Journal of Trace Elements in Medicine and Biology* 127661. https://doi.org/10.1016/j.jtemb.2025.127661.

Anas, M. et al. (2024). Manganese oxide nanoparticles elicit cadmium tolerance in wheat seedlings by ionomic and phenomic adjustment, regulation of AsA–GSH pathway, cellular thickness recovery, and antioxidant modulation [10.1039/D4EN00490F]. *Environmental Science: Nano* 11 (11): 4588–4605. https://doi.org/10.1039/D4EN00490F.

Anas, M. et al. (2025). Mitigating heavy metal pollution in agriculture: A multi-omics and nanotechnology approach to safeguard global wheat production. *Journal of Hazardous Materials Advances* 17: 100584. https://doi.org/10.1016/j.hazadv.2024.100584.

Ashitha, A., & Rakhimol, K. (2021). Fate of the conventional fertilizers in environment. In Controlled release fertilizers for sustainable agriculture (pp. 25–39). Elsevier.

Aulakh, M. S., & Malhi, S. S. (2005). Interactions of nitrogen with other nutrients and water: Effect on crop yield and quality, nutrient use efficiency, carbon sequestration, and environmental pollution. *Advances in agronomy*, 86, 341–409.

Cavagnaro, T. R. (2016). Soil moisture legacy effects: impacts on soil nutrients, plants and mycorrhizal responsiveness. *Soil Biology and Biochemistry*, 95, 173–179.

Chandra, M. S., Lal, M., Naresh, R., Yadav, S., Kumar, R., Kumar, R., & Lavanya, N. (2019). Role of polymer coated fertilizers (PCFS) an advance technology for improving nutrient use efficiency and crop productivity: A review. *International Journal of Chemical Studies*, 7(6), 2667–2679.

Chien, S., Prochnow, L., & Cantarella, a. H. (2009). Recent developments of fertilizer production and use to improve nutrient efficiency and minimize environmental impacts. *Advances in agronomy*, 102, 267–322.

Cui, L., Li, D., Wu, Z., Xue, Y., Xiao, F., Gong, P., Zhang, L., Song, Y., Yu, C., & Du, Y. (2022). Effects of combined nitrification inhibitors on soil nitrification, maize yield and nitrogen use efficiency in three agricultural soils. *Plos one*, 17(8), e0272935.

Da Silveira, F., Da Silva, S. L. C., Machado, F. M., Barbedo, J. G. A., & Amaral, F. G. (2023). Farmers' perception of the barriers that hinder the implementation of agriculture 4.0. *Agricultural Systems*, 208, 103656.

Dawar, K., Hussain, A., Roomi, I., Abbas, A., Ali, A., Hussain, A., & Akram, W. Chapter-2 From Scraps to Saplings: Transforming Vegetable Production with Kitchen Waste.

Ebbisa, A. (2022). Nutrient Use Efficiency (NUE): Strategies to Improve Phosphorus. Sustainable Crop Production: Recent Advances, 3.

Fan, M., Shen, J., Yuan, L., Jiang, R., Chen, X., Davies, W. J., & Zhang, F. (2012). Improving crop productivity and resource use efficiency to ensure food security and environmental quality in China. *Journal of experimental botany*, 63(1), 13–24.

Faridvand, S., Amirnia, R., Tajbakhsh, M., El Enshasy, H. A., & Sayyed, R. (2021). The effect of foliar application of magnetic water and nano-fertilizers on phytochemical and yield characteristics of fennel. *Horticulturae*, 7(11), 475.

GAWDE, M. S. H. (2024). NUTRIENT RELEASE PATTERN OF DIFFERENT BRIQUETTES FORTIFIED WITH HUMIC ACID IN BRINJAL (Solanum melongena L.) IN LATERITIC SOILS OF KONKAN DR. BALASAHEB SAWANT KONKAN KRISHI VIDYAPEETH].

Gorai, T., Yadav, P. K., Choudhary, G. L., & Kumar, A. (2021). Site-specific crop nutrient management for precision agriculture—A review. *Curr. J. Appl. Sci. Technol*, 40, 37–52.

Gregory, P. J. (1994). Root growth and activity. *Physiology and determination of crop yield*, 65–93.

Gruhn, P., Goletti, F., & Yudelman, M. (2000). Integrated nutrient management, soil fertility, and sustainable agriculture: current issues and future challenges. *Intl Food Policy Res Inst*.

Gu, B., Chen, B., & Zhang, X. (2023). Managing nitrogen to achieve Sustainable Development Goals in China.

Hossain, A., Bhatt, R., Sarkar, S., Barman, M., Majumder, D., Saha, S., Islam, M. T., Maitra, S., & Meena, R. S. (2021). Cost-effective and eco-friendly agricultural technologies in rice-wheat cropping systems for food and environmental security. Sustainable intensification for agroecosystem services and management, 69–96. Research Square.

Iqbal, M., Umar, S., & Mahmooduzzafar, f. (2019). Nano-fertilization to enhance nutrient use efficiency and productivity of crop plants. *Nanomaterials and plant potential*, 473–505. Springer.

Islam, S. (2025). Agriculture, food security, and sustainability: a review. *Exploration of Foods and Foodomics*, 3, 101082.

Karthik, A., & Maheswari, M. U. (2021). Smart fertilizer strategy for better crop production. *Agricultural Reviews*, 42(1), 12-21.

Khan, A., Singh, A. V., Gautam, S. S., Agarwal, A., Punetha, A., Upadhayay, V. K (2023). Microbial bioformulation: A microbial assisted biostimulating fertilization technique for sustainable agriculture. *Frontiers in Plant Science* 14: 1270039.

Khariri, R. B. A. (2016). Evaluation of selected coated urea on nitrogen use efficiency of rice PhD Thesis, Universiti Putra Malaysia, Serdang, Malaysia].

Kumar, M.S. et al. (2018). Role of bio-fertilizers towards sustainable agricultural development: A review. *Journal of Pharmacognosy and Phytochemistry* 7 (6): 1915–1921.

Lam, S.K. et al. (2022). Next-generation enhanced-efficiency fertilizers for sustained food security. *Nature Food* 3 (8): 575–580.

Madhupriyaa, D. et al. (2024). Efficacy of chelated micronutrients in plant nutrition. *Communications in Soil Science and Plant Analysis* 55 (22): 3609–3637.

ME Trenkel, T. (2021). Slow-and controlled-release and stabilized fertilizers: An option for enhancing nutrient use effiiency in agriculture. International Fertilizer Industry Association (IFA).

Mgendi, G. (2024). Unlocking the potential of precision agriculture for sustainable farming. *Discover Agriculture* 2 (1): 87.

Munir, A. et al. (2024). Advancing agriculture: Harnessing smart nanoparticles for precision fertilization. *BioNanoScience* 14 (4): 3846–3863. https://doi.org/10.1007/s12668-024-01597-5.

Mustafa, A. et al. (2022). Improving crop productivity and nitrogen use efficiency using sulfur and zinc-coated urea: A review. *Frontiers in Plant Science* 13: 942384.

Nagpal, S. et al. (2021). Microbial bioformulations: Revisiting role in sustainable agriculture. In: *Biofertilizers* (ed. A. Rakshit, V.S. Meena, M. Parihar, et al.), 329–346. Elsevier, Elsevier.

Nakachew, K. et al. (2024). Review on enhancing the efficiency of fertilizer utilization: Strategies for optimal nutrient management. *Open Agriculture* 9 (1): 20220356.

Nandy, P. et al. (2025). Impact of foliar spray of zinc in nano form on lentil grown under residual fertility of preceding rice. *International Journal of Plant Production* 19 (1): 15–30.

Oyetunji, O. et al. (2022). A comprehensive review on enhancing nutrient use efficiency and productivity of broadacre (arable) crops with the combined utilization of compost and fertilizers. *Journal of Environmental Management* 317: 115395.

Pahalvi, H. N. et al. (2021). Chemical fertilizers and their impact on soil health. *Microbiota and biofertilizers, Vol 2: Ecofriendly tools for reclamation of degraded soil environs*, 1–20.

Pandao, M.R. et al. (2024). Soil health and nutrient management. *International Journal of Plant and Soil Science* 36 (5): 873–883.

Panpatte, D. G. et al. (2016). Nanoparticles: The next generation technology for sustainable agriculture. *Microbial inoculants in sustainable agricultural productivity: Vol. 2: Functional applications*, 289–300.

Patel, K. et al. (2023). Site specific nutrient management in precision agriculture. *Agriculture and Food* 44090: 262–264.

Penuelas, J. et al. (2023). A better use of fertilizers is needed for global food security and environmental sustainability. *Agriculture & Food Security* 12 (1): 1–9.

Pratyusha, P. et al. (2023). Site specific nutrient management. *Rajat Yadav* 63: 149.

Pretty, J. (2018). Intensification for redesigned and sustainable agricultural systems. *Science* 362 (6417): eaav0294.

Qiong, W. et al. (2021). Effects of different types of slow-and controlled-release fertilizers on rice yield. *Journal of Integrative Agriculture* 20 (6): 1503–1514.

Rahman, G. M., et al. (2020). Biochar and organic amendments for sustainable soil carbon and soil health. *Carbon and nitrogen cycling in soil*, 45–85.

Rasul, G. and Thapa, G.B. (2004). Sustainability of ecological and conventional agricultural systems in Bangladesh: An assessment based on environmental, economic and social perspectives. *Agricultural Systems* 79 (3): 327–351.

Raza, A. et al. (2025). Harnessing metabolomics for enhanced crop drought tolerance. *The Crop Journal* 13 (2): 311–327. https://doi.org/10.1016/j.cj.2025.01.001.

Ren, C. et al. (2022). Socioeconomic barriers of nitrogen management for agricultural and environmental sustainability. *Agriculture, Ecosystems & Environment* 333: 107950.

Rudnick, J. et al. (2023). Understanding barriers to adoption of sustainable nitrogen management practices in California. *Journal of Soil and Water Conservation* 78 (4): 347–363.

Sachan, D.S. et al. (2023). Assessing grain yield and achieving enhanced quality in maize by next generation fertilizer: A review. *International Journal of Environment and Climate Change* 13 (8): 626–637.

Saha, L. and Bauddh, K. (2020). Sustainable agricultural approaches for enhanced crop productivity, better soil health, and improved ecosystem services. In: *Ecological and practical applications for sustainable agriculture* (ed. K. Bauddh, S. Kumar, R.P. Singh, and J. Korstad), 1–23. Springer.

Sairam, M. et al. (2023). Evaluation of precision nutrient tools and nutrient optimization in maize (Zea mays L.) for enhancement of growth, productivity and nutrient use efficiency. *Research on Crops* 24 (4): 666.

Salim, N. and Raza, A. (2020). Nutrient use efficiency (NUE) for sustainable wheat production: A review. *Journal of Plant Nutrition* 43 (2): 297–315.

Sarangthem, I. et al. (2023). The book of abstract: International conference on natural farming for revitalizing environment and resilient agriculture (NF-RERA, 2023). *Pub: College of Agriculture, CAU, Imphal* 374 (3): 498–501.

Senapaty, M.K. et al. (2023). IoT-enabled soil nutrient analysis and crop recommendation model for precision agriculture. *Computers* 12 (3): 61.

Shah, F. and Wu, W. (2019). Soil and crop management strategies to ensure higher crop productivity within sustainable environments. *Sustainability* 11 (5): 1485.

Shenoy, V. and Kalagudi, G. (2005). Enhancing plant phosphorus use efficiency for sustainable cropping. *Biotechnology Advances* 23 (7–8): 501–513.

Singh, S. (2025). Optimizing nutrient uptake: Nano-fertilizers and soil amendments. In: *Food and industry 5.0: Transforming the food system for a sustainable future* (ed. P.K. Dutta, A. Hamad, A.K. Haghi, and P.K. Prabhakar), 229–248. Springer.

Singh, S. et al. (2024). Smart fertilizer technologies: An environmental impact assessment for sustainable agriculture. *Smart Agricultural Technology* 8: 100504.

Smith, V. et al. (2021). New plant breeding technologies: An assessment of the political economy of the regulatory environment and implications for sustainability. *Sustainability* 13 (7): 3687.

Sohail, D. (2024). Advancements in precision fertilization technologies and controlled-dispersion fertilizers for sustainable rice cultivation. *International Journal of Agriculture and Sustainable Development* 6 (1): 29–42.

Solanki, M.K. et al. (2023). Microbial resilience in plant nutrient management towards sustainable farming. *Frontiers in Microbiology* 14: 1280811.

Sukprasert, K. and Phadungkit, W. (2024). Integrating sustainable development frameworks into agricultural policies: A policy analysis perspective. *International Journal of Contemporary Issues in Social Sciences* 11 (2): 1–12.

Tang, H. et al. (2022). The critical role of arbuscular mycorrhizal fungi to improve drought tolerance and nitrogen use efficiency in crops. *Frontiers in Plant Science* 13: 919166.

Tantalaki, N. et al. (2019). Data-driven decision making in precision agriculture: The rise of big data in agricultural systems. *Journal of Agricultural & Food Information* 20 (4): 344–380.

Teklu, E. and Hailemariam, T. (2009). Agronomic and economic efficiency of manure and urea fertilizers use on vertisols in Ethiopian highlands. *Agricultural Sciences in China* 8 (3): 352–360.

Tiemann, T. and Douxchamps, S. (2023). Opportunities and challenges for integrated smallholder farming systems to improve soil nutrient management in Southeast Asia. *World Development Sustainability* 3: 100080.

Wang, Y. and Wu, W.-H. (2015). Genetic approaches for improvement of the crop potassium acquisition and utilization efficiency. *Current Opinion in Plant Biology* 25: 46–52.

Wu, W. and Ma, B. (2015). Integrated nutrient management (INM) for sustaining crop productivity and reducing environmental impact: A review. *Science of the Total Environment* 512: 415–427.

Yadav, M. et al. (2017). Strategies for improving nitrogen use efficiency: A review. *Agricultural Reviews* 38 (1): 29–40.

Yadav, M.R. et al. (2023). Mechanistic understanding of leakage and consequences and recent technological advances in improving nitrogen use efficiency in cereals. *Agronomy* 13 (2): 527.

6

Global Perspectives: The Role of Next Generation Fertilizers in Feeding the World

Faisal Nadeem[1,2], *Sundas Abbas*[3], *Ali Hasnain*[4], *Hina Jabeen*[5], *Tayyaba Samreen*[6], *Rubab Sarfaraz*[7], *Saif Ullah Masood*[4], *Rahib Hussain*[8], *Rashid Mahmood*[1], *Faiza Waseem*[1], *and Wajih ul Hassan*[4]

[1] *Department of Soil Science, Faculty of Agricultural Sciences, University of the Punjab, Lahore, Pakistan*
[2] *Soil and Water Testing Laboratory for Research, Gujranwala, Pakistan*
[3] *Regenerative Climate Smart Agriculture, SAWiE (Sustainable Agriculture Water and Intelligence Ecosystem), Faisalabad, Pakistan*
[4] *Department of Agronomy, Faculty of Agricultural Sciences, University of the Punjab, Lahore, Pakistan*
[5] *Knowledge Center, SAWiE (Sustainable Agriculture Water and Intelligence Ecosystem), Faisalabad, Pakistan*
[6] *Institute of Soil and Environmental Sciences, University of Agriculture, Faisalabad, Pakistan*
[7] *Institute of Agriculture and Life Sciences, Gyeongsang National University, Jinju, South Korea*
[8] *Sustainable Development Study Centre, Government College University Lahore, Lahore, Pakistan*

6.1 Introduction: The Global Challenge of Feeding a Growing Population

A considerable surge from 7.8 to 9.7 billion, reaching up to 11.2 billion by 2050 and 2100, respectively, is expected in the world's population (Randive et al., 2021; Sadigov, 2022). In comparison to the annual average growth rate of around 1.1%, there is only a 2.8% increase in food production per year (Byrnes & Bumb, 2017; Gu et al., 2021). The developing countries with limitations of importing or producing food for a growing population are facing serious threats of food insecurity in the future. Since the 1970s, the lowest food grain reserves were recorded in 1996, and the rate of increase in food production has slowed. It was estimated that the food grain production needed to be increased from about 2 billion tons to over 3 billion tons per year by the year 2020 (Byrnes & Bumb, 2017). Considering the limitations of arable land, the enhancement in food production either requires an enhancement in the efficiency of fertilizers (Randive et al., 2021) or increments in their doses.

Since the green revolution, a continuous increase in the utilization of synthetic fertilizers has been observed to ensure food security on the globe. In 2019, the fertilizer market attained a value of US$155.8 billion with the expectations of registering a compound annual growth rate (CAGR) of 3.8% during 2019–2024 (Williams, 2019). Such scenarios always demand fertilizer production industries to increase the fertilizer plant capacity, considering its practical economics (UNEP, 2016b). In contrast to the reliance of food security on nitrogen fertilizers, it accounts, approximately 5% of global greenhouse gas (GHG) emissions (Gao & Cabrera, 2023). The thrust of ensuring food security, the use of synthetic fertilizers has tripled in the last three decades, which subsequently contributed to agricultural emissions. Given the circumstances and availability of resources, various scientific and technological measures were adopted by the developed countries during the 20th century to limit fertilizer consumption. These adopted innovations had no dramatic impact other than the increased ability of synthetic nitrogen fertilizer production (Byrnes & Bumb, 2017).

Serious environmental concerns are associated with the excessive use of chemical fertilizers, a significant portion of which remains unused and deteriorates soil health. This unused portion of chemical fertilizers accumulates in the environment, affecting important regulators of ecosystems. The widespread incorporation of traditional fertilizers brings associated challenges of low fertilizer efficiency due to inefficient nutrient utilization and environmental risks, which call for the urgency of adopting more sustainable, targeted, and intelligent solutions, such as next-generation fertilizers (NGFs), to achieve sustainable agriculture production.

NGFs refer to innovative and scientifically advanced nutrient formulations that go beyond traditional chemical fertilizers. Their design is thought to reduce environmental harm and improve nutrient use efficiency to support sustainable agricultural production. NGFs may include biofertilizers (contain microbes, e.g., Rhizobia and Azotobacter), nanofertilizers (nanoscale carriers deliver nutrients efficiently at the cellular level), controlled-release fertilizers (CRFs) or slow-release fertilizers (SRFs) (nutrients are released gradually over time to reduce waste), and stabilized fertilizers (inhibit chemical processes, like urea hydrolysis and nitrification, that lead to nutrient loss). Bionanofertilizers, engineered through nano- and biotechnological innovations, are the alternatives to chemical/synthetic fertilizers, promising to improve nutrient uptake without soil health deterioration. Having the nanoscale constituents and beneficial microbes, they offer benefits of seed germination enhancement, soil quality improvement, nutrient use efficiency enhancement, and residual pesticide degradation. The research community has welcomed NGFs as enhanced-efficiency fertilizers (EEFs), having the potential of upsurging nitrogen use efficiency and providing agronomic benefits through plant growth stimulation as indicated by the Association of American Plant Food Control Officials (AAPFCO). They are engineered to release nutrients in synchronization with crop demand, aiming to address food security challenges and environmental constraints. In 2018, USD 2.37 billion was the value of the global CRF market, which by 2026, is predicted to be worth USD 3.862 billion, having a 6.37% CAGR.

Focusing on the low-income countries, NGFs offer a transformative solution to their agricultural challenges. In developing regions of the globe where farmers often have limited access to quality inputs and where soils are degraded, NGFs can be adopted as a cost-effective and environmentally friendly approach to address crop productivity challenges. NGFs also help to conserve natural resources, such as water, along with the minimization of agricultural emissions by minimizing leaching and volatilization losses, aligning with the sustainable agricultural goals. Their ability to enhance yields on marginal lands further supports food security in areas where agricultural expansion is constrained by poor soil health and climate variability.

The successful deployment of NGFs in low-income countries, however, often depends on robust international collaboration. Global partnerships help to disseminate proven scientific technologies, enabling the adaptation of NGFs to local agroecological conditions. International development programs and public-private initiatives may enable low-income nations to access funding, training, and infrastructure necessary to integrate NGFs into their mainstream farming systems. Collaborative research initiatives can support the co-development of region-specific formulations that are both effective and affordable. Moreover, harmonized international regulatory frameworks can help streamline the approval and distribution of innovative fertilizers across borders, reducing delays and enhancing farmer access. Multilateral climate action programs also contribute by embedding NGFs within broader strategies for climate-resilient agriculture, thereby addressing environmental vulnerabilities to secure food production. Ultimately, the synergy between advanced fertilizer technologies and global cooperation holds the key to empowering smallholder farmers, sustaining natural resources, and building resilient food systems in low-income regions.

6.2 Current Fertilizer Use Around the World

Fertilizers are described as chemical formulations consisting of minerals, commonly known as essential plant nutrients (macronutrients and/or micronutrients), which enable plants to complete their life cycle. These fertilizers enrich soils with essential plant nutrients, supporting qualitative as well as quantitative crop production. By 2050, the 9.7 billion of the world's population will require ~70% more food production. Consequently, it becomes critical to effectively use fertilizers for the uplift of agricultural production on available arable lands. In a surge to meet agriculture production goals, the fertilizer demand is forecasted to rise from 200 million tons (by 2024) to 324 million tons (by 2050). It includes 1.2%, 1.7%, and 1.8% annual growth in nitrogen, phosphates, and potash, respectively (Drescher et al., 2011; Quinn, 2020).

6.2.1 Synthetic Fertilizers

The differential geological conditions and intensity of exploration make world reserves vary in estimation. Russia led the production of NPK fertilizer with a production increment of 5.2% in 2019 compared to 2018. The production of ammonia (82% N), the richest source of nitrogen, is expected to rise by a CAGR of 4.4% in terms of revenue during 2025 (Grand View Research, 2019). The occurrence of phosphate rock, a precursor for phosphatic fertilizers (UNEP IFA, 2001; Geissler et al., 2019), is found to be 72% in Western Sahara and Morocco, 5% in China, and 3% in Syria and Algeria (Randive et al., 2021). The reserves of phosphate rock are estimated to be 69 billion tons globally, out of which 240 million tons were mined during 2019 (Pistilli, 2020; Randive et al., 2021). China leads the world in the production of phosphatic fertilizers, followed by the United States of America, India, Russia, and Morocco at present (USGS, 2019; FAO, 2020a, b). On the other hand, the estimation for global potash reserves was 5.8 billion tons in 2018, including 34.3% and 20.6% shares of Russia and Canada, respectively (Randive et al., 2021; USGS, 2020). In terms of total production, 26%, 19%, and

16% are the contributions of Canada, Russia, and Belarus, respectively, as major potash producers in 2019 (FAO, 2020b). The Canadian fertilizer industry contributed $23 billion to the country's economy through an 18.6% increase in potash production during 2017–2018 (Fertilizer Canada, 2018). The fourfold increase in fertilizer use since 1960 has played a pivotal role in driving the Green Revolution.

During 2018–2019, 188 million tons of fertilizer were consumed worldwide (Randive et al., 2021). In this scenario, China experienced an increase of more than eightfold in grain production at the cost of 55 times more consumption of nitrogen fertilizers. The estimates of phosphatic fertilizer in the global market for cropland are forecasted to be 22–27 million tons per year by 2050, having an additional 4–12 million tons per year production for grassland (Mogollon et al., 2018; Bindraban et al., 2020). Singapore, Egypt, China, Netherlands, the United Kingdom, Japan, Germany, India, the United States, Italy, and Canada consumed about 34 708 kg, 675 kg, 519 kg, 270 kg, 248 kg, 228 kg, 198 kg, 174 kg, 143 kg, 125 kg, and 94 kg fertilizers per hectare, respectively, in 2019 (World Bank, 2020). The usage of inorganic fertilizers, both nutrient-wise and fertilizer-wise, for the years 2020–2022 is given in Tables 6.1 and 6.2, respectively, for different regions of the world.

Table 6.1 Inorganic fertilizer use by nutrient (thousand tons).

	Nitrogen		Phosphorus (P_2O_5)		Potassium (K_2O)	
Regions	**2020**	**2022**	**2020**	**2022**	**2020**	**2022**
Global	81 024.7	108 058.2	32 450.2	41 854.5	21 744.6	35 454.8
Africa	2 464.6	4 129.7	927.6	1 654.4	469.6	920.0
Americas	17 433.3	25 784.9	8 152.3	12 482.1	8 247.4	14 081.9
Asia	46 723.3	62 628.2	17 698.1	23 088.3	7 864.5	16 342.6
Europe	13 213.8	13 138.7	4 092.6	3 111.5	4 814.1	3 742.1
Oceania	1 189.7	2 376.8	1 579.5	1 518.3	349.0	368.1

Table 6.2 Inorganic fertilizer use per cropland area by nutrient (kg per hectare).

	Nitrogen (N)		Phosphorus (P_2O_5)		Potassium (K_2O)	
Regions	**2020**	**2022**	**2020**	**2022**	**2020**	**2022**
Global	51.3	65.4	21.0	26.0	13.7	21.7
Africa	10.4	13.3	3.9	5.3	2.0	2.9
Americas	40.4	60.9	21.9	33.7	20.7	35.8
Asia	83.1	108.4	31.5	40.0	14.0	28.3
Europe	38.5	41.6	12.4	10.2	14.3	12.0
Oceania	34.2	52.9	31.5	24.8	7.6	6.5

The maximum limit for the application of phosphorus (P) and nitrogen (N) is proposed to be 6.2 million tons and 62 million tons per year, respectively (Moberg et al., 2015). In contrast, 14 million tons and 150 million tons are recent consumption rates of N and P per year, which are rising continuously, rendering the risk of a destabilized environmental system on the planet Earth (Rockström et al., 2009; Steffen et al., 2015). Asia accounts for over 60% of global nitrogen fertilizer consumption, while Sub-Saharan Africa uses less than 5%, reflecting significant access and economic barriers. The demand for K fertilizers is expected to increase by 3.3% from 2013 (30.0 million tons) to 2014 (31.04 million tons). The regional trends of consumption vary greatly from the highest in South and East Asia to the lowest in Africa (Kotschi, 2015). Nigeria reported about 52% per year of average annual growth in the nitrogen fertilizer production, whereas Guinea consumed about 48.4% nitrogen fertilizer per year (FAO, 2020b).

6.2.2 Next-Generation Fertilizers

The ecosystem and human health have suffered irreversible harm resulting from the considerable loss of fertilizer in agricultural systems. NGFs (Figure 6.1), also known as EEFs, are categorized as nanofertilizers, SRFs, and/or CRFs (Liu & Lal, 2015; Shaviv & Mikkelson, 1993; Trenkel, 1997; Wei et al., 2020).

6.2.3 Nanofertilizers

Literature describes nanofertilizers as nanoparticles that function as plant nutrients or carriers of nutrients (Liu & Lal, 2015). Based on plant requirements and carrier support, nanofertilizers can be categorized into five basic classes: (i) macronutrient-based,

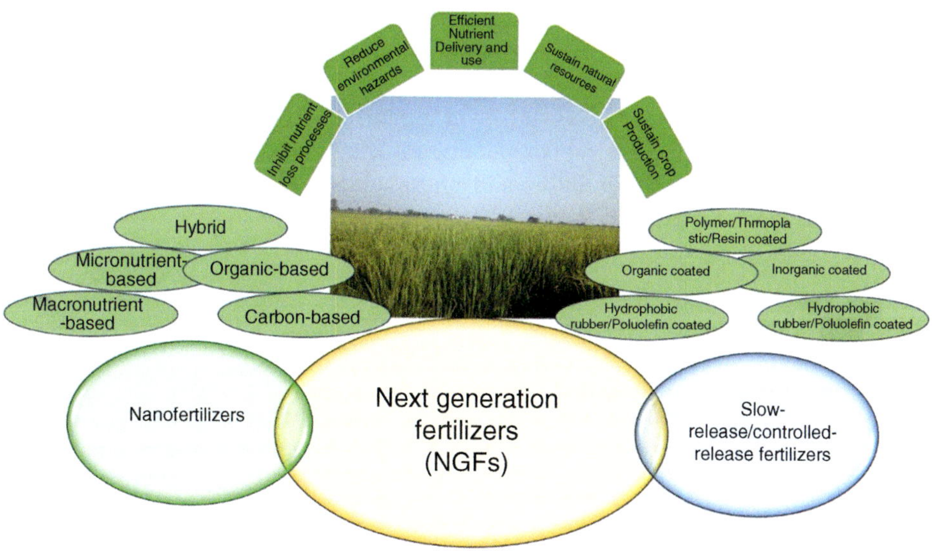

Figure 6.1 Next-generation fertilizers (NGFs) and their role in modern agriculture.

(ii) micronutrient-based, (iii) organic nanofertilizers, (iv) hybrid nanofertilizers, and (v) carbon-based nanofertilizers. The need for macronutrients will rise in tandem with the demand for food. Compared to conventional fertilizers, nanofertilizers can improve the efficiency of macronutrient uptake with fewer amounts because of their high volume-to-surface ratio (Afreen et al., 2018). The productivity can be enhanced more through nitrogen-based nanofertilizers than traditional fertilizers by enhancing chlorophyll biosynthesis in leaves and promoting aboveground and belowground biomass production (Verma et al., 2023). Likewise, grain and straw yield are increased by phosphorus-based nanofertilizers in wheat (Poudel et al., 2023). Similarly, anthocyanin formation in petals, biomass accumulation, and water uptake in daffodil plants vary notably under potassium-based nanofertilizer application. In the same context, the growth, mineral accumulation, and stress-responsive attributes are improved by potassium-based nanofertilizers in alfalfa plants (Asgari et al., 2018; El-Sharkawy et al., 2017). Nano-calcium fertilizer has led to a considerable biomass increase of almost 15% in peanuts. Wheat observes a boost in antioxidant enzyme activity and photosynthetic pigments, which subsequently strengthens photosynthesis through the application of nano-calcium carbonate (Gao et al., 2023; Liu et al., 2005). Water-stressed sesame plants show enhancements in physiological processes and total yield when supplied with nano-magnesium and chitosan (Khordadi et al., 2019). A boost in the stress resistance helps to attain better growth and improve the nutritional quality of plants through nano-sulfur application (Yuan et al., 2021). Iron-based nanofertilizer, applied in low doses, improved rice development; however, at larger concentrations, growth is adversely affected through reductions in root system architecture, including length, diameter, surface area, and volume of roots (Li et al., 2021). Soybean plants accumulate lignin in roots at the cost of reduced root growth under copper nanoparticles (Nair & Chung, 2014). Pomegranate yield is increased by applying boron nanofertilizers, while fruit quality is improved by using zinc nanofertilizers. Similarly, mung bean plants increase their height and pod count when fed with boron-based nanofertilizers (Davarpanah et al., 2016; Ibrahim & Al Farttoosi, 2019). Nickel nanofertilizer at low concentration impacted chlorophyll biosynthesis through the regulation of chlorophyll a and b pigments (Tsygvintsev et al., 2018). Zinc nanoparticles, such as zinc oxide (ZnO), improve physical (shoot and root length) and biochemical attributes (chlorophyll and protein biosynthesis, and phosphatase enzyme activity) of various plant species (Lin & Xing, 2008; Raliya & Tarafdar, 2013).

6.2.4 Slow-Release Fertilizers (SRFs)

The phenomenon of improving soil fertility and health by SRFs (Liu et al., 2014) is obvious from the higher rice yield under the application of neem cake-coated urea (NCU), a slow-release chemical, in comparison to prilled urea in India (Budhar et al., 1991; Singh & Singh, 1994). Neem-extract (nimin) coated urea (NICU) requires less nitrogen than prilled urea for rice at 30 kg/ha (De et al., 1992). In field experiments, Geethadevi et al. (1991) found that NCU produced higher rice yields than prilled urea; however, the highest yield was obtained from urea super granules.

6.2.5 Controlled-Release Fertilizers

The CRFs, which contain the same amount of nitrogen as conventional fertilizers or only half of it, have been shown to improve economic production and quality on commodity crops like maize and wheat (Ma et al., 2006). The practice of applying a combined dose of urea and controlled-release urea at a conventional or 30% lesser rate enhances grain production, increases or preserves protein content, and dramatically decreases nitrate accumulation because of better N-use efficiency in winter wheat (Yang et al., 2005). A 1.5-fold decrease in the frequency of application is observed in young Valencia orange trees when supplied with CRFs. This implies that a citrus plant nutrition program that combines soluble and CRFs is a cost-effective and successful approach (Zekri & Koo, 1991a, b). Tachibana (2007) provides the % decrease in the suggested N rate for a variety of Japanese crops when CRFs are used in place of traditional fertilizers. In addition to reducing potential nitrate leaching losses, increased N-use efficiency saves labor and energy expenses. Similar positive outcomes were confirmed by Shao et al. (2007), where CRFs were applied to apple plants, leading to encouraging tree development, yield, and quality. In Chinese field tests, 10–40% higher rice yield is observed using CRFs instead of urea (Song et al., 2005). CRFs enhanced rice output by 15% even with a 1/3 less N.

The agricultural intensification is required to meet the 100–110% increased crop demand for sustained global food production (Tilman et al., 2011). On the other hand, the escalation in fertilizer prices has decreased their economic efficiency dramatically. Furthermore, the issue of global climate change, owing to fertilizer-associated agricultural emissions, is an unprecedented challenge for agriculture as well as the fertilizer sector. This scenario is further worsened by the limited geological availability of raw material for fertilizer manufacturing and associated market competition. Historically, the regional fluctuation in fertilizer consumption existed in contrast to their constantly rising production. Western Europe and North America experienced both rise and decline in consumption, whereas a rapid increase in consumption was observed in countries of Eastern and Southern Asia. It indicates a rapidly growing share (60%) of Asia in the consumption of global fertilizers and thus influences the global fertilizer trade flow and competition.

The climate change-influenced global crop yield reductions were estimated to be 3.8% for maize and 5.5% for wheat. It is obvious from the increased release of nitrous oxide in Japan from the use of N fertilizers (Hayashi et al., 2021). It highlights the necessity of exploring fertilization systems for improvements in crop production. It also shows the importance of population-based calculations of NPK consumption rates to monitor agroecosystem performance.

6.3 The Need for Sustainable Fertilizer Solutions

Environmentally friendly and renewable resource-based, more sustainable and effective agricultural methods are desperately needed to feed the growing population under climate change. They should aim to provide economical food at the cost of efficient water, energy, and nutrient use (Urso & Gilbertson, 2018). The agricultural systems, across the globe, are dominated by the unprecedented use of N-, P-, and K-based chemical fertilizers, the long-term use of which has proved detrimental at environmental, economic, and social fronts (Culman et al., 2023; Naseem et al., 2024). Their unchecked use has degraded soils and accumulated nutrients in freshwater bodies through runoff, causing eutrophication and increased GHG emissions (Canatoy et al., 2024). Moreover, the production of synthetic N fertilizers is an energy-consuming process, mostly relying on fossil fuels. Similarly, P reservoirs are also finite, nonrenewable, and concentrated in only a few countries (Kumar and Kumar, 2023). Hence, in the recent scenario, the shift toward sustainable fertilizer solutions has become essential.

The primary aim of sustainable fertilizers is to reduce environmental damage while maintaining crop productivity and soil health. Hence, various alternative approaches such as plant and animal waste-based compost, vermicompost, biochar, biofertilizers containing beneficial microorganisms, and recycled nutrients from wastewater or food residues can be used (Sarfraz et al., 2020). Adopting sustainable fertilizer strategies can support climate-smart agriculture, enhance soil health, and contribute to circular economy principles. It also ensures long-term food security by reducing dependency on volatile global markets for synthetic fertilizer components. As global population and food demand continue to rise, the need for more resilient, regenerative, or next-generation fertilization practices becomes increasingly urgent (Babcock-Jackson et al., 2023).

6.4 Next-Generation Fertilizers: A Global Game Changer

The global agricultural landscape is evolving rapidly under the pressure of meeting the food demands of a growing population while ensuring environmental sustainability at the same time. Although fertilizers are a necessity for crop production; however, their conventional use brings significant drawbacks for soil, water, and atmospheric resources. In this context, NGFs emerge as a revolutionary innovation in terms of nutrient delivery optimization, crop productivity enhancement, and environmental constraints mitigation. Further, NGFs represent a major shift in global efforts to create sustainable food systems (Savci, 2012). Agriculture uses about 70% of all the fresh water in the world and accounts for nearly 40% of all land use (FAO, 2017). Fertilizers, particularly those with nitrogen, are the key to higher yields, yet they are also wasteful: plants typically take up less than half the nutrients. The remainder seeps into the surrounding environment, where it creates other issues like eutrophication and groundwater pollution. This inefficiency is economically impossible and environmentally detrimental as the global food demand increases (Zhang et al., 2015). The objective of NGFs is to overcome these constraints through the exploitation of advances like nanotechnology, bioengineering, and controlled-release technologies to deliver execute nutrient application in line with crop development stages and soil types.

6.4.1 Environmental Impact

NGFs are attributed as new-generation products based on fertilizer, which contain special substances and elements that eliminate the shortcomings of conventional fertilizers. CRFs are coated with polymers or resins to release nutrients slowly in coordination with plant uptake, leading to less wastage and more efficiency (MET Trenkel, 2021). To minimize losses of nitrogen through various processes, stabilized fertilizers have additives like nitrification inhibitors and urease inhibitors, whereas compounds like nitropyrin are added to stabilized fertilizers in an effort to mitigate leaching and discharge (Subbarao et al., 2006). Biofertilizers utilize beneficial microbes as soil conditioners to enhance nutrient availability naturally. Nanofertilizers use nanotechnology to enhance the solubility of nutrients, absorption, and site-specific delivery at the cellular level, enabling the nanoscale elements, for example, zinc and iron, to be better absorbed by plants (DeRosa et al., 2010). Collectively, these innovative fertilizers help solve some of the major problems confronting agriculture, including nutrient inefficiency and environmental and soil degradation, and they encourage higher productivity and sustainability. All of these NGF types solve major agricultural issues, which are nutrient use inefficiency, environmental damage, and soil erosion, and allow increasing productivity and resilience (Figure 6.2).

The traditional fertilizers are adding to the problems of the environment. The nutrient runoff eutrophicates lakes and rivers, generates dead zones, and causes biodiversity losses (Smith et al., 1999). The burden on the environment can be substantially decreased by the innovations related to NGFs. Stabilized fertilizers and CRFs minimize nitrogen volatilization and leaching, and hence minimize emissions of GHGs. Biofertilizers enhance

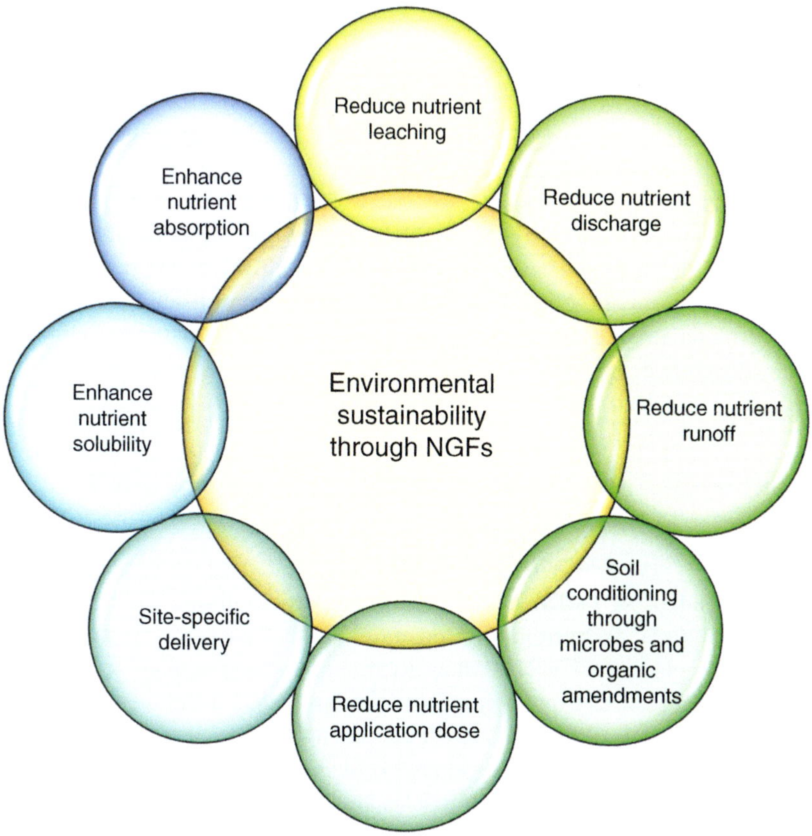

Figure 6.2 Environmental improvement by the use of next-generation fertilizers (NGFs).

soil aggregate stability, microbial diversity, and decrease chemical input dependency. With the use of nanofertilizers, fertilization can be controlled, preventing excesses and erosion. Enhancing nutrient absorption and minimizing wastage, NGFs are aligned with the practices defined as climate-smart agriculture.

6.4.2 Economic Benefits

Although NGFs might be more expensive than conventional fertilizers, yet they offer notable long-term economic benefits through enhanced nutrient use and crop productivity. Indicatively, studies conducted by the International Fertilizer Association (IFA) demonstrate that CRFs can minimize fertilizer applications by 20–30% without compromising yield (IFA, 2021). This improvement in efficiency has been converted to reduced input costs, such as reduced use of fertilizers and reduced labor demand. Second, NGFs are beneficial in enhancing the quality as well as quantity parameters of crops, hence raising the commercial worth of goods. NGFs also create new market opportunities, since they spur innovation and job creation in the agricultural technology and biotechnology sectors. Collectively, NGFs are not only beneficial to enhance the profitability of the farmers but also align with sustainable production in agriculture.

6.4.3 Social and Food Security Implications

Providing nutrients to 10 billion people on the globe without increasing agricultural land is a great challenge; however, NGFs can contribute dramatically to solving the issue. By making production sustainable intensity, NGFs can contribute to per-hectare yield increases, relieving the pressure of forest land or grassland conversions to cropland. Biofertilizers and CRFs can enhance food security and livelihoods in developing regions where soil fertility is low and resources are limited (Vanlauwe et al., 2010). Moreover, training and availability of NGFs can assist small-hold farmers to expand their output sustainably to assist in poverty alleviation and rural advancement. Large-scale usage of NGFs is futuristically aligned with the Sustainable Development Goals (SDGs), such as zero hunger, clean water, and climate action, devised by the United Nations.

6.5 Overcoming Barriers to Adoption in Low-Income Regions

6.5.1 Barriers to Adoption

A farmer's decision to adopt NGFs depends on social, economic, cultural, and behavioral factors (Wreford et al., 2017). Apart from huge potential, NGFs have many obstacles hindering their universal implementation. NGFs may not be affordable to smallholder farmers of low-income countries because of higher initial investment than conventional fertilizers (Sachan et al., 2023; Shaviv & Mikkelsen, 1993) due to higher costs and difficulties in finding comparable alternatives (Fragomeli et al., 2024). Such circumstances compel smallholder farmers to use suboptimal fertilizers, which lead to low yields (Khundi et al., 2025). Second, a lot of farmers are not familiar with NGF products nor do they have any training that would help them to apply them efficiently. Irregularities in policies and the absence of uniform guidelines further curtail the progress of the market. Moreover, NGFs differ in their effectiveness based on soil type, crop type, and local climate, which necessitates their customization.

In developing countries, accessibility to NGFs is a challenge for farmers who have limited credits, agricultural inputs, and access to extension services (Krivanek et al., 2007). The economic challenges are the main hindrance to adopting NGFs, as it is difficult to afford their expenses to comply with regulations. The fluctuations in farm economics also affect the acceptance of NGFs negatively in developing countries (Goodwin & Gouldthorpe, 2013). Furthermore, poor infrastructure, terms of payments, lack of insurance, and complicated requirements of certification also become negative factors compromising NGF adaptation (Boys & Fraser, 2019). International competition is another challenge as small-scale farmers have to abide by the food safety-related regulations (Goodwin & Gouldthorpe, 2013). That's why even if they have a high production level, it would be wasted due to limited market access (Mpandeli & Maponya, 2014).

6.5.2 Strategies for Overcoming Barriers

The technology alone cannot play a role in producing more food with low environmental consequences. The economic and social factor-based policies are required that could ease the decision-making process of farmers (Davidson et al., 2015). In Pakistan, 80% of farmers have small farm holdings, and due to financial constraints, they are unable to apply the recommended fertilizer dose. Subsidies provided by the government on fertilizer purchase are their only motivation to use recommended fertilizer doses that could increase farm output and improve their livelihoods (Ali et al., 2019). Similarly, the subsidized availability of fertilizers in Ghana resulted in sector-based improvements in agriculture and the food industry, which led to GDP growth. The subsidized production of sorghum, rice, and maize yielded 4.5%, 3.8%, and 8.3% more gains, respectively, in 2017. In similar terms, the gains were recorded to be three, six, and four times more in 2020 than in 2017 in the sorghum, rice, and maize sectors, respectively (Iddrisu et al., 2020). Extension services could have a widespread positive influence on the farming community to accept the transition from traditional fertilizers to NGFs (Becerra-Encinales et al., 2024). There are examples of Bangladesh and Ethiopia having a positive impact of extension services enabling farmers to better crop cultivation and adopt marketable crops (Buehren et al. 2017; Rahman & Connor, 2022). Additionally, public and private sectors should work together to develop data that strengthens markets and provides farmers with the latest market and agronomic information (Ariga et al., 2019).

Community participation is crucial to improving fertilizer use in low-income regions. In Nepal, only 14% of farmers adopted fertilizers without community-led extension programs (Kumar et al., 2020). In Ethiopia, cooperative membership raised adoption by 23% (Tadesse, 2014), whereas farmer organizations boosted yields by 16% in Pakistan (Khan et al., 2022). Lack of regulation on fertilizer quality undermines trust, which discourages adoption (Michelson et al., 2023). Similarly, policies bundling extension advice with subsidies have a stronger effect than isolated subsidy programs (Takahashi et al., 2020). Fertilizer pricing policies proved more impactful when coupled with advisory services and measured using value–cost ratios (Assefa et al., 2025). Spatially differentiated nitrogen policies outperformed flat-rate subsidies, reducing excess use by up to 20% (Snapp et al., 2023). The use of digital agronomy tools underperformed in Sub-Saharan Africa due to policy gaps in input credit and agro-dealer networks (Alawode, 2025). Moreover, a combination of fertilizer subsidies and soil testing improves both adoption and environmental conditions (Sane et al., 2021). Therefore, there is a need to have concerted efforts from all the stakeholders to facilitate NGF adoption through combined efforts.

6.6 The Role of International Collaboration in Advancing Fertilizer Technology

Fertilizers have been used for ages to maintain soil fertility, food security, and agricultural productivity. The Green Revolution brought a trend for the widespread use of synthetic fertilizers, allowing for previously unimaginable increases in crop yields. However, such intensification also entailed considerable environmental costs, such as nutrient leaching into the groundwater, atmospheric volatilization, and contamination of surface waters, threatening ecosystems and human health. Climate change and an increasing global population by 2050 to over 10 billion people only aggravate these problems. Hence, a switch to more efficient and environmentally friendly fertilizer application technologies is also warranted for this dilemma. This transition cannot be foreseen unilaterally; though, for it to happen, strong international cooperation is necessary to share ideas, harmonize laws, and adapt solutions to different agro-ecological setups (Shahwar et al., 2023).

6.6.1 Collaborations Within the Existing Framework of NGFs

The surge in material sciences holds the capability to make CRFs for a determined delivery system for nutrients. CRFs control by coating the fertilizer granules, thereby protecting them from the microenvironments of soil. A consequence of such coating is the diminished loss by leaching, volatilization, and denitrification. Nevertheless, South Asia and Sub-Saharan Africa are still experiencing the prevalence of traditional fertilizers, whereas CRFs are in extensive use in developed countries. It is equally important to have slow-release or stabilized nitrogen fertilizers that contain urease and nitrification inhibitors to curtail nitrogen losses. There is still limited adoption of SRFs, where nitrogen losses remain an economic and environmental bane. With nanoscale carriers to improve nutrient encapsulation, mobility, and uptake efficiency, nanofertilizers are at the forefront of fertilizer technology. While developments and pilot applications of nanofertilizers have been led by a few nations, especially China and India, insurmountable barriers such as regulatory frameworks, risk assessment tests from the ecological standpoint, and gaps in reach to the farmer community have captured and hindered the commercial scaling-up of the technology in its use. To accommodate the myriad of physicochemical characteristics of soils worldwide, international research collaborations will have to optimize these formulations based on the local conditions, validate the long-term implications, and set up policy recommendations that will guide their safe and equitable adoption (Jakhar et al., 2022). Promotion of the technology argues for the need for international cooperation through joint field trials, adapted research, and regional policy dialogues, to ensure gains realized in temperate systems are replicated across the climatic zones (Jariwala et al., 2022; Thapa et al., 2015).

The plant growth-promoting rhizobacteria (PGPR) fix nitrogen, suppress pathogens, and produce siderophores and plant hormones to enhance soil nutrient status and promote plant growth through biofertilizer application. Brazil has been a pioneer in introducing the use of inoculants at large scales to underpin conventional agriculture toward maintaining soil conditions and crop productivity (Shahwar et al., 2023). However, the successful adoption of such biofertilizer technology demands the use of locally predominant strains and relevant customized inoculation approaches, which will call for international microbial culture collections and cross-boundary genetic research and collaborative follow-on farmer training programs. This cooperative approach will fast-track the research and dissemination of effective biofertilizer technologies among farmers in developing countries where synthetic inputs remain expensive or ecologically unsustainable (Shahwar et al., 2023).

Recent progress observed in foliar fertilization, particularly the application of chelated formulations, signifies yet another promising aspect to be pursued in NGF strategies. These chelated fertilizers are known to enhance nutrient uptake by stabilizing essential micronutrients and enhancing their translocation within plants, unlike classical inorganic foliar sprays. Nevertheless, several factors modulate the effects of the chelating agents, including their molecular structures and binding strengths, which may vary with climatic and soil conditions. Therefore, international cooperation on this subject is necessary to test, adapt, and optimize the use of chelated foliar fertilizers across different agroecosystems to ensure these benefits can be delivered fairly reliably to smallholder farmers on nutrient-deficient lands in Africa and Asia, where trace element deficiencies are commonplace (Niu et al., 2021).

Biochar now comprises an innovative soil amendment that fulfills agronomic and environmental needs. Biochar, an organic residue pyrolysis product, brings the promise of developing soil structure, minimizing nutrient leaching, incorporating carbon in soils, and reducing agricultural emissions (Oni et al., 2019). Besides, integrating biochar into rural bio-economies provides avenues for the farmers to derive extra income from local biomass residues. For all its advantages, the global uptake of biochar remains poorly synchronized worldwide, and even where biowastes are available, the projects in question often suffer from technology costs and local market systems. Partnerships promoting the dissemination of best practices regarding biochar production technologies, adapting pyrolysis systems for region-biomas types, and establishing market incentives would project biochar as a mainstream intervention for the maintenance of agricultural soils (Oni et al., 2019).

Microbial fertilizers containing plant growth-promoting microorganisms (PGPM) are justified as ecologically healthy substitutes to synthesized agrochemicals. These bio-inoculants improve nutrient availability via nitrogen fixation, phosphorus solubilization, phytohormone synthesis, and suppression of soilborne pathogens. The major bottleneck remains the inconsistent quality and stability of commercialized microbial products, which, in turn, damages the trust of farmers and hampers the market. As a remedy to this situation, a collective research effort is needed from local and international aspirants for the standardization of mass multiplication protocols, development of stable formulations, and identification of strains based on their efficacy under various field conditions. By implementing coordinated international quality control measures further aided by regional microbial banks, the microbial fertilizer technologies can be tailored to local crop-soil systems, ultimately enhancing their acceptability and leading to measurable productivity gains worldwide (Stamenković et al., 2018). Thus, a collaborative transnational approach is required in the quest to deliver fertilizer innovations that score high on resilience, productivity, and ecological balance to uphold efforts of global food production (Jakhar et al., 2022; Jariwala et al., 2022; Shahwar et al., 2023; Thapa et al., 2015).

6.7 The Future of Fertilizer Innovation: Global Prospects and Trends

It is believed that the NGFs industry will expand rapidly due to the rapidly evolving technologies and the emergence of data-driven agriculture. Using modern precision agriculture technologies, including GPS, drones, and remote sensing, fertilizers can be applied to a certain piece of land precisely, resulting in successful optimization of input use and reducing waste. This process is also aided by digital agriculture platforms, which deliver real-time data analytics to enable farmers to adapt and optimize nutrient management plans. The manufacturing of personalized biofertilizers, through the innovative developments at the forefront of synthetic and microbial engineering, can address the unique nutrient requirements of crops under various soil types. Moreover, artificial intelligence (AI) is becoming a more significant factor as it is represented in decision support systems determining the most effective time and dose of NGF application. Collectively, these innovations are changing how fertilizers are created, delivered, and used, and NGF is becoming a pillar of the next-generation smart and sustainable farm systems. Traditional approaches to increase agricultural production will no longer be able to secure food in underdeveloped countries (Elhassani et al, 2023; El Idrissi et al., 2023). It calls for a shift toward devising a responsible "plant nutrition paradigm" inclusive of fertilizer product advancements within the framework of the local economy.

The application of CRF and nanofertilizer is being pioneered in China and India (Li & Li, 2024). The Chinese government is promoting CRFs with farmer education programs and incentives (Asghar & Kazerooni, 2024). European countries are focusing on sustainable farming practices, with a preference for bio-based nutrient input and organic fertilizers. The consumption of organic-based fertilizers is being driven by the EU's Common Agricultural Policy (Tröster, 2023). To maximize nutrient delivery, fertilizer application is being combined with precision agriculture in the United States. Environmental discharge has decreased, and fertilizer use efficiency has increased with the use of sensors in soil analysis (Fixen et al., 2015). African nations are investigating the manufacturing and application of biofertilizers using domestic organic materials, such as manure and compost. The biofertilizers have shown promise of increasing crop yields, soil fertility, and soil health in small-scale agricultural systems in Kenya and Nigeria (Gama & Millinga, 2016). Governments and the commercial sector around the world are realizing how important sustainable nutrient management is. The US Department of Agriculture has made significant investments in precision agriculture and EEFs (McFadden et al., 2023). Policies that promote renewable energy utilization in fertilizer production have been put in place by the European Union. Research is being done to develop plant need-based fertilizers to enhance efficiency and minimize the ecological impact (Barłóg et al., 2022). Additionally, the chemical fertilizer-generated GHG emissions may be reduced through bionanofertilizers (Tarafdar, 2021). Scientific and technological development is necessary for optimizing fertilizer use in diverse agricultural systems in different countries around the world. However, the literature lacks enough studies to support the utilization and efficiency of all these categories of fertilizer products. The research gap in data, technology, and knowledge of NGFs needs a more comprehensive assessment (Maaz et al., 2025). The worldwide adoption of new technologies requires farmer training and infrastructure improvements (Batan, 2025). In the near future, nanotechnology will provide a sustainable solution to the maximization and enhancement of the world's food production, safety, and security (Arora et al., 2024).

6.8 Feeding the World in a Changing Climate

6.8.1 Fertilizer Solutions for Climate-Resilient Agriculture

Agricultural productivity depends on land and crop management practices influenced by biotic and abiotic modulators. Climate conditions affect all the crop productivity parameters; for example, wheat yield decreases from about 4.1% to 6.4% with a 1°C increase in global temperature (Liu et al., 2016). Droughts, floods, and temperature-related heat and cold stresses negatively impact crop productivity globally (Anderson et al., 2020).

Historically, the conventional chemical fertilizers have long been associated with environmental consequences (Shanmugavel et al., 2023). Such climatic anomalies demand the exploration of more resilient alternatives to conventional fertilizer, which can ensure the supply of nutrients for sustainable crop production (Abdelsalam et al., 2019).

Biochar, a plant or animal waste-based pyrolyzed product, typically contains about 85% carbon (Fakhar et al., 2025). Its use can strategically incorporate carbon into soils and minimize N_2O emissions (Mahmood et al., 2024). Biochar may prove beneficial for soils in terms of its both mineralizable and recalcitrant components of carbon. Furthermore, biochar can retain NO_3 inside its pores, which may regulate N loss processes (nitrification and denitrification) in soils (Borchard et al., 2019). Biofertilizers constitutively contain bacteria that can fix nitrogen and solubilize phosphorus along with arbuscular mycorrhizal fungi for natural soil conditioning and improved nutrient foraging. These fertilizers are usually rich in beneficial bacteria and fungi that improve nutrient uptake, nitrogen fixation, and organic matter decomposition. They may prove effective in minimizing dependency on traditional fertilizers. Organic fertilizers, such as vermicompost, bone meal, and poultry manure, have the potential to enhance soil physical conditions, sustain essential plant nutrients for longer periods of time, and retain soil moisture (Shu et al., 2022) under variable environmental conditions. Some biofertilizers are marine algae-based (liquid seaweed extracts and powdered seaweed fertilizers), enriched with natural growth promoters such as microelements, organic matter, vitamins, fatty acids, and other minerals (Karthik & Jayasri, 2023).

A steady and sustained nutrient supply to plant roots is ensured by CRFs through gradual release (Shanmugavel et al., 2023). This attribute helps in improving nutrient efficiency and reducing the frequency of fertilizer application, subsequently minimizing environmental impact. Nanoparticles of <100 nm size can be effectively utilized for nutrient management in the wake of environmental sustainability. They can encapsulate nutrients and mobilize them for better agricultural productivity (Jakhar et al., 2022). They are useful to reduce the quantity of fertilizer required, minimize environmental pollution, and improve crop productivity.

6.8.2 Integrating Fertilizer Use with Climate Adaptation Strategies

To maximize agricultural productivity and to overcome food security issues, integrating fertilizer practices with efficient nutrient use and land management strategies can be a good option under continuously altering climate conditions. Adoption of 4Rs (right source, right rate, right time, and right place) can perform better to reduce nutrient losses during droughts or heavy rains. It can further help to optimize plant uptake when resources are limited. Research on nutrient-based genetic innovations in crop varieties should be considered. Crop varieties selected through climate resilience and nutrient efficiency screening can provide a better option to cope with climate change issues. A combination of organic amendments and biofertilizers can minimize the nutrient loss, replenish soil fertility, increase plant health, and avoid soil degradation (Altieri et al., 2012). These amendments can further help to buffer soil zones against extreme temperatures and climate change. Precision agriculture tools may be incorporated into fertilizer application strategies. For example, GIS can enable farmers to refrain from fertilizer overuse. The analytical capabilities of GIS can help assess the variable parameters negatively influencing crop production in changing agroclimatic conditions (Yadav et al., 2010). Moreover, animal raising and crop cultivation on the same farm within the same farming system can help smallholder farmers close nutrient loops and reduce dependency on external fertilizers (Sekaran et al., 2021). Animal manure returning to the field can serve as organic fertilizer, while crop residue can serve as livestock feed. Introducing leguminous fodders can minimize nitrogen inputs, thus reducing the dependency on inorganic or chemical fertilizers.

6.9 Conclusion

The concept of NGFs is exciting for the research and farming community as it promises to meet sustainability targets and reduce carbon footprints. The entire transformation of fertilizers, from their enormous chemical applications and impact into systems that entail minute amounts of careful applications under stringent biological control, is indeed a paradigm change in its own right in the perspectives of agroecosystem stewardship. Policy frameworks, particularly in regions such as the European Union and parts of Asia, are evolving to incentivize the use of low-emission and high-efficiency fertilizers, which signals a regulatory shift that will shape global fertilizer production and trade. The convergence of climate-smart agriculture goals with sustainable fertilizer use will further upscale the idea of NGFs to meet global environmental commitments as per the SDGs. Indeed, the full promise of NGFs, in developing regions in particular, will require a great deal of international cooperation. Any major breakthrough in this area will need to be shared among farming communities across the globe. The future of fertilizer innovation calls to combine biotechnology, nanotechnology, and information technology for the creation of alternative fertilizers (NGFs) that can promise sustainable crop production without compromising the environment. Briefly, policy support and extension services are crucial for scaling these practices. Farmers need access to capacity-building trainings, adaptive technologies, and risk-sharing mechanisms (e.g., insurance and subsidies) to confidently shift toward integrated, climate-responsive nutrient strategies. The future of NGFs is increasingly tied to a vision of agriculture that is not only more productive but also regenerative, inclusive, and resilient to global challenges.

References

Abdelsalam, N. R., Kandil, E. E., Al-Msari, M. A., Al-Jaddadi, M. A., Ali, H. M., Salem, M. Z., & Elshikh, M. S. (2019). Effect of foliar application of NPK nanoparticle fertilization on yield and genotoxicity in wheat (Triticum aestivum L.). *Science of the Total Environment*, 653, 1128–1139.

Afreen, S., Omar, R. A., Talreja, N., Chauhan, D., & Ashfaq, M. (2018). Carbon-based nanostructured materials for energy and environmental remediation applications. In R. Prasad & E. Aranda (Eds.), *Approaches in bioremediation: The new era of environmental microbiology and nanobiotechnology* (pp. 369–392). Cham: Springer.

Alawode, A. (2025). Evaluating agricultural subsidy reforms and their effects on smallholder farmer income and efficiency. *International Journal of Advance Research Publication and Reviews*, 2(5), 180–201.

Ali, A., Rahut, D. B., & Imtiaz, M. (2019). Affordability linked with subsidy: Impact of fertilizers subsidy on household welfare in Pakistan. *Sustainability*, 11(19), 5161.

Altieri, M. A., Funes-Monzote, F. R., & Petersen, P. (2012). Agroecologically efficient agricultural systems for smallholder farmers: Contributions to food sovereignty. *Agronomy for Sustainable Development*, 32(1), 1–13.

Anderson, R., Bayer, P. E., & Edwards, D. (2020). Climate change and the need for agricultural adaptation. *Current Opinion in Plant Biology*, 56, 197–202.

Ariga, J., Mabaya, E., Waithaka, M., & Wanzala-Mlobela, M. (2019). Can improved agricultural technologies spur a green revolution in Africa? A multicountry analysis of seed and fertilizer delivery systems. *Agricultural Economics*, 50, 63–74.

Arora, P. K., Tripathi, S., Omar, R. A., Chauhan, P., Sinhal, V. K., Singh, A., & Singh, V. P. (2024). Next-generation fertilizers: The impact of bionanofertilizers on sustainable agriculture. *Microbial Cell Factories*, 23(1), 254.

Asgari, S., Moradi, H., & Afshari, H. (2018). Evaluation of some physiological and morphological characteristics of narcissus tazatta under BA treatment and nano-potassium fertilizer. *Journal of Chemical Health Risks*, 4(4), 63–70.

Asghar, R., & Kazerooni, E. (2024). Current trends in the use of controlled-release fertilizers for sustainable agriculture. *International Journal of Chemical and Biochemical Sciences*, 25(17), 119160.

Assefa, T. W., Berhane, G., Abate, G. T., & Abay, K. A. (2025). Fertilizer demand and profitability amid global fuel-food-fertilizer crisis: Evidence from Ethiopia. *Food Policy*, 133, 102785.

Babcock-Jackson, L., Konovalova, T., Krogman, J. P., Bird, R., & Diaz, L. L. (2023). Sustainable fertilizers: Publication landscape on wastes as nutrient sources, wastewater treatment processes for nutrient recovery, biorefineries, and green ammonia synthesis. *Journal of Agricultural and Food Chemistry*, 71(22), 8265–8296.

Barłóg, P., Grzebisz, W., & Łukowiak, R. (2022). Fertilizers and fertilization strategies mitigating soil factors constraining efficiency of nitrogen in plant production. *Plants*, 11(14), 1855.

Batan, A. (2025). Assessing the role of public-private partnerships in advancing sustainable development goals through agri-tech and renewable energy solutions. *Transactions on Social Innovation, Digital Inclusion, and Ethical AI*, 15(2), 1–11.

Becerra-Encinales, J. F., Bernal-Hernandez, P., Beltrán-Giraldo, J. A., Cooman, A. P., Reyes, L. H., & Cruz, J. C. (2024). Agricultural extension for adopting technological practices in developing countries: A scoping review of barriers and dimensions. *Sustainability*, 16(9), 3555.

Bindraban, P. S., Dimkpa, C. O., & Pandey, R. (2020). Exploring phosphorus fertilizers and fertilization strategies for improved human and environmental health. *Biology and Fertility of Soils*, 56(3), 299–317.

Borchard, N., Schirrmann, M., Cayuela, M. L., Kammann, C., Wrage-Mönnig, N., Estavillo, J. M., & Novak, J. (2019). Biochar, soil and land-use interactions that reduce nitrate leaching and N_2O emissions: A meta-analysis. *Science of the Total Environment*, 651, 2354–2364.

Boys, K. A., & Fraser, A. M. (2019). Linking small fruit and vegetable farmers and institutional foodservice operations: Marketing challenges and considerations. *Renewable Agriculture and Food Systems*, 34(3), 226–238.

Budhar, M. N., Ramaswami, C., & Kareem, A. A. (1991). Relative efficiency of coated N fertilizers in lowland rice.

Buehren, N., Goldstein, M., Molina, E., & Vaillant, J. (2017). The impact of strengthening agricultural extension services: Evidence from Ethiopia. *World Bank Policy Research Working Paper*, (8169).

Byrnes, B. H., & Bumb, B. L. (2017). Population growth, food production and nutrient requirements. In Z. Rengel (Ed.), *Nutrient use in crop production* (pp. 1–27). CRC Press.

Canatoy, R. C., Cho, S. R., Galgo, S. J. C., Kim, P. J., & Kim, G. W. (2024). Reducing ammonia volatilization in rice paddy: The importance of lower fertilizer rates and soil incorporation. *Frontiers in Environmental Science*, 12, 1479712.

Culman, S. W., Fulford, A., Camberato, J., & Steinke, K. (2023). *Tri-state fertilizer recommendations for corn, soybeans, wheat, and alfalfa*. Ohio State University, College of Food, Agricultural, and Environmental Sciences.

Davarpanah, S., Tehranifar, A., Davarynejad, G., Abadía, J., & Khorasani, R. (2016). Effects of foliar applications of zinc and boron nano-fertilizers on pomegranate (Punica granatum cv. Ardestani) fruit yield and quality. *Scientia Horticulturae*, 210, 57–64.

Davidson, E. A., Suddick, E. C., Rice, C. W., & Prokopy, L. S. (2015). More food, low pollution (Mo Fo Lo Po): A grand challenge for the 21st century. *Journal of Environmental Quality*, 44(2), 305–311.

De, G. C., Das, S., & Pal, D. (1992). Efficiency of neem-extract-coated urea on transplanted summer rice (Oryza sativa).

DeRosa, M. C., Monreal, C., Schnitzer, M., Walsh, R., & Sultan, Y. (2010). Nanotechnology in fertilizers. *Nature Nanotechnology*, 5(2), 91–91.

Drescher, A., Glaser, R., Richert, C., & Nippes, K. R. (2011). Demand for key nutrients (NPK) in the year.

El Idrissi, A., Dardari, O., Metomo, F. N. N. N., Essamlali, Y., Akil, A., Amadine, O., & Zahouily, M. (2023). Effect of sodium alginate-based superabsorbent hydrogel on tomato growth under different water deficit conditions. *International Journal of Biological Macromolecules*, 253, 127229.

El-Sharkawy, M. S., El-Beshsbeshy, T. R., Mahmoud, E. K., Abdelkader, N. I., Al-Shal, R. M., & Missaoui, A. M. (2017). Response of alfalfa under salt stress to the application of potassium sulfate nanoparticles. *American Journal of Plant Sciences*, 8(8), 1751–1773.

Fakhar, A., Galgo, S. J. C., Canatoy, R. C., Rafique, M., Sarfraz, R., Farooque, A. A., & Khan, M. I. (2025). Advancing modified biochar for sustainable agriculture: A comprehensive review on characterization, analysis, and soil performance. *Biochar*, 7(1), 8.

FAO. (2017). *The future of food and agriculture—trends and challenges*. Rome: Food and Agriculture Organization of the United Nations.

FAO. (2020a). *World fertilizer trends and outlook to 2022*. Rome: Food and Agriculture Organization of United Nations.

FAO (2020b). Global trends, statistics and insights for fertilizer. Published by NationMaster. https://www.nationmaster.com/nmx/sector/fertilizer

Fertilizer Canada (2018). Annual report value add+. https://fertilizercanada.ca/wp-content/uploads/2018/08/fc_annualreport2018_en_vf2-web.pdf

Fixen, P., Brentrup, F., Bruulsema, T., Garcia, F., Norton, R., & Zingore, S. (2015). Nutrient/fertilizer use efficiency: Measurement, current situation and trends. *Managing Water and Fertilizer for Sustainable Agricultural Intensification*, 270, 1–30.

Fragomeli, R., Annunziata, A., & Punzo, G. (2024). Promoting the transition towards agriculture 4.0: A systematic literature review on drivers and barriers. *Sustainability*, 16(6), 2425.

Gama, J., & Millinga, M. L. (2016). Latest developments in organic agriculture in Africa. In *The world of organic agriculture* (p. 174). Research Institute of Organic Agriculture FiBL, IFOAM – Organics International.

Gao, Y., & Cabrera Serrenho, A. (2023). Greenhouse gas emissions from nitrogen fertilizers could be reduced by up to one-fifth of current levels by 2050 with combined interventions. *Nature Food*, 4(2), 170–178.

Gao, Y., Chen, S., Li, Y., & Shi, Y. (2023). Effect of nano-calcium carbonate on morphology, antioxidant enzyme activity and photosynthetic parameters of wheat (Triticum aestivum L.) seedlings. *Chemical and Biological Technologies in Agriculture*, 10(1), 31.

Geethadevi, T., Srinivas, N., Swamygowda, S. N., & Kenchaiah, R. B. (1991). Nitrogen loss studies in wetland rice using modified urea materials.

Geissler, B., Mew, M. C., & Steiner, G. (2019). Phosphate supply security for importing countries: Developments and the current situation. *Science of the Total Environment*, 677, 511–523.

Goodwin, J. N., & Gouldthorpe, J. L. (2013). Small farmers, big challenges: A needs assessment of Florida small-scale farmers' production challenges and Training needs. *Journal of Rural Social Sciences*, 28(1), 3.

Grand View Research (2019). Nitrogenous fertilizer market size, share and trends analysis report by product (urea, ammonium nitrate), by application (cereals and grains, oilseeds & pulses), by region, and segment forecasts, 2019–2025. https://www.grandviewresearch.com/industryanalysis/nitrogenous-fertilizer-market#:~:text=Some%20of%20the%20key%20market,Limited%2C%20and%20Coromandel% 20International%20Ltd

Gu, D., Andreev, K., & Dupre, M. E. (2021). Major trends in population growth around the world. *China CDC Weekly*, 3(28), 604.

Hayashi, K., Shibata, H., Oita, A., Nishina, K., Ito, A., Katagiri, K., & Winiwarter, W. (2021). Nitrogen budgets in Japan from 2000 to 2015: Decreasing trend of nitrogen loss to the environment and the challenge to further reduce nitrogen waste. *Environmental Pollution*, 286, 117559.

Ibrahim, N. K., & Al Farttoosi, H. A. K. (2019). Response of mung bean to boron nanoparticles and spraying stages (Vigna Radiata L.). *Plant Arch*, 19(2), 712–715.

Iddrisu, A. M., Gafa, D., Emini, C. A., & Beaumais, O. (2020). *Implications of the fertilizer-subsidy programme on income growth, productivity, and employment in Ghana*. Partnership for Economic Policy Working Paper, (2020-05).

IFA. (2021). *Fertilizer and nutrient efficiency report*. International Fertilizer Association.

Jakhar, A. M., Aziz, I., Kaleri, A. R., Hasnain, M., Haider, G., Ma, J., & Abideen, Z. (2022). Nano-fertilizers: A sustainable technology for improving crop nutrition and food security. *NanoImpact*, 27, 100411.

Jariwala, H., Santos, R. M., Lauzon, J. D., Dutta, A., & Wai Chiang, Y. (2022). Controlled release fertilizers (CRFs) for climate-smart agriculture practices: A comprehensive review on release mechanism, materials, methods of preparation, and effect on environmental parameters. *Environmental Science and Pollution Research*, 29(36), 53967–53995.

Karthik, T., & Jayasri, M. A. (2023). Systematic study on the effect of seaweed fertilizer on the growth and yield of Vigna radiata (L.) R. Wilczek (Mung bean). *Journal of Agriculture and Food Research*, 14, 100748.

Khan, N., Ray, R. L., Kassem, H. S., Ihtisham, M., Siddiqui, B. N., & Zhang, S. (2022). Can cooperative supports and adoption of improved technologies help increase agricultural income? Evidence from a recent study. *Land*, 11(3), 361.

Khordadi, V. J., Fanoodi, F., Masuod, S. J., Rezvan, S., & Damavandi, A. (2019). Physiological response of sesame (Sesamum indicum L.) to application of chitosan and magnesium-nano fertilizers under irrigation cut-off in a sustainable agriculture system.

Khundi, Q., Jiang, Y., Sun, Y., & Rui, Y. (2025). Nanofertilizers for sustainable African agriculture: A global review of agronomic efficiency and environmental sustainability. *Nanomaterials*, 15(5), 390.

Kotschi, J. (2015). Adverse impacts of mineral fertilizers in tropical agriculture. In T. Koberich (Ed.), *A soiled reputation. A study commission by Heinrich Böll Foundation, WWF Germany* (pp. 1–60). Heinrich Böll Foundation.

Krivanek, A. F., Groote, H. D., Gunaratna, N. S., Diallo, A. O., & Friesen, D. (2007). Breeding and disseminating quality protein maize (QPM) for Africa.

Kumar, D. and Kumar, D. (2023). *Phosphate Rock: An Industry in Transition*. ISBN 0323959830, 9780323959834, Elsevier, pp 298. https://books.google.com.pk/books?id=2C2gEAAAQBAJ&source=gbs_navlinks_s

Kumar, A., Takeshima, H., Thapa, G., Adhikari, N., Saroj, S., Karkee, M., & Joshi, P. K. (2020). Adoption and diffusion of improved technologies and production practices in agriculture: Insights from a donor-led intervention in Nepal. *Land Use Policy*, 95, 104621.

Li, X., & Li, Z. (2024). Global trends and current advances in slow/controlled-release fertilizers: A bibliometric analysis from 1990 to 2023. *Agriculture*, 14(9), 1502.

Li, M., Zhang, P., Adeel, M., Guo, Z., Chetwynd, A. J., Ma, C., & Rui, Y. (2021). Physiological impacts of zero valent iron, Fe_3O_4 and Fe2O3 nanoparticles in rice plants and their potential as Fe fertilizers. *Environmental Pollution*, 269, 116134.

Lin, D., & Xing, B. (2008). Root uptake and phytotoxicity of ZnO nanoparticles. *Environmental Science & Technology*, 42(15), 5580–5585.

Liu, R., & Lal, R. (2015). Potentials of engineered nanoparticles as fertilizers for increasing agronomic productions. *Science of the Total Environment*, 514, 131–139.

Liu, X. M., Zhang, F. D., Zhang, S. Q., He, X. S., Wang, R. F., Feng, Z. B., & Wang, Y. J. (2005). Responses of peanut to nano-calcium carbonate. *Journal of Plant Nutrition and Fertilizers*, 11(3), 385–389.

Liu, G., Zotarelli, L., Li, Y., Dinkins, D., Wang, Q., & Ozores-Hampton, M. (2014). Controlled-release and slow-release fertilizers as nutrient management tools: HS1255/HS1255, 10/2014. *EDIS*, 2014(8), 1–6.

Liu, B., Asseng, S., Müller, C., Ewert, F., Elliott, J., Lobell, D. B., & Zhu, Y. (2016). Similar estimates of temperature impacts on global wheat yield by three independent methods. *Nature Climate Change*, 6(12), 1130–1136.

Ma, L., Zhang, M., Chen, J. Q., Kong, F. M., & Yang, Y. C. (2006). Effect of coated controlled release nitrogen fertilizer on yield increase of corn. *Phosphate and Compound Fertilizer*, 21(4), 12–14.

Maaz, T. M., Dobermann, A., Lyons, S. E., & Thomson, A. M. (2025). Review of research and innovation on novel fertilizers for crop nutrition. *npj Sustainable Agriculture*, 3(1), 25.

Mahmood, R., Arooj, A., & Sarfraz, R. (2024). Potential limitations of biochar. In R. Sarfaraz & C. Rensing (Eds.), *Biochar-solid carbon for sustainable agriculture* (pp. 126–138). Bentham Science Publishers.

McFadden, J., Njuki, E., & Griffin, T. (2023). Precision agriculture in the digital era: Recent adoption on US farms.

MET Trenkel. (2021). *Slow-and controlled-release and stabilized fertilizers: An option for enhancing nutrient use effiiency in agriculture*. International Fertilizer Industry Association (IFA).

Michelson, H., Gourlay, S., Lybbert, T., & Wollburg, P. (2023). Purchased agricultural input quality and small farms. *Food Policy*, 116, 102424.

Moberg, F., Höök, K., Salomonsson, L., Mastroianni, K., Lundberg, J., Karlberg, L., & Rundgren, G. (2015). How to feed nine billion within the planet's boundaries: The need for an agroecological approach. *Moberg, F., Ed.*

Mogollón, J. M., Beusen, A. H. W., Van Grinsven, H. J. M., Westhoek, H., & Bouwman, A. F. (2018). Future agricultural phosphorus demand according to the shared socioeconomic pathways. *Global Environmental Change*, 50, 149–163.

Mpandeli, S., & Maponya, P. (2014). Constraints and challenges facing the small scale farmers in Limpopo Province, South Africa. *Journal of Agricultural Science*, 6(4), 135.

Nair, P. M. G., & Chung, I. M. (2014). A mechanistic study on the toxic effect of copper oxide nanoparticles in soybean (Glycine max L.) root development and lignification of root cells. *Biological Trace Element Research*, 162, 342–352.

Naseem, M., Iqbal, S., Malik, H., Awais, M., Jehan, S., & Jabeen, S. (2024). Environmental implications of biochar. In R. Sarfaraz & C. Rensing (Eds.), *Biochar-solid carbon for sustainable agriculture* (pp. 109–125). Bentham Science Publishers.

Niu, J., Liu, C., Huang, M., Liu, K., & Yan, D. (2021). Effects of foliar fertilization: A review of current status and future perspectives. *Journal of Soil Science and Plant Nutrition*, 21(1), 104–118.

Oni, B. A., Oziegbe, O., & Olawole, O. O. (2019). Significance of biochar application to the environment and economy. *Annals of Agricultural Sciences*, 64(2), 222–236.

Pistilli, M. (2020). *10 top phosphate countries by production*. Investing News Networ.

Poudel, A., Singh, S. K., Jiménez-Ballesta, R., Jatav, S. S., Patra, A., & Pandey, A. (2023). Effect of nano-phosphorus formulation on growth, yield and nutritional quality of wheat under semi-arid climate. *Agronomy*, 13(3), 768.

Quinn, R. (2020). Global fertilizer outlook–6 global fertilizer demand bright. https://www.dtnpf.com/agriculture/web/ag/crops/article/2020/01/03/

Rahman, M. M., & Connor, J. D. (2022). Impact of agricultural extension services on fertilizer use and farmers' welfare: Evidence from Bangladesh. *Sustainability*, 14(15), 9385.

Raliya, R., & Tarafdar, J. C. (2013). ZnO nanoparticle biosynthesis and its effect on phosphorous-mobilizing enzyme secretion and gum contents in Clusterbean (Cyamopsis tetragonoloba L.). *Agricultural Research*, 2(1), 48–57.

Randive, K., Raut, T., & Jawadand, S. (2021). An overview of the global fertilizer trends and India's position in 2020. *Mineral Economics*, 34(3), 371–384.

Rockström, J., Steffen, W., Noone, K., Persson, Å., Chapin, F. S., III, Lambin, E., & Foley, J. (2009). Planetary boundaries: Exploring the safe operating space for humanity. *Ecology and Society*, 14(2), 32.

Sachan, D. S., Reddy, K. J., Saini, Y., Rai, A. K., Singh, O., & Laxman, T. (2023). Assessing grain yield and achieving enhanced quality in maize by next generation fertilizer: A review. *International Journal of Environment and Climate Change*, 13(8), 626–637.

Sadigov, R. (2022). Rapid growth of the world population and its socioeconomic results. *The Scientific World Journal*, 2022(1), 8110229.

Sane, M., Hajek, M., Nwaogu, C., & Purwestri, R. C. (2021). Subsidy as an economic instrument for environmental protection: A case of global fertilizer use. *Sustainability*, 13(16), 9408.

Sarfraz, R., Yang, W., Wang, S., Zhou, B., & Xing, S. (2020). Short term effects of biochar with different particle sizes on phosphorous availability and microbial communities. *Chemosphere*, 256, 126862.

Savci, S. (2012). An agricultural pollutant: Chemical fertilizer. *International Journal of Environmental Science and Development*, 3(1), 73.

Sekaran, U., Lai, L., Ussiri, D. A., Kumar, S., & Clay, S. (2021). Role of integrated crop-livestock systems in improving agriculture production and addressing food security—a review. *Journal of Agriculture and Food Research*, 5, 100190.

Shahwar, D., Mushtaq, Z., Mushtaq, H., Alqarawi, A. A., Park, Y., Alshahrani, T. S., & Faizan, S. (2023). Role of microbial inoculants as bio fertilizers for improving crop productivity: A review. *Heliyon*, 9(6), e16134.

Shanmugavel, D., Rusyn, I., Solorza-Feria, O., & Kamaraj, S. K. (2023). Sustainable SMART fertilizers in agriculture systems: A review on fundamentals to in-field applications. *Science of the Total Environment*, 904, 166729.

Shao Lei, S. L., Zhang Min, Z. M., Chen XueSen, C. X., & Wang LiXia, W. L. (2007). Effects of controlled release nitrogen fertilizer on yield and nitrogen content of soil and apple tree.

Shaviv, A., & Mikkelsen, R. L. (1993). Controlled-release fertilizers to increase efficiency of nutrient use and minimize environmental degradation—a review. *Fertilizer Research*, 35(1), 1–12.

Shu, X., He, J., Zhou, Z., Xia, L., Hu, Y., Zhang, Y., & Wang, C. (2022). Organic amendments enhance soil microbial diversity, microbial functionality and crop yields: A meta-analysis. *Science of the Total Environment*, 829, 154627.

Singh, G., & Singh, O. P. (1994). Effect of slow release urea materials on rice and succeeding wheat.

Smith, V. H., Tilman, G. D., & Nekola, J. C. (1999). Eutrophication: Impacts of excess nutrient inputs on freshwater, marine, and terrestrial ecosystems. *Environmental Pollution*, 100(1–3), 179–196.

Snapp, S., Sapkota, T. B., Chamberlin, J., Cox, C. M., Gameda, S., Jat, M. L., & Govaerts, B. (2023). Spatially differentiated nitrogen supply is key in a global food–fertilizer price crisis. *Nature Sustainability*, 6(10), 1268–1278.

Song FuPeng, S. F., Zhang Min, Z. M., Shi YanXi, S. Y., & Hu YingYing, H. Y. (2005). Releasing characteristics of controlled-release nitrogen fertilizer and its effects on rice yield.

Stamenković, S., Beškoski, V., Karabegović, I., Lazić, M., & Nikolić, N. (2018). Microbial fertilizers: A comprehensive review of current findings and future perspectives. *Spanish Journal of Agricultural Research*, 16(1), e09R01.

Steffen, W., Richardson, K., Rockström, J., Cornell, S. E., Fetzer, I., Bennett, E. M., & Sörlin, S. (2015). Planetary boundaries: Guiding human development on a changing planet. *Science*, 347(6223), 1259855.

Subbarao, G. V., Ito, O., Sahrawat, K. L., Berry, W. L., Nakahara, K., Ishikawa, T., & Rao, I. M. (2006). Scope and strategies for regulation of nitrification in agricultural systems—challenges and opportunities. *Critical Reviews in Plant Sciences*, 25(4), 303–335.

Tachibana, M. (2007). *Chissoasahi fertilizer Co. Ltd: Personal report. Aglukon (1992): Plantacote-the system of crop-specific plant nutrition.* Dusseldorf, Germany: Aglukon Spezialdunger GmbH.

Tadesse, M. (2014). Fertilizer adoption, credit access, and safety nets in rural Ethiopia. *Agricultural Finance Review*, 74(3), 290–310.

Takahashi, K., Muraoka, R., & Otsuka, K. (2020). Technology adoption, impact, and extension in developing countries' agriculture: A review of the recent literature. *Agricultural Economics*, 51(1), 31–45.

Tarafdar, J. C. (2021). Bionanofertilizer: A new invention for global farming. In A. Rakshit, V. S. Meena, M. Parihar, H. B. Singh, & A. K. Singh (Eds.), *Biofertilizers* (pp. 347–358). Woodhead Publishing.

Thapa, R., Chatterjee, A., Johnson, J. M., & Awale, R. (2015). Stabilized nitrogen fertilizers and application rate influence nitrogen losses under rainfed spring wheat. *Agronomy Journal*, 107(5), 1885–1894.

Tilman, D., Balzer, C., Hill, J., & Befort, B. L. (2011). Global food demand and the sustainable intensification of agriculture. *Proceedings of the National Academy of Sciences*, 108(50), 20260–20264.

Trenkel, M. E. (1997). *Controlled-release and stabilized fertilizers in agriculture* (Vol. 11). Paris: International Fertilizer Industry Association.

Tröster, M. F. (2023). Assessing the value of organic fertilizers from the perspective of EU farmers. *Agriculture*, 13(5), 1057.

Tsygvintsev, P. N., Goncharova, L. I., Manin, K. V., & Rachkova, V. M. (2018). Morphophysiological features of wheat (Triticum aestivum L.) seedlings upon exposure to nickel nanoparticles.

UNEP (2016b). Food systems and natural resources. A report of the working group on food systems of the international resource panel Westhoek H, Ingram J, Van Berkum S, Özay L and Hajer M p 164. https://www.resourcepanel.org/sites/default/files/documents/

UNEP/IFA. (2001). *Environmental aspects of phosphate and potash mining*. Paris: United Nations Environment Programme and International Fertilizer Industry Association.

Urso, J. H., & Gilbertson, L. M. (2018). Atom conversion efficiency: A new sustainability metric applied to nitrogen and phosphorus use in agriculture. *ACS Sustainable Chemistry & Engineering*, 6(4), 4453–4463.

USGS (2019). Mineral commodity summaries 2019. U.S. Geological Survey. https://prd-wret.s3-us-west-2.amazonaws.com/assets/palladium/production/atoms/files/mcs2019_all.pdf

USGS National Minerals Information Center (2020). Mineral commodity summaries 2020.

Vanlauwe, B., Bationo, A., Chianu, J., Giller, K. E., Merckx, R., Mokwunye, U., & Sanginga, N. (2010). Integrated soil fertility management: Operational definition and consequences for implementation and dissemination. *Outlook on Agriculture*, 39(1), 17–24.

Verma, K. K., Song, X. P., Degu, H. D., Guo, D. J., Joshi, A., Huang, H. R., & Li, Y. R. (2023). Recent advances in nitrogen and nano-nitrogen fertilizers for sustainable crop production: A mini-review. *Chemical and Biological Technologies in Agriculture*, 10(1), 111.

Wei, X., Chen, J., Gao, B., & Wang, Z. (2020). Role of controlled and slow release fertilizers in fruit crop nutrition. In A. K. Srivastava & C. Hu (Eds.), *Fruit crops* (pp. 555–566). Elsevier.

Williams, L.A. (2019) 'SNACK ATTACK! Hand-held entrees, portable snacks taking over consumption trends-bite by bite', *Prepared Foods*, 188(10), 15+, available: https://link.gale.com/apps/doc/A605833406/AONE?u=anon~63b3361f&sid=googleScholar&xid=bd6f0e91 [accessed 27 Nov 2025

World Bank (2020). Fertilizer consumption (% of fertilizer production). https://datacatalog.worldbank.org/search?search_api_views_fulltext_op=AND&query=fertilizer&nid=&sort_by=search_api_relevance&sort_order=DESC

Wreford, A., Ignaciuk, A., & Gruère, G. (2017). Overcoming barriers to the adoption of climate-friendly practices in agriculture.

Yadav, D. S., Chander, S., & Selvaraj, K. (2010). Agro-ecological zoning of brown planthopper [Nilaparvata lugens (Stal)] incidence on rice (Oryza sativa L.). *Journal of Scientific and Industrial Research*, 69(11), 818–822.

Yang, W. Y., He, M. R., Wang, Y. J., Wang, X., Zhang, B., & Wu, C. P. (2005). Effect of controlled-release urea combined application with urea on nitrogen utilization efficiency of winter wheat. *Journal of Plant Nutrition and Fertilizers*, 11(5), 627–633.

Yuan, H., Liu, Q., Guo, Z., Fu, J., Sun, Y., Gu, C., & Dhankher, O. P. (2021). Sulfur nanoparticles improved plant growth and reduced mercury toxicity via mitigating the oxidative stress in Brassica napus L. *Journal of Cleaner Production*, 318, 128589.

Zekri, M., & Koo, R. C. (1991a). Comparative effects of controlled-release and soluble fertilizers on young'Valencia'orange trees.

Zekri, M., & Koo, R. C. (1991b). Evaluation of controlled-release fertilizers for young citrus trees. *Journal of the American Society for Horticultural Science*, 116(6), 987–990.

Zhang, X., Davidson, E. A., Mauzerall, D. L., Searchinger, T. D., Dumas, P., & Shen, Y. (2015). Managing nitrogen for sustainable development. *Nature*, 528(7580), 51–59.

7

Enhancing Crop Resilience to Biotic and Abiotic Stresses Through Advanced Fertilizer Technologies

Muhammad Nauman Khan[1], Hansa Gul[2], Nasir Assad[3], Arshad Ali[3], Sher Ullah[3], Sofia Nazar[4], Haris Ahmed Khan[5], and Shah Fahad[6]

[1] Department of Botany, University of Chakwal, Punjab, Pakistan
[2] Department of Zoology, University of Mianwali, Mianwali, Pakistan
[3] Institute of Chemistry, University of Sargodha, Sargodha, Pakistan
[4] Department of Zoology, University of Sargodha, Sargodha, Pakistan
[5] Department of Biotechnology, University of Mianwali, Mianwali, Pakistan
[6] Department of Agronomy, Abdul Wali Khan University Mardan, Mardan, Khyber-Pakhtunkhwa, Pakistan

7.1 Introduction

Currently, farming has to deal with supplying enough food for a world where population rates are higher, as well as with various stresses in the environment and from living things. Changes in the climate, damage to land, and unsustainable farming practices have aggravated the problems with drought, temperature extremes, and harmful organisms that endanger crops worldwide. These difficulties make it tough for agroecosystems to function well, resulting in a loss of 30–50% of crops yearly (Lesk et al., 2016).

An increasing number of nations rely on intensive farming with high levels of chemical fertilizer application for feeding their populations. Over the years, chemical fertilizers have greatly helped to boost crop yields, causing many areas to rely less on imports. Yet, their excessive and careless application has caused considerable harm to the environment. The residue from fertilizers enters the soil, streams, and aquifers and causes contamination of water sources (Savci and Development, 2012). These chemical fertilizers affect the composition of nutrients, decrease the soil's ability to retain water, alter microbial balance, and increase soil salinity, all of which erode long-term soil health and fertility.

Savci's study shows that chemical fertilizers pose a high risk of contaminating the environment. Unabsorbed nutrients, notably nitrogen and phosphorus, are carried into water sources where they cause explosive algal growth and deoxygenation of the water (Savci & Development, 2012). This results in a downward spiral where soil fertility diminishes, plants grow less capable of defending themselves, and ever-higher input amounts are needed to maintain yields—a trend that is economically and environmentally unsustainable (Foley et al., 2011).

There has been growing interest in using organic and sustainable methods as alternative ways of producing food. Organic farming practices depend on natural inputs and favor soil diversity, minimal nutrient loss, and cleaner products. Yet, organic agriculture cannot provide sufficient food for the world but shows great promise for certain crops and regions. Additionally, it provides insights that can help mainstream agriculture reduce its reliance on risky inputs (Glick, 2012).

Nevertheless, total abandonment of synthetic nutrient products is impractical due to extensive worldwide food needs. Advanced, ecologically friendly fertilizer technologies are now drawing greater attention from the agricultural industry. These new fertilizers are developed to deliver the productivity advantages of conventional fertilizers without the negative impact on the environment. The range of advanced fertilizers addresses these issues by utilizing specially engineered formulations that promote efficient nutrient distribution and minimize environmental impact.

They are being developed to both nourish plants and strengthen their ability to adapt to environmental challenges. Several types of nutritional supplements, such as those containing a combination of micronutrients, plant growth-promoting rhizobacteria (PGPR), and biostimulants, have the ability to promote the activation of antioxidative mechanisms and help plants cope better in the face of stress (Mahanty et al., 2017). Moreover, given that abiotic stresses are expected to intensify due to climate change, these defense pathways and metabolic activities are all the more important.

Oerke (2006) estimates that these problems cost the agriculture industry between 25% and 40% of its total production every year. Heavy use of pesticides to combat biotic stress can be expensive and endangers both people and the environment. Integrated stress management practices that include improved crop varieties, natural enemies, and careful fertilizer application are rapidly becoming recognized as the most viable way forward (Oerke, 2006).

Effective fertilization plays an essential part in the development of integrated stress management. Applying the right fertilizers can build plant health and boost defense mechanisms against both living and nonliving threats. In particular, potassium has been shown to improve resistance to certain diseases and promote better water use by plants. Calcium strengthens the cell wall and guards against diseases, while silicon prevents wilting under dry conditions or salty soil. Blending these insights with advanced fertilizer formulation strategies and innovative delivery systems is transforming how we ensure that crops get exactly what they require, no more and certainly no less, as seasons change and challenges arise. Such delivery systems help lower waste and mitigate greenhouse gases, enabling farmers to achieve higher productivity through sustainable intensification strategies anywhere in the world (Pandey et al., 2017).

However, fertilizer application should be carefully coordinated with efforts to maintain and improve soil health. Soils with high levels of organic substances and a diverse population of microbes can better resist stressors, store nutrients, and maintain long-term productivity. The combined application of organic amendments, cover cropping, reduced tillage, and biologically active fertilizers promotes the recovery of degraded soils and enhances their capacity to withstand climate stress (Bitew & Alemayehu, 2017).

Fertilizers continue to play a major role in maintaining food security, but their application should become smarter and more environmentally friendly over time. Fertilizers increasingly play a more strategic role in helping plants overcome stress and preserving productive ecosystems. This chapter examines how fertilizers can help improve plants' ability to withstand biotic stresses and contribute to a more durable and environmentally friendly food production (Suhag, 2016).

7.2 Understanding Biotic Stresses

Biotic stress occurs when organisms living on or around plants hinder their normal development and reduce yields. Examples of biotics that cause stress are viruses, bacteria, fungi, nematodes, insects, and arachnids, as well as weeds that compete for available resources (Figure 7.1). Biotic stresses originate from living organisms and sometimes contribute to substantial reductions in crop production both before and following harvest.

Plants are hurt by pathogens such as fungi, bacteria, and viruses when these organisms penetrate the tissues and interfere with important bodily functions. Plants are susceptible to damage caused by pathogenic organisms that enter and interfere with the plant's normal functions. Fungal diseases account for some of the greatest losses in plants by causing a range of symptoms like wilting, brown spots, decay of tissues, and clogging of roots. Infections by bacteria frequently lead to cankers, galls, and soft rots, whereas viruses often result in mosaic patterns, stunted growth, and yellowing of leaves. A single species of fungus, such as the rice blast pathogen (*Magnaporthe oryzae*), is responsible for enormous economic damage on an annual basis. These pathogens penetrate plants via susceptible areas such as wounds or openings and release chemicals that damage and destroy host tissue (Dean et al., 2012; Oerke, 2006).

Besides depriving plants of their resources, insects and pests can spread hundreds of plant diseases through their movements. These insects suck the sap from plants, making them

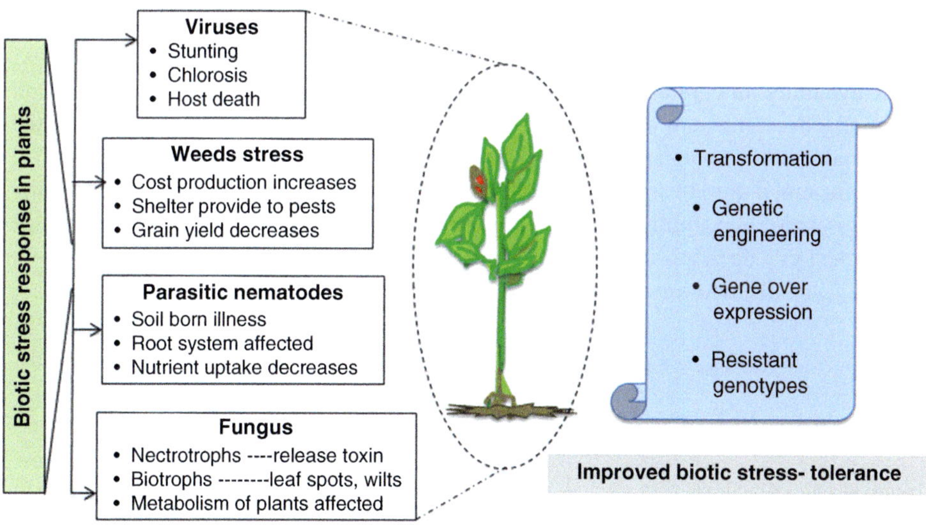

Figure 7.1 Different types of biotic stress in plants.

vulnerable and helping to spread viruses that result in broad-based disease outbreaks. Infestations of root-knot nematodes damage the root system and cause decreases in water and nutrient absorption, leading to poor plant growth and reduced yield. Crop interactions with pests and weeds can result in severe damage, since these organisms compete for essential natural resources. Severe weed infestations could lower yields by half (Oerke, 2006). Weeds also provide habitats that allow pests and diseases to multiply and move to different crops. Weed invasions are especially challenging in monoculture farming systems, since there is little plant variety to discourage their growth. Losses caused by biotic stressors greatly affect the agricultural industry. It is reported that up to 40% of crops are lost to pests and diseases around the world each year. In developing countries, these losses are often higher due to limited access to effective pest management tools and resistant cultivars. Furthermore, chemical control methods such as pesticides and fungicides, while effective, carry environmental and health risks, including soil and water contamination, nontarget species harm, and development of resistance in pest populations (Glick, 2012, 2014).

Plants have developed complex systems to counter biotic stresses, although they do not possess an adaptive immune system. They comprise anatomical resistance that comprises thickened cell walls, waxy cuticles, and trichomes, and biochemical resistances that include phytoalexins, pathogenesis-related (PR) proteins, and the hypersensitive response that restricts the spread of pathogens by causing programmed death of host cells. These defense responses seem to be coordinated by molecular signaling pathways mediated by such hormones as salicylic acid, jasmonic acid, and ethylene.

Resistance to genetic factors by means of breeding programs has been one pillar in the biotic stress management. Production of resistant strains will decrease the use of chemical methods and make them sustainable. Faster methods of incorporating resistance into crops by using modern methods such as marker-assisted selection and genetic engineering have modernized this process. Other growth regulators like auxin and cytokinins can also be used in the priming of seeds to increase plant vigor and tolerance to stress (Hafeez et al., 2023).

However, managing biotic stress remains challenging due to evolving pathogen virulence, pest resistance to chemicals, and changing climatic conditions that affect pest and disease dynamics. Integrated pest management (IPM), which combines biological control, cultural practices, resistant varieties, and judicious chemical use, is increasingly promoted to sustainably mitigate biotic stresses (Savary et al., 2019).

In conclusion, biotic stresses significantly limit crop productivity worldwide. Addressing these stresses requires a multifaceted approach that integrates advanced breeding, improved management practices, and sustainable use of agrochemicals to safeguard food security in a changing global environment.

7.3 Understanding Abiotic Stresses: Drought, Salinity, Temperature Extremes

Abiotic stresses are those that a nonparticipating plant undergoes, that is, drought, salinity, high temperatures, low temperatures, and harmful factors. All these pressures reduce the production of crops and disrupt the functioning of plants at various levels. As the global population passed 7 billion and is predicted to go up even further in 2020, there is increasing demand for agricultural production to meet the needs of everyone. Furthermore, at the same time as industrialization and urbanization, the environment is suffering more damage, which makes it even tougher for crops to grow properly. Among abiotic stressors, drought is the most damaging and greatly reduces the amount of food that can be grown. Less rainfall, inconsistent rainfall, and stronger evaporation, influenced by climate change, are the reasons it occurs (Glick, 2012). If drought is severe and lasts for a long time, it lessens plant metabolism by slowing photosynthesis and stopping growth. While plants try to conserve water by slowing down growth and reducing activity during drought, the results can be permanent damage and a lack of crops if there is not enough rain. When soil does not hold enough moisture, plants find it difficult to take up nutrients, disturbing their cells, which results in lower yield at harvest (Gull et al., 2019).

Agriculture in arid and semiarid areas is often limited by the stress caused by high salinity. Weathering of rocks is one natural way for soil to become salty, but the process is sped up when we use salty water to irrigate and there is poor drainage. Too much salt in the soil makes water difficult for roots to take up and can destroy cells in the body of the plant. The concentration of sodium and chloride ions can change the amount of nutrients and make the soil less fertile. As a result, the ability of photosynthesis to work properly, the enlargement of plant tissues, and the vitality of the crop are all lowered, which, in turn, causes crops to produce fewer yields. Salt stress can make soil less capable of storing water and make the soil more salty (Fahad et al., 2017).

Agricultural activities can be severely affected by sharp changes in temperature, whether the weather is hot or cold. When plants are subjected to cold, their cell walls are harmed and enzymatic actions are interrupted, leading to things like sterility in flowers of both rice and maize. Some plants are able to respond to colder temperatures to survive them. Also, it is important to note that many paleo-agriculture crops are still at risk, especially when they are growing the most. Increasing global warming causes more instances of heat stress, and this results in proteins becoming unstable and both membranes and cells becoming damaged. It also pushes out more water from drying out the soil, creating a worse impact of drought. Hot weather decreases both the number of viable pollen grains and the number of seeds, resulting in poorer and smaller crop yields (Umar et al., 2021).

Chemical fertilizers are necessary for high yields, but their continuous use may cause abiotic stresses to become worse. Using too many chemical fertilizers results in water and soil pollution, dirtier soil, and problems with soil nutrients. Residual chemicals in the soil get into groundwater and encourage unwanted algal growth in water bodies, harming both nature and humans. For this reason, efforts like organic farming are being recognized due to their effects on soil quality and capacity to withstand several unfavorable stresses in the environment (Hossain et al., 2022).

Plants deal with abiotic stress by using different physiological and molecular ways such as osmotic adjustment, production of stress proteins, and activating antioxidant actions. Scientists apply current biological discoveries and technological tools to breed drought- and salt-tolerant crop varieties. Advanced fertilizer methods help plants withstand stress by ensuring better absorption of needed nutrients (Raza et al., 2025).

Overall, drought, salt in the soil, and high or low temperatures are the biggest challenges to food production worldwide. Solving these issues means using better crop types, careful soil handling, and using less fertilizer to help plants deal with these challenges and maintain an agricultural system that lasts (Figure 7.2).

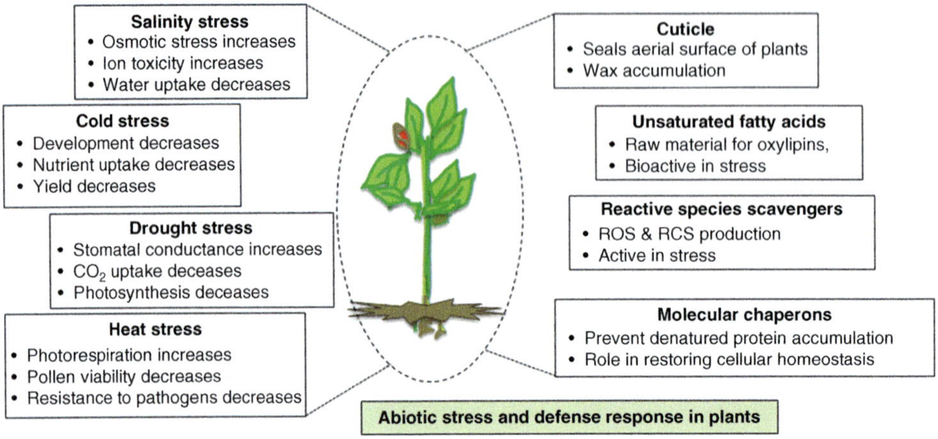

Figure 7.2 Abiotic stress and defense in plants.

7.4 Mechanisms of Stress Tolerance in Plants: The Role of Fertilizers

Nitrogen (N), phosphorus (P), and potassium (K), known as NPK, are basic to modern agriculture and are vital for proper growth, good development, and increased plant yields. In addition to providing basic nutrients, macronutrients greatly affect the tolerance of plants to stresses such as heat, sunlight, disease, and insects. Nitrogen helps with many tasks such as being used in amino acids, nucleic acids, and chlorophyll, as well as regulating several things like the way stomata open and close, the efficiency of photosynthesis, and how stress-responsive proteins are made (Xu et al., 2012). Under biotic stress, nitrogen has been shown to enhance the expression of PR genes and promote systemic acquired resistance (SAR), thus strengthening plant immunity (Sun et al., 2020).

Phosphorus supports energy transfer through ATP, provides signals via cell messengers, and is included in phospholipids, which strengthen cell membranes. When there is drought or high salt levels, phosphorus helps the plant's roots to go deeper and support finding both water and nutrients. A lack of phosphorus results in less lignin in the plant's tissues, which makes it easier for diseases and pathogens to affect the plant. The process of making nucleotides and coenzymes, which are key during stress on the body, is better with phosphorus (Khan et al., 2023a).

Potassium ensures that photosynthesis, protein synthesis, and fluid control happen in the plant. It manages the amount of opening and closing in plant pores that control water loss during drought and heat stress. Potassium also helps to make more secondary metabolites and antioxidant enzymes like catalase (CAT), peroxidase (POD), and superoxide dismutase (SOD), which look for reactive oxygen species (ROS) and stop damage caused by stress. If plants lack potassium, they become more likely to contract powdery mildew and blight (Pandey & Mahiwal, 2020).

Nevertheless, using conventional NPK fertilizers too much and improperly has resulted in various environmental and agronomic problems. Nitrogen and phosphorus that flow into lakes and rivers cause eutrophication and areas with low oxygen, which greatly impact both water quality and the diversity of aquatic life. When nitrogen fertilizer is applied in large amounts, it changes the pH of the soil and lowers the number of helpful microbes, leading to eventual soil infertility. Zhang et al. (2015) state that the worldwide use of nitrogen fertilizers on cereals is just 33%, so the other 67% is expected to end up in the environment as greenhouse gases, mainly nitrous oxide (N_2O), which harms the climate (Zhang et al., 2015).

In addition, repeated use of synthetic fertilizers can make soil less healthy, lower its organic material, disturb the soil's bacteria and microorganisms, and cause crops to face more disease and stress from weather changes. They have led to more efforts to find better and greener solutions, like biofertilizers, nanofertilizers, and biostimulant-containing formulations, which boost plant nutrition absorption and make plants more stress-resistant. All in all, farmers still need traditional fertilizers in agriculture, though they cause concerns for the environment through their stress tolerance benefits. It is important to find smarter ways to fertilize that support plant nutrition and help them defend themselves against all types of stress.

7.5 Next-Generation Fertilizers for Biotic Stress Management

NGFs go beyond regular NPK fertilizers by including advanced materials and living agents that help nutrients be used better and also boost plants' resistance to diseases, pests, and nematodes. These fertilizers involve nanofertilizers, biofertilizers, slow-release or controlled-release fertilizers (CRFs), and products with microbial inoculants or plant biostimulants that engage with the plant immune system or the soil around the roots to help the plant resist. Among the most promising innovations are nanofertilizers, which utilize nanoparticles—materials with dimensions typically less than 100 nanometers (Assad et al., 2025d; Ghaffar et al., 2023; Jabbar et al., 2023). These particles can be synthesized through physical, chemical, or green (plant-based) methods and are designed to carry essential nutrients in a controlled-release manner (Khan et al., 2023b; Roheen et al., 2025). Due to their high surface area and reactivity, nanoparticles enhance nutrient uptake, reduce losses, and improve stress tolerance (Hashmi et al., 2024; Hussain et al., 2025). Different metal and metal oxide nanoparticles, such as gold (Au) (Assad et al., 2025c), silver (Ag) (Kanwal et al., 2025), and zinc (ZnO) (Assad et al., 2024), have shown remarkable potential in improving plant immunity and nutrient efficiency (Assad et al., 2025a). In agriculture, nanoformulations such as nano-urea, nano-zinc, and nano-silica are increasingly recognized for their role in boosting plant resistance to both biotic and abiotic stresses while minimizing environmental impact (Assad et al., 2025a, b). They control the release of nutrients over a suitable period that causes less waste and provides more prolonged growth. Moreover, some nanoparticles, such as zinc oxide (ZnO), iron oxides, copper oxide(CuO), and silver nanoparticles, are antimicrobial (Gula et al., 2025), which are helpful to reduce the count of fungal and bacterial diseases in crops. The application of zinc nanofertilizers has been found to boost different genes and antioxidants in the plant and, as a consequence, raise the resistance of the plant to bacterial blight and rust. Similarly, CuO nanoparticles synthesized using green methods have been demonstrated to exhibit not only biocidal but also nutrient effects, most often on fungal pathogens (Pascoli et al., 2018).

Nutrient fixation and plant defense happen because of the biofertilizers containing Rhizobium, Azospirillum, *Pseudomonas fluorescens*, *Trichoderma harzianum*, and arbuscular mycorrhizal fungi. When activated on plant roots, the microbes produce substances like antimicrobials, siderophores, and enzymes known as chitinases and beta-glucanases that have the effect of retarding the growth of a pathogen (Vessey & Soil, 2003).

P. fluorescens produces the compound DAPG, which helps prevent harmful fungi and improves root growth. Moreover, mycorrhizal fungi can strengthen plant resistance against root diseases by modifying how roots are formed and increasing their production of secondary metabolites. CRFs and polymer-coated urea can help plants grow well by giving them nutrients at the same time as they are needed. With these fertilizers, buildups of nitrogen are prevented, saving the plants from excess production of fungus and helping them thrive with fewer diseases (Du Jardin, 2015; Rico et al., 2011).

Additionally, fertilizers that contain seaweed extracts, humic substances, protein hydrolysates, and plant signaling molecules have been shown to increase a plant's defense system and the production of phytoalexins. These substances switch on pathways such as JA and SA, which are important for the plant's protection against disease and pest attacks. Recent findings suggest that biostimulants can influence stress hormones and antioxidant systems, improving how crops respond to various stresses (Calvo et al., 2014). By including NGFs in sustainable crop management, farmers can cut back on using chemical pesticides while staying true to climate-smart agriculture. By using fertilizers, soil health is maintained as they help reduce nutrient leaks, greenhouse gases, and aid in defending crops in areas where weather is uncertain and resources are limited.

Overall, advanced fertilizers use several approaches to fight biotic stress by boosting plant nutrient use and resistance at the same time. The use of nanotechnology, beneficial microbes, and biostimulants by NGFs is guiding agriculture toward a sustainable and productive future. As shown in Table 7.1, different types of fertilizers—including conventional NPK and advanced formulations like nanofertilizers and biofertilizers—play distinct roles in enhancing plant tolerance to both abiotic and biotic stresses through various physiological and biochemical mechanisms.

7.6 Fertilizer Innovations for Abiotic Stress Resistance

In modern times, farming has to cope with more difficulties because of drought, salinity, and extreme temperatures. Due to this, using improved fertilizers helps crops cope with difficult environmental factors and continue to thrive. A traditional fertilizer can be less exact and efficient, causing nutrients to wash away, harm the environment, and prevent plants from using them well during stressful times. Advances in fertilizer research are focused on creating "smart" fertilizers that improve plant nutrient use and increase their ability to tolerate stress (Dimkpa & Bindraban, 2016).

Nanofertilizers are an important innovation because they allow nutrients to be applied at the nanoscale and make them easily absorbed by crops. Nanofertilizers play a role in boosting photosynthesis, enzymes, and the management of water in the plant under stress situations. The use of nano-zinc, iron, and copper on plants has led to notable improvements in their defense systems against ROS-caused oxidative damage during drought and salt

Table 7.1 Comparison of traditional and next-generation fertilizers based on their mechanisms, stress mitigation roles, benefits, and environmental implications in plant stress tolerance.

Fertilizer type	Mechanism of action	Stress type managed	Key benefits	Environmental concerns/advantages
Nitrogen (N)	Regulates stomatal function, photosynthesis, and synthesis of proteins and chlorophyll, activates PR genes and SAR pathways	Heat and biotic stress (pathogens)	Enhances immunity, growth, and metabolic function	Overuse leads to eutrophication, GHG emissions (N_2O), and soil acidification
Phosphorus (P)	Enhances ATP production, signal transduction, root development, and lignin synthesis	Drought, salinity, and pathogens	Aids in energy metabolism, strengthens tissues, and supports water/nutrient uptake	Loss in water bodies causes eutrophication
Potassium (K)	Regulates stomatal opening, enzyme activation, and ROS scavenging via SOD, CAT, and POD	Drought, heat, and fungal diseases	Improves water use efficiency, stress enzyme production, and disease resistance	Excessive use reduces soil microbial health
Nanofertilizers (e.g., ZnO and CuO)	Controlled nutrient release, antimicrobial action, and gene regulation	Fungal and bacterial infections	Sustained nutrient availability, disease suppression, and improved immunity	Reduces nutrient leaching and waste
Biofertilizers (e.g., AMF, Rhizobium, and P. fluorescens)	Produce antimicrobial compounds, fix nutrients, and enhance secondary metabolites	Pathogens and soil-borne diseases	Boosts plant defense, root growth, and soil health	Eco-friendly and enhances microbial biodiversity
Controlled-release fertilizers (CRFs)	Slow nutrient release matches plant demand	General stress and fungal overgrowth	Prevents nutrient buildup and improves plant health	Reduces nutrient runoff and minimizes excessive growth of pathogens
Seaweed extracts and biostimulants	Induce signaling pathways (JA and SA), promote phytoalexin and antioxidant production	Pest and disease attack	Strengthens innate immunity and enhances stress hormone balance	Biodegradable and promotes resilience
Traditional NPK (overuse scenario)	Direct nutrient supply but often inefficient	Temporary relief from nutrient stress	Boosts immediate growth	Causes environmental degradation, GHG emissions, and microbial imbalance
Next-generation fertilizers (NGFs)	Combine nanotech, microbes, CRFs, and biostimulants to enhance nutrient use and immunity	Multiple biotic stresses	Holistic stress management, reduced pesticide need, and enhanced crop yield	Align with climate-smart agriculture; minimize environmental impact

stress (Afzal et al., 2024; Faryal et al., 2022). Nano-urea is notable because it helps plants use nitrogen more effectively and cuts ammonia losses, especially in hot conditions.

Biostimulant-enriched fertilizers are another important advancement. They include substances like humic substances, amino acids, seaweed extracts, and helpful microorganisms, which help address plant stress by balancing hormone levels in plants, improving the shape of roots, and boosting the uptake of nutrients. Example of how amino acid-based fertilizers assist plants in handling and regulating water in their tissues under drought conditions (Calvo et al., 2014). CRFs permit nutrients to be distributed as plants require them, which reduces both losses and harm to the environment. CRFs show high effectiveness when plants are under stress and their metabolism changes. Even if the roots are stressed from drought or salinity, both essential and minor nutrients are available because they are slowly and continuously released by the material.

Biofertilizers containing stress-tolerant PGPR also represent an innovative strategy. These microbes can fix nitrogen, solubilize phosphorus, and produce stress-alleviating phytohormones like indole-3-acetic acid (IAA) and abscisic acid (ABA), which help plants maintain growth and water balance under adverse conditions (Saeed et al., 2022; Wang et al., 2016).

For the most part, progress in fertilizers for abiotic stress resistance is key to keeping global food security steady. Thanks to progress in nanotechnology, microbiology, and biochemistry, these new fertilizers make plants more able to survive in different environments and less reliant on standard chemical products.

7.7 Fertilizer Coatings and Delivery Systems for Stress Management

How nutrients are given to crops is a major factor in determining how effective and efficient fertilizers work, especially when the plants are facing stress. New fertilizer coatings and modern ways of applying them are promising methods to resolve the issues of traditional fertilizer use and enhance plants' ability to survive droughts, high salt, and extreme weather.

With a coating of polymer, sulfur, or resin, these fertilizers are built to release their nutrients over time, ensuring a longer and more sustained feeding. The way nutrients are slowly released means they are available when plants need them most in stressful conditions. One way to do this is to coat urea with a polymer, which can minimize the loss of nitrogen through leaching and volatilization and increase its efficiency under water-short conditions. By ensuring that nitrogen is properly available, these systems support photosynthesis and protein production in stressed plants. Because of this, the plants can maintain their yield. In addition to delivering nitrogen slowly, sulfur-coated fertilizers add sulfur, which plants use to make stress-related amino acids. They contribute to the formation of antioxidant enzymes, which reduce the impact of high salinity and temperature changes on the organisms (Mustafa et al., 2022).

Microencapsulation is another cutting-edge technique in which nutrients are enclosed within biodegradable materials that degrade under specific environmental triggers. This delivery method can protect sensitive nutrients such as micronutrients or biostimulants from degradation in extreme pH or oxidative environments, enhancing their effectiveness during stress. For instance, encapsulated micronutrients such as zinc and iron are more efficiently taken up by crops under salinity stress compared to conventional forms, as they remain stable and bioavailable in adverse soil conditions (Mustafa et al., 2022, Shen et al., 2023).

New developments in smart fertilizers, featuring sensors or pH-controlled release, make it possible to give plants nutrients on demand based on the present soil or plant state. Though they are not fully developed yet, these systems show promise and will likely drive improvements in agriculture in areas facing stress. As a result, proper use of fertilizer coatings and application methods helps crops make better use of nutrients and withstand stress. With these innovations, plants get the nutrients they need when they need them, which lessens abiotic stress and supports sustainable farming.

7.8 The Condition of Soil and Stress Resistance

When oil health is good, agricultural yields can increase and are more likely to withstand abiotic stress. Plants are better able to resist dry conditions, salt problems, and extreme temperature changes when soils are strong and healthy. Combining soil health methods with the use of fertilizers can make crops better able to withstand stress. Soil organic matter (SOM) greatly helps plants adapt to stress caused by abiotic factors. It makes it easier for plants to hold water, maintain healthy soil structure, and remove harmful chemicals, all important in extreme conditions. Adding organic fertilizers to soil increases SOM and speeds up biological actions, which strengthen the soil ecosystem. They enhance the number of roots and help nutrients go back into the system, which plants need to survive under stress. When chemical fertilizers and organic fertilizers are combined in a strategy referred to as integrated nutrient management (INM), the results for improving soil quality and crop growth have been very good, especially when crops face stress. By following the guidelines of INM, crop nutrients are used more constructively, and less harmful chemical fertilizers are required, which helps to preserve the properties of the soil. As an illustration, blending urea with compost in drought-like environments helps the plant take up nitrogen and keep the soil strong and wet (Singh et al., 2024). A plant's ability to fight stress largely depends on the role of AMF and PGPR in soil microbial communities. Using fertilizers that are enriched with bio-inoculants can attract useful microbes, which lead to greater nutrient uptake and healthy hormones in the plants. Thanks to these microbes, plants can use more phosphorus, microelements, and water more efficiently (Wahab et al., 2023).

All these factors, including pH and salinity, can affect the amounts of nutrients soil can hold. In seawater, Na^+ and Cl^- tend to build up and block the entry of nutrient elements like potassium and calcium into cells. If used as a soil amendment, gypsum ($CaSO_4$) replaces excess sodium ions and brings ionic balance back to sodic soils. If fertilizers with calcium and magnesium are given to the soil at the right time, they can conduct a better root function under stress (Chhabra et al., 2021). Micronutrient fertilization helps plants maintain their activity level when they experience stress. For enzymes to function and for chlorophyll to be produced, zinc, boron, and iron are important. Providing essential nutrients to the leaves or directly into the soil in degraded areas can make plants healthier and improve their response to stresses such as by producing osmolytes and ROS scavengers (Oerke, 2006).

Overall, strong and healthy soil is vital, and proper fertilization plays a big role in helping plants survive during stressful conditions. Applying organic matter restoration, inoculating with microbial materials, and choosing appropriate chemical fertilization promote a sustainable farming system able to handle climate-related stresses.

7.9 Applying Fertilizer Technologies Alongside Other Pressure-Reduction Methods

To manage stress well in agriculture, fertilizers should be combined with additional agronomic and biological techniques to address issues such as drought, salinity, and excessive temperature. Even though fertilizer increases nutrients for plants, it rarely makes them resilient enough to survive stressful conditions. Discovering solutions like choosing drought-resistant plants, more precise irrigation, mulching, or introducing biological helpers means plants can do better and help the environment. A suitable strategy in arid and semiarid zones is to blend CRFs with both deficit and drip irrigation approaches.

Through CRFs, nutrients are less likely to be carried away by rain or evaporation, and water irrigation helps provide water to plants in their growing area. Because of this, farmers use water and nutrients more effectively than they would alone, a key benefit in times of drought. Likewise, biofertilizers applied with organic or mineral fertilizers have demonstrated their use in strengthening plants. Fertilizers made from helpful microbes such as Azospirillum, Rhizobium, and Pseudomonas support healthy soil and growth of plant roots, allowing better access to food and water under pressure (Anand & Practices, 2023; Vessey & soil, 2003).

Fertilizers enriched with gypsum or having sulfur on urea, together with compost, are useful for soils with high salt content. They enhance the structure of the soil and give a better balance of ions to avoid sodium toxicity and help minerals become more available. PGPR applied with fertilizers additionally helps reduce the production of ethylene, manages stress-related genes, and enhances how plants use light during photosynthesis (Egamberdieva et al., 2017). The use of nanotechnology in fertilizers, together with precision agriculture technologies like remote sensing and GIS mapping, creates new possibilities. These fertilizers deliver nutrients and chemicals that reduce stress to each site where they are applied. By combining them with stress-resistant plants, these fertilizers form a reliable way to fight climate-related stress (Kah et al., 2018).

By adopting this approach, farmers are putting into practice what climate-smart agriculture stands for: efforts to raise productivity and to keep the land safe and able to withstand extreme weather conditions. But successful use of these technologies relies on training farmers, having good policies in place, and considering the cost-effectiveness to encourage wider adoption.

7.10 Future Directions: Fertilizer Research Dedicated to Enhancing Resistance to Stresses

Scientists are now paying more attention to creating fertilizer technologies that are climate-proof, eco-friendly, and economical. In today's changing world, fertilizers should not only feed the soil but also increase plants' resistance to stress, protect the quality of soil, and reduce the harm to the environment. Research on developing smart fertilizers that react appropriately to plants and the environment looks very promising. They are made up of fertilizers that provide nutrients when the pH level, temperature, or moisture is right. For example, moisture-triggered fertilizers supply nutrients when the field receives irrigation or rain, so nutrients are not lost and water is used wisely in areas where there is a drought (Sun et al., 2020).

Also, scientists focus on nanofertilizers that deliver essential nutrients with exact control at the level of cells. It has been found that small quantities of zinc, copper, and iron can help plants survive stressful conditions by improving their enzymes, their defenses against oxidative damage, and by strengthening their root system. Furthermore, crafting micronutrients within biodegradable materials like chitosan or alginate ensures the nutrients dissolve slowly and need less fertilizer (Co-operation & Nanomaterials, 2012).

Managing stress in the future is expected to rely more on the use of biological fertilizers. Thanks to new findings in microbial genomics and synthetic biology, we can now build customized groups of microorganisms that help fix nitrogen, free stored phosphorus, and produce essential hormones when plants are under stress. Applying biostimulants with organic or chemical fertilizers can improve the efficiency of the process. Using AI and data analytics in fertilizer application is totally transforming how farmers handle nutrients. By using IoT soil sensors, drones, and machine learning, farmers are able to see soil nutrients, notice when their crops are stressed, and get timely guidance on fertilizer use (Klerkx & Rose, 2020; Reddy et al., 2024).

Rules and guidelines to support the study and use of innovative fertilizers are being set up as well. Organizations and international groups are spending money on eco-efficient fertilizers and encouraging teams from the public and private sectors to work together and produce more for use by farmers. All in all, future studies on fertilizer will focus primarily on finding products that feed crops, enhance stress tolerance, maintain healthy soils, and help the environment.

7.11 Conclusion

The growing prevalence of biotic and abiotic stresses in agriculture presents a critical challenge to global food security. Traditional farming methods and conventional fertilizers alone are no longer sufficient to meet the nutritional demands of a growing population under rapidly changing climatic conditions. This chapter presents the power of advanced fertilizer technologies in increasing the resilience of plants and the sustainability of crop yields. The new generation of fertilizers (nanofertilizers, biostimulants, biofertilizers, and controlled-release formulations) enables more precise provision of nutrients to the plant, minimizes their losses to the environment, and enhances the physiological and biochemical defenses of the plant. Their insertion into crop management strategies has been proven to improve plant growth, even yield stability and resistance to adversities in crop production such as pests, diseases, drought, and salinity tolerance. Furthermore, the technologies in fertilizer coating and delivery mechanism also result in the ability to maximize nutrient uptake during periods of stress so that plants in need obtain the necessary assistance at the most appropriate time. The concept of soil health becomes a defining variable in the establishment of stress response within the plant. The use of advanced fertilizers in combination with innovative agricultural activities such as organic amendments, inoculation of microorganisms, and INM would strengthen the soil and increase microbial diversity and overall soil fertility in the long run. The combination of smart fertilization and precision farming with agronomy interventions such as crop rotation, mulching, and efficient irrigation is an example of a system approach in attaining stress-resilient farming. In the distant perspective, future technologies in fertilizer should be developed with the perception of encroachment, organic assimilation, and cyber intelligence. Artificial intelligence and sensor-enabled systems can turn fertilizing into a smarter process where nutrients are added at the time when they are really needed. It will be through capacity-building, research, and farmer awareness that these innovations can be seen to have achieved their potential in the field. Novelty in fertilizer can eventually become one of the pillars of a climate-resilient, productively oriented, ecologically friendly agriculture, provided it is anchored in a system of sustainable practices. This is the way ahead, with integration, innovation, and responsible implementation.

References

Afzal, M., Ullah, S., Assad, N., Naeem-Ul-Hassan, M., Kanwal, M., Mubashar, B., Shahzad, R., Nauman Khan, M., Javed, M. A., & Wahab, S. (2024). Photocatalytic degradation of organic dyes and biological potentials of biogenic zinc oxide nanoparticles synthesized using the polar extract of cyperus scariosus R. Br.(Cyperaceae). *Green Processing and Synthesis*, 13, 20240038.

Anand, A. J. M. A.-T., & Practices. (2023). Slow release and coated nitrogen fertilizer for enhancing nutrient use efficiency and increasing crop production. In S. Gulaiya, A. Sharma, P. Kochale, P. Sharma, & S. B. Agrawal (Eds.), *Modern Agronomy-Theory and Practices* (p. 254–267). Laxmibai colony Gwalior, M.P.: Navi Publications.

Assad, N., Abbas, A., Ur Rehman, M. F., & Naeem-Ul-Hassan, M. (2024). Photo-catalytic and biological applications of phyto-functionalized zinc oxide nanoparticles synthesized using a polar extract of equisetum diffusum d. *RSC Advances*, 14, 22344–22358.

Assad, N., Ahmad, H., Abbas, A., Ur Rehman, M. F., Naeem-Ul-Hassana, M., Sher, M., Riaz, T., Laila, M. B., Zahra, I., & Ullah, F. (2025a). Green synthesis of selenium nanoparticles using equisetum diffusum: characterization, antibacterial potential, effects on plant growth. *IEEE Transactions On Nanobioscience*, 24(4), 443–453.

Assad, N., Ansa, Laila, M. B., Ullah, S., Adnan, M., Wahab, S., Kaplan, A., Khan, M. N., & Ullah, B. (2025b). Nanotechnology in oilseed crops: a new frontier for abiotic stress adaptation. In A. A. H. Abdel Latef (Ed.), *Oilseed Crops Under Abiotic Stress. Sustainability Sciences in Asia and Africa*. Singapore: Springer.

Assad, N., Laila, M. B., Hassan, M. N. U., Rehman, M. F. U., Ali, L., Mustaqeem, M., Ullah, B., Khan, M. N., Iqbal, M., & Ercişli, S. (2025c). Eco-friendly synthesis of gold nanoparticles using equisetum diffusum D. Don. with broad-spectrum antibacterial, anticancer, antidiabetic, and antioxidant potentials. *Scientific Reports*, 15, 19246.

Assad, N., Naeem-Ul-Hassan, M., Ajaz Hussain, M., Abbas, A., Sher, M., Muhammad, G., Assad, Y., & Farid-Ul-Haq, M. (2025d). Diffused sunlight assisted green synthesis of silver nanoparticles using cotoneaster nummularia polar extract for antimicrobial and wound healing applications. *Natural Product Research*, 39, 2203–2217.

Bitew, Y., & Alemayehu, M. J. A. J. P. S. (2017). Impact of crop production inputs on soil health: a review. *Asian Journal of Plant Sciences*, 16, 109–131.

Calvo, P., Nelson, L., & Kloepper, J. W. (2014). Agricultural uses of plant biostimulants. *Plant and Soil*, 383, 3–41.

Chhabra, R., Chhabra, R. J. S.-A. S, Perspectives, M. W. G., & Management, S. (2021). Nutrient management in salt-affected soils. 349–429.

Co- Operation, O. F. E., & Nanomaterials, D. J. S. O. M. (2012). *Guidance on Sample Preparation And Dosimetry For The Safety Testing Of Manufactured Nanomaterials.* Oecd Paris. Environment Directorate, Organisation for Economic Co-Operation and Development, Paris, 2012.

Dean, R., Van Kan, J. A., Pretorius, Z. A., Hammond-Kosack, K. E., Di Pietro, A., Spanu, P. D., Rudd, J. J., Dickman, M., Kahmann, R., & Ellis, J. J. M. P. P. (2012). The top 10 fungal pathogens. *Molecular Plant Pathology*, 13, 414–430.

Dimkpa, C. O., & Bindraban, P. S. (2016). Fortification of micronutrients for efficient agronomic production: a review. *Agronomy for Sustainable Development*, 36, 7.

Du Jardin, P. J. S. H. (2015). Plant biostimulants: definition, concept, main categories and regulation. *Scientia Horticulturae*, 196, 3–14.

Egamberdieva, D., Wirth, S. J., Alqarawi, A. A., Abdallah, E. F., & Hashem, A. J. F. I. M. (2017). Phytohormones and beneficial microbes: essential components for plants to balance stress and fitness. *Frontiers in Microbiology*, 8, 2104.

Fahad, S., Bajwa, A. A., Nazir, U., Anjum, S. A., Farooq, A., Zohaib, A., Sadia, S., Nasim, W., Adkins, S., & Saud, S. J. F. I. P. S. (2017). Crop production under drought and heat stress: plant responses and management options. *Frontiers in Plant Science*, 8, 1147.

Faryal, S., Ullah, R., Khan, M. N., Ali, B., Hafeez, A., Jaremko, M., & Qureshi, K. A. (2022). Thiourea-capped nanoapatites amplify osmotic stress tolerance in Zea Mays L. by conserving photosynthetic pigments, osmolytes biosynthesis and antioxidant biosystems. *Molecules*, 27, 5744.

Foley, J. A., Ramankutty, N., Brauman, K. A., Cassidy, E. S., Gerber, J. S., Johnston, M., Mueller, N. D., O'connell, C., Ray, D. K., West, P. C., Balzer, C., Bennett, E. M., Carpenter, S. R., Hill, J., Monfreda, C., Polasky, S., Rockström, J., Sheehan, J., Siebert, S., ... Zaks, D. P. M. (2011). Solutions for a cultivated planet. *Nature*, 478, 337–342.

Ghaffar, S., Abbas, A., Naeem-Ul-Hassan, M., Assad, N., Sher, M., Ullah, S., Alhazmi, H. A., Najmi, A., Zoghebi, K., & Al Bratty, M. (2023). Improved photocatalytic and antioxidant activity of olive fruit extract-mediated Zno nanoparticles. *Antioxidants*, 12, 1201.

Glick, B. R. (2012). Plant growth-promoting bacteria: mechanisms and applications. *Scientifica*, 2012, 963401.

Glick, B. R. J. M. R. (2014). Bacteria with acc deaminase can promote plant growth and help to feed the world. *Microbiological Research*, 169, 30–39.

Gula, H., Khan, H. A., Nasreen, Z., Assad, N., Turab, S. A., & Hanif, M. J. I. T. O. N. (2025). Exploring the antibacterial, anticoagulant, and hemolytic potential of Green-synthesized Fe_2 O_3 nanoparticles by cucurbita pepo pulp.

Gull, A., Lone, A. A., & Wani, N. U. I. (2019). Biotic and abiotic stresses in plants. In A. B. de Oliveira (Ed.), *Abiotic and biotic stress in plants*. Intechopen.

Hafeez, A., Ali, B., Javed, M. A., Saleem, A., Fatima, M., Fathi, A., Afridi, M. S., Aydin, V., Oral, M. A., & Soudy, F. A. J. P. (2023). Plant breeding for harmony between sustainable agriculture, the environment, and global food security: an era of genomics-assisted breeding. *Planta*, 258, 97.

Hashmi, S. S., Ibrahim, M., Adnan, M., Ullah, A., Khan, M. N., Kamal, A., Iqbal, M., Kaplan, A., Assad, N., & Elshikh, M. S. (2024). Green synthesis of silver nanoparticles from olea europaea l. extracted polysaccharides, characterization, and its assessment as an antimicrobial agent against multiple pathogenic microbes. *Open Chemistry*, 22, 20240016.

Hossain, M. E., Shahrukh, S., & Hossain, S. A. (2022). Chemical fertilizers and pesticides: impacts on soil degradation, groundwater, and human health in Bangladesh. In V. P. Singh, S. Yadav, K. K. Yadav, & R. N. Yadava (Eds.), *Environmental Degradation: Challenges and Strategies for Mitigation. Water Science and Technology Library* (Vol. 104). Cham: Springer.

Hussain, S. B., Zahra, S., Hussain, S., Mahmood, H., Haq, A. U., Shafique, M., Gill, A. R., Ali, A., Javed, M. A., & Khan, M. N. (2025). Characterization and comparative antibacterial activities of gold nanoparticles synthesized by tagetes Patula L.(Asteraceae) flower extract. *Microscopy Research and Technique*, 88, 1869–1880.

Jabbar, A., Abbas, A., Assad, N., Naeem-Ul-Hassan, M., Alhazmi, H. A., Najmi, A., Zoghebi, K., Al Bratty, M., Hanbashi, A., & Amin, H. M. (2023). A highly selective Hg 2+ colorimetric sensor and antimicrobial agent based on green synthesized silver nanoparticles using equisetum diffusum extract. *RSC Advances*, 13, 28666–28675.

Kah, M., Kookana, R. S., Gogos, A., & Bucheli, T. D. J. N. N. (2018). A critical evaluation of nanopesticides and nanofertilizers against their conventional analogues. *Nature Nanotechnology*, 13, 677–684.

Kanwal, M., Sher, M., Abbas, A., Akhtar, S., Siddique, A. B., Ul Hasan, M. N., Assad, N., Alhazmi, H. A., & Amin, H. M. (2025). Dual colorimetric sensing of Hg (Ii) And Fe (Iii) using sulfanilamide-stabilized silver nanoparticles and evaluating their photodegradation and antibacterial properties. *Journal of Water Process Engineering*, 75, 107981.

Khan, F., Siddique, A. B., Shabala, S., Zhou, M., & Zhao, C. J. P. (2023a). Phosphorus plays key roles in regulating plants' physiological responses to abiotic stresses. *Plants*, 12, 2861.

Khan, Z. U. R., Assad, N., Naeem-Ul-Hassan, M., Sher, M., Alatawi, F. S., Alatawi, M. S., Omran, A. M., Jame, R. M., Adnan, M., & Khan, M. N. (2023b). Aconitum lycoctonum L. (Ranunculaceae) mediated biogenic synthesis of silver nanoparticles as potential antioxidant, anti-inflammatory, antimicrobial and antidiabetic agents. *BMC Chemistry*, 17, 128.

Klerkx, L., & Rose, D. J. G. F. S. (2020). Dealing with the game-changing technologies of agriculture 4.0: how do we manage diversity and responsibility in food system transition pathways? *Global Food Security*, 24, 100347.

Lesk, C., Rowhani, P., & Ramankutty, N. (2016). Influence of extreme weather disasters on global crop production. *Nature*, 529, 84–87.

Mahanty, T., Bhattacharjee, S., Goswami, M., Bhattacharyya, P., Das, B., Ghosh, A., & Tribedi, P. (2017). Biofertilizers: a potential approach for sustainable agriculture development. *Environmental Science and Pollution Research*, 24, 3315–3335.

Mustafa, A., Athar, F., Khan, I., Chattha, M. U., Nawaz, M., Shah, A. N., Mahmood, A., Batool, M., Aslam, M. T., & Jaremko, M. J. F. I. P. S. (2022). Improving crop productivity and nitrogen use efficiency using sulfur and zinc-coated urea: a review. *Frontiers in Plant Science*, 13, 942384.

Oerke, E.-C. J. T. J. O. A. S. (2006). Crop losses to pests. *The Journal of Agricultural Science*, 144, 31–43.

Pandey, G. K., & Mahiwal, S. (2020). *Role Of potassium in plants*. Springer.

Pandey, P., Irulappan, V., Bagavathiannan, M. V., & Senthil-Kumar, M. J. F. I. P. S. (2017). Impact of combined abiotic and biotic stresses on plant growth and avenues for crop improvement by exploiting physio-morphological traits. *Frontiers in Plant Science*, 8, 537.

Pascoli, M., De Lima, R., & Fraceto, L. F. J. F. I. C. (2018). Zein nanoparticles and strategies to improve colloidal stability: a mini-review. *Frontiers in Chemistry*, 6, 6.

Raza, A., Khare, T., Zhang, X., Rahman, M. M., Hussain, M., Gill, S. S., Chen, Z. H., Zhou, M., Hu, Z., Varshney, R. K. J. J. O. S. A., & Environment. (2025). Novel strategies for designing climate-smart crops to ensure sustainable agriculture and future. *Food Security*, 4, E70048.

Reddy, G. V., Reddy, M. V. K., Spandana, K., Subbarayudu, Y., Albawi, A., Chandrashekar, R., Singla, A., & Praveen, P. (2024). Precision farming practices with data-driven analysis and machine learning-based crop and fertiliser recommendation system. E3s Web Of Conferences. Edp Sciences, 01078. 29 March 2024

Rico, C. M., Majumdar, S., Duarte-Gardea, M., Peralta-Videa, J. R., Gardea-Torresdey, J. L. J. J. O. A., & Chemistry, F. (2011). Interaction of nanoparticles with edible plants and their possible implications in the food chain. *Journal of Agricultural and Food Chemistry*, 59, 3485–3498.

Roheen, T., Bibi, M., Rehman, M. F. U., Assad, N., Gondal, H. Y., Nadeem, M., Atif, F. A., Batool, F., Perveen, N., & Qadir, R. (2025). Enhanced antibacterial and anti-inflammatory efficiency of serratiopeptidase immobilized on CMC-silver nanoparticles. *Journal of Basic Microbiology*, 65(11), E70060.

Saeed, S., Ullah, A., Ullah, S., Noor, J., Ali, B., Khan, M. N., Hashem, M., Mostafa, Y. S., & Alamri, S. J. L. (2022). Validating the impact of water potential and temperature on seed germination of wheat (Triticum Aestivum L.) via hydrothermal time model. *Life*, 12, 983.

Savary, S., Willocquet, L., Pethybridge, S. J., Esker, P., Mcroberts, N., & Nelson, A. (2019). The global burden of pathogens and pests on major food crops. *Nature Ecology & Evolution*, 3, 430–439.

Savci, S. J. I. J. O. E. S., & Development. (2012). An agricultural pollutant: chemical fertilizer. *Journal of Environmental Science and Development*, 3, 73.

Shen, M., Liu, S., Jiang, C., Zhang, T., & Chen, W. J. E.-E. (2023). Recent advances in stimuli-response mechanisms of nano-enabled controlled-release fertilizers and pesticides. *Eco-Environment & Health*, 2, 161–175.

Singh, N., Sachan, K., Bp, M., Panotra, N., Katiyar, D. J. A. J. O. S. S., & Nutrition, P. (2024). Building soil health and fertility through organic amendments and practices: a review. *Asian Journal of Soil Science and Plant Nutrition*, 10, 175–197.

Suhag, M. J. I. A. R. J. S. E. T. (2016). Potential of biofertilizers to replace chemical fertilizers. *International Advanced Research Journal in Science, Engineering and Technology*, 3, 163–167.

Sun, Y., Wang, M., Mur, L. A. J., Shen, Q., & Guo, S. J. I. J. O. M. S. (2020). Unravelling the roles of nitrogen nutrition in plant disease defences. *International Journal of Molecular Sciences*, 21, 572.

Umar, O. B., Ranti, L. A., Abdulbaki, A. S., Bola, A. U. L., Abdulhamid, A. K., Biola, M. R., & Victor, K. O. (2021). Stresses in plants: biotic and abiotic. In A. MUR. (Ed.), *Current trends in wheat research*. Intechopen.

Vessey, J. K. J. P., & Soil. (2003). Plant growth promoting rhizobacteria as biofertilizers. *Plant and Soil*, 255, 571–586.

Wahab, A., Muhammad, M., Munir, A., Abdi, G., Zaman, W., Ayaz, A., Khizar, C., & Reddy, S. P. P. J. P. (2023). Role of arbuscular mycorrhizal fungi in regulating growth, enhancing productivity, and potentially influencing ecosystems under abiotic and biotic stresses. *Plants*, 12, 3102.

Wang, W., Bai, X., Dong, Y., Chen, W., Song, Y., Tian, X. J. J. O. S. S., & Nutrition, P. (2016). Effects of application of exogenous no on the physiological characteristics of perennial ryegrass grown in cd-contaminated soil. *Journal of Soil Science and Plant Nutrition*, 16, 731–744.

Xu, G., Fan, X., & Miller, A. J. J. A. R. O. P. B. (2012). Plant nitrogen assimilation and use efficiency. *Annual Review of Plant Biology*, 63, 153–182.

Zhang, X., Davidson, E. A., Mauzerall, D. L., Searchinger, T. D., Dumas, P., & Shen, Y. (2015). Managing nitrogen for sustainable development. *Nature*, 528, 51–59.

8

Balancing the Soil: Enhancing Fertilizer Use Efficiency and Soil Health

Yasir Assad[1], *Muhammad Nauman Khan*[2], *Hansa Gul*[3], *Zahida Nasreen*[3], *Nasir Assad*[4], *Iram Zahra*[4], *and Alevcan Kaplan*[5]

[1] *Department of Zoology, Hazara University Mansehra, Mansehra, Pakistan*
[2] *Department of Botany, University of Chakwal, Punjab, Pakistan*
[3] *Department of Zoology, University of Mianwali, Mianwali, Pakistan*
[4] *Institute of Chemistry, University of Sargodha, Sargodha, Pakistan*
[5] *Department of Pharmaceutical Botany, Dicle University, Diyarbakır, Turkey*

8.1 Introduction

The rapid modernization of agriculture, coupled with increasingly intensive farming practices, has created a global paradox: the very fertilizers that have fueled unprecedented yield gains are now contributing to soil degradation and environmental instability (Saeed et al., 2022).

While nitrogen- and phosphorus-based fertilizers remain indispensable for meeting the food demands of a growing global population, their excessive and inefficient use has triggered profound ecological repercussions. Current estimates indicate that nitrogen use efficiency (NUE) in major cereal systems remains at a suboptimal 30–50%, meaning that more than half of applied nitrogen is lost to the atmosphere as nitrous oxides or leached into waterways, exacerbating greenhouse gas emissions, eutrophication, and biodiversity loss (Congreves et al., 2021). Over 60% of global agricultural soils are already experiencing fertility decline, and nearly 35% show severe compaction, jeopardizing the long-term viability of existing production systems.

Historically, fertilizer use has been credited with contributing to at least half of the global crop yield gains achieved during the 20th century. For example, maize yields would be expected to drop by about 40% in the absence of nitrogen fertilization, while multi-decade trials show a 40–57% reduction in wheat yields without fertilizer inputs. Similarly, the combined application of nitrogen, phosphorus, and potassium (NPK) has been shown to boost rice yields by 19–41% and rapeseed yields by 61–76% (Krasilnikov et al., 2022; Yousaf et al., 2017).

With forecasts predicting a global population of 9.7 billion by 2050 and 10.4 billion by 2100, expanding agricultural land at the expense of forests, shrublands, or marginal ecosystems is neither economically sustainable nor ecologically responsible (Barłóg, 2023a). The only viable pathway forward is to maximize yields from the existing agricultural footprint while safeguarding soil health (Hafeez et al., 2025). Achieving this requires a paradigm shift from input-intensive strategies toward precision nutrient management, where fertilizer use efficiency (FUE) becomes as critical a metric as total yield (Barłóg, 2023a).

Integrated nutrient management (INM) offers a promising solution, combining mineral fertilizers with organic amendments to improve nutrient retention, enhance soil structure, and foster beneficial microbial communities. For example, vermicompost at $5\,t\,ha^{-1}$ can increase soil macroaggregate formation by 40%, while poultry manure-derived dissolved organic carbon can mobilize phosphorus by chelating soil calcium (Liu et al., 2021). Beyond physical and chemical benefits, these approaches reshape rhizosphere ecology—integrated regimes have been shown to increase arbuscular mycorrhizal fungal biomass nearly threefold compared to chemical-only treatments, strengthening nutrient cycling and plant resilience (Xing et al., 2025).

The microbial dimension of this strategy is particularly compelling. Metagenomic studies reveal that integrated fertilization fosters diverse and functionally significant microbial consortia, including nitrogen-fixing *Bradyrhizobium* and phosphate-solubilizing *Pseudomonas*. These communities not only enhance nutrient availability but also display heightened metabolic activity, with integrated systems showing up to a 34% increase in substrate-induced respiration rates. Such biological synergies can maintain crop yields while reducing synthetic nitrogen inputs by up to 40% and simultaneously sequester substantial amounts of soil organic carbon (Xing et al., 2025).

This chapter synthesizes global research and field evidence to explore the intricate link between fertilizer efficiency and soil health. Drawing on insights from soil chemistry, microbiology, and agronomy, it outlines strategies for optimizing nutrient use while safeguarding the ecological integrity of farmlands. By bridging molecular-scale understanding of microbial nutrient transformations with field-level agronomic practices, we present a framework for sustainable intensification of agricultural systems that nourish both humanity and the planet.

8.2 Understanding Soil Health: Key Principles and Components

Soil health is a multifaceted concept that extends beyond mere fertility to encompass the physical, chemical, and biological integrity of the soil ecosystem. At its chemical core, soil health is governed by the maintenance of a balanced nutrient profile and an optimal pH. Soil pH acts as a master variable, controlling the solubility and bioavailability of virtually all plant nutrients. Most crops thrive in a slightly acidic to neutral pH range of 6.0–7.5, where essential macronutrients like NPK, as well as secondary and micronutrients, are most available for plant uptake. However, modern agricultural practices, particularly the long-term application of nitrogen-based fertilizers, can lead to soil acidification (Amin et al., 2024). This not only renders the soil infertile over time but also causes crops to become unresponsive to further fertilization, as the acidic conditions lock up essential nutrients and can even induce deficiencies of phosphorus and potassium (Panta & Parajulee, 2021).

Beyond chemical balance, soil health is fundamentally biological, driven by the activity of a diverse community of microorganisms. Soil pH directly influences these organisms; for example, the population and activity of bacteria responsible for decomposing organic matter decline significantly in highly acidic soils. This biological activity is critical because the decomposition of organic matter is the primary mechanism through which many essential micronutrients—such as zinc, boron, and manganese—are released into the soil in plant-available forms. Therefore, a decline in microbial function due to poor chemical balance directly translates into a less fertile soil, even if macronutrients are supplied synthetically (Panta and Parajulee, 2021).

The challenge in modern agriculture often arises from an imbalanced approach to nutrient management. An over-reliance on NPK fertilizers, while boosting short-term yields, frequently neglects the full spectrum of secondary and micronutrients required for sustained plant health. This "unbalanced use" has been shown to progressively deplete the soil of these other essential elements, creating hidden deficiencies that limit crop potential. Furthermore, such mismanagement can lead to broader soil degradation, including nutrient losses through runoff and negative impacts on soil structure (Selim, 2018). Addressing these interconnected challenges requires moving beyond simple fertilizer application toward integrated strategies that restore and maintain the soil's complex chemical and biological equilibrium.

8.3 The Impact of Fertilizers on Soil Health

Fertilizers play a pivotal role in sustaining agricultural productivity, but their effects on soil health are shaped largely by the type used and the manner of their application. When applied responsibly, fertilizers can replenish essential nutrients and support crop growth. However, prolonged and excessive reliance on synthetic chemical fertilizers often undermines soil vitality, leading to chemical, physical, and biological degradation. In contrast, microbial fertilizers offer a restorative approach, working with the soil's natural ecosystem to rebuild fertility and resilience (Figure 8.1).

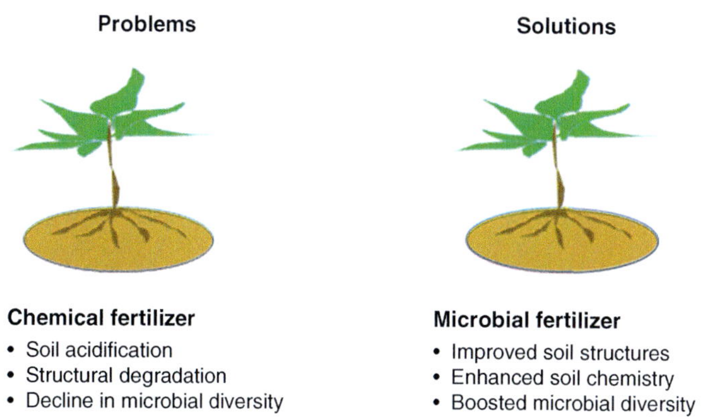

Figure 8.1 The impact of fertilizers on soil health.

The long-term overuse of chemical fertilizers is linked to several well-documented issues. Excessive application can cause soil salinization and acidification, lowering pH to levels harmful for plant roots and beneficial microorganisms. These chemical imbalances alter the soil's physical properties, promoting compaction, reducing porosity, and restricting air and water flow. Compacted soil not only impedes root penetration but also creates localized anaerobic zones that further stress plants. Biologically, such conditions disrupt the native microbial community, suppressing beneficial species and favoring opportunistic or pathogenic organisms, ultimately weakening soil function and plant health (Wei et al., 2024).

Microbial fertilizers directly counter these problems by introducing diverse populations of beneficial microorganisms that enhance soil structure, chemistry, and biological activity. Fungi and actinomycetes, for example, bind soil particles into stable aggregates using hyphae and extracellular polysaccharides, improving porosity, aeration, and water infiltration. Many of these microbes also secrete organic acids, such as malic and citric acid, which help regulate soil pH, offsetting the acidifying effects of chemical fertilizers while increasing nutrient availability, particularly phosphorus.

The benefits of microbial fertilizers extend beyond structural and chemical restoration to the functional enrichment of the soil microbiome:

- **Nutrient Cycling**: They include microorganisms capable of nitrogen fixation, phosphorus solubilization, and potassium mineralization, thus improving the bioavailability of key macronutrients.
- **Pathogen Suppression**: Beneficial microbes outcompete harmful organisms for resources and, in many cases, produce antimicrobial compounds that reduce the incidence of soil-borne diseases. This effect can be augmented by nanoparticles synthesized from plant extracts, which have shown potent activity against multidrug-resistant bacterial strains (Gul et al., 2025; Hussain et al., 2025).
- .**Pollutant Degradation**: Certain strains break down pesticide residues and other contaminants, detoxifying the soil and supporting healthier root systems (Liu et al., 2022a). Phyto-functionalized nanoparticles also offer a promising avenue for degrading organic pollutants through photocatalysis (Assad et al., 2024; Ghaffar et al., 2023).

By rebuilding microbial diversity and function, microbial fertilizers not only repair damage caused by conventional practices but also create a more resilient and self-sustaining soil ecosystem. This makes them an essential component of INM strategies aimed at improving fertilizer efficiency, maintaining long-term productivity, and safeguarding soil health.

8.4 Fertilizer Use Efficiency: Maximizing Nutrient Uptake and Minimizing Waste

FUE is an important indicator to determine the ability of a fertilizer; it is used to enhance crop yield capacity and the potential of the soil-plant system. The world population is expected to increase to 9.7 billion annually by the year 2050, thus necessitating the need to ensure that the current agricultural land yields are maximized. Nevertheless, the productivity of the most important nutrient, nitrogen, is still dangerously low. About 30–50% of the placed nitrogen is estimated to be recovered through the harvest of crops, and this percentage is indicative of an enormous farming inefficiency in the world (Congreves et al., 2021). The unused excess flows into the environment, causing the so-called nitrogen cascade of causes and effects, such as the release of greenhouse gases, ammonia volatilization, ozone depletion, eutrophication of water bodies, and degeneration of ecosystems. Hence, an efficient agronomic optimization of FUE is not only a means of economic viability and ecological sustainability but also the very necessity of existence.

The increase of FUE meets a profound comprehension of intricate relationships between the genotype of plants, environment, and management systems. The most important soil variables that control nutrient acquisition are soil composition, moisture levels, compaction, heat, salinity, organic matter, and, most importantly, soil pH. The genetics of the plant are also critical; that is, in the case of crop genotypes, modified root system architecture (RSA) may appear in reaction to a deficiency of nutrients, and, in response to potassium starvation, crop genotypes may lose the ability to probe the soil. Most importantly, there is the aspect of balanced fertilization. Supplying enough of such supporting nutrients as phosphorus (P), potassium (K), magnesium (Mg), and sulfur (S) is also key to the efficiency of nitrogen. A lack of any of them might cause a bottleneck that could restrict the potential of the plant to use the nitrogen absorbed by it (Barłóg, 2023b).

A comprehensive framework for implementing balanced fertilization across a timeline is the Soil Fertility Clock (SFC) concept (Grzebisz et al., 2022). The SFC operates on three core assumptions: (i) the soil must be managed to a critical fertility level sufficient for the most nutrient-sensitive crop in a given rotation; (ii) less sensitive crops in the rotation provide a timeframe to restore this critical nutrient level; and (iii) the availability of any single nutrient must not become a limiting factor for nitrogen uptake by any crop grown. This long-term, rotation-based perspective moves beyond simplistic single-season application rates, promoting a holistic strategy for sustaining both soil health and FUE over time (Barłóg, 2023b).

8.5 Advances in Fertilizer Technology for Improved Efficiency

Owing to the great demand for increased FUE, there has been considerable advancement in the technology of fertilizers. Such controlled-release fertilizers (CRFs) are more like a significant breakthrough in comparison with traditional highly soluble fertilizers since they are designed in a way that optimizes the availability of nutrients to the plant when they are needed and reduces environmental losses.

Among the most notable of those developments are the inventions of CRFs and slow-release fertilizers (SRFs). Such products aim at postponing the availability of nutrients upon their application. One of the strategies that the CRFs will adopt is they add a protective polymer layer to conventional fertilizer granules. One of the examples is research on winter oilseed rape, which revealed that when calcium ammonium nitrate (CAN) coated with vegetable oil-based polymers and used as a 1:2 ratio with conventional CAN in spring fertilizer, it proved to be an effective method. Such management provided enough quickly accessible nitrogen to initiate early plant regeneration and lead to subsequent slower and steady release that was in line with the requirements of the crop at the later stages of its growth, so that it could drive yield and minimize N losses (Lu et al., 2024).

A similar tool aims at the biochemical routes of nitrogen loss by means of the application of urease and nitrification inhibitors. When urea comes into contact with soil, the enzyme urease speeds up its processing to the unstable ammonia gas. Nitrification is the process by which stable ammonium is turned into highly leachable nitrate. The processes are blocked by inhibitors. For example, a survey on the production of corn in Brazil revealed that the use of enhanced-efficiency fertilizers with urease inhibitor (NBPT + Duromide) or nitrification inhibitor resulted in a substantial ammonia volatilization reduction as compared to regular urea use. Losses were minimized in ammonium sulfate and ammonium nitrate, as they decreased ammonia volatilization by 84% and 80%, respectively, compared to urea, in addition to increasing the yield of corn grain (Cassim et al., 2022).

Lastly, another frontier includes soil amendments such as biochar. Biochar is a durable carbon-dense substance that enhances soil structure and nutrient retention and sequesters carbon (Palansooriya et al., 2019). It gets better when used with other organic inputs. In tomatoes grown in a greenhouse, using vermicompost at a rate of 3% was the most economic and effective treatment to achieve all measurements of enhancing photosynthesis, increasing the water use efficiency (WUE), and finally the tomato yield. It was revealed in the study that although biochar also exerted positive effects, the synergistic effect of high-quality organic matter through vermicomposting was an excellent stimulator of plant physiological activity (Liu et al., 2022b). Together, these advanced technologies give farmers a powerful set of tools to improve nutrient management and construct more productive, resiliency-based agricultural systems.

8.6 Integrated Nutrient Management (INM): Optimizing Fertilizer and Organic Inputs

A combination strategy of organic amendments and mineral fertilizers has become an essential step towards soil health management, maintenance of crop productivity, and lowering the harm done to the Earth through the development of the so-called INM. Unlike long-term application of chemical fertilizers without any inclusion of organic fertilizers, INM reverses microbial imbalance and restores soil functionality as well as increases agroecosystem resilience (Xing et al., 2025).

One of the processes behind these advantages is an increased stimulation of heterogeneous and functional microbial communities of the soil. To illustrate, the replacement of half of the mineral nitrogen with organic sources like sheep manure has a considerable positive effect on the microbial metabolic pathways, where richness, dominance, and evenness grow by 12–15%, whereas the yield of oats jumps by as much as 15% (Zhu et al., 2020). It has also been demonstrated that long-term organic replacement of paddy soils not only enriches desirable taxa (such as nitrogen-fixing Bradyrhizobium and phosphate-solubilizing Burkholderia as well as key organisms such as Acidobacteria, Proteobacteria, and Ascomycota). These microbial communities enhance the breakdown of complex organic matter, for example, cellulose and lignin, which enhances the nutrient availability (facilitated by the increase of nutrient availability to plants) (Xing et al., 2025). The practical consequences of these microbial changes are heavyweight. INM is always related to high levels of soil enzyme activation, such as urease (38.3%), 7-glucosidase

(122.4%), invertase (51.9%), and cellulase (122.4%), showing much ability to mineralize and recycle nutrients. Such biochemical gains can be directly transferred to the performance of the crop, with published results showing that lettuce biomass under INM regimes has risen by 75% and rice productivity by 1520% (Liu et al., 2021). Beyond yield gains, INM contributes to environmental sustainability by increasing soil organic carbon stocks by over 100%, stabilizing soil pH, and reducing harmful leaf nitrate and lead concentrations by 27% and 46%, respectively (Xing et al., 2025).

Critically, INM is also central to the alleviation of the negative impacts of drivers of global change. The degradation of soil microbial networks due to decreased precipitation and oversupply of nitrogen can be overcome by multi-nutrient plans to restore diversity and benefit healthy microbial networks to make the soil more resistant to changing ecological functions. By maintaining a robust and healthy soil microbiome, INM not only maintains the productivity of agrarian economies but also promotes long-term environmental well-being (Yang et al., 2021). The key contribution areas of INM are summarized in Table 8.1.

Table 8.1 Key contribution areas of integrated nutrient management (INM) in sustainable agriculture.

Contribution area	Key points	References
Soil health improvement	Enhances soil physical (structure, bulk density), chemical (nutrient balance, pH stabilization), and biological (microbial activity) properties; increases organic matter and cation exchange capacity.	Selim (2020)
Nutrient-use efficiency	Balances macro- and micronutrient supply; minimizes nutrient losses via leaching, volatilization, and runoff; synchronizes nutrient release with crop demand.	Selim (2020)
Water-use efficiency	Improves water retention and infiltration; increases economic water use and drought resistance through better root systems and organic matter content.	Selim (2020)
Residual benefits	Long-term application of organic and inorganic fertilizers improves soil organic matter and nutrient reserves, with residual effects on subsequent crops. Positive impacts were observed on yield stability over multiple seasons.	Selim (2020), Wu and Ma (2015)
Environmental protection	Reduces greenhouse gas emissions by lowering excessive nitrogen application; minimizes nitrate leaching and phosphorus runoff; promotes carbon sequestration through increased soil organic matter.	Wu and Ma (2015)
Stress Resistance	Enhances resilience to climate variability and drought by improving soil structure, water retention, and root development; supports plant defense mechanisms through better nutrient balance.	Wu and Ma (2015)
Adaptability	Can be adjusted to diverse agroecosystems and cropping systems; compatible with smallholder and large-scale farming; adaptable to varying resource availability and climatic conditions.	Selim (2020), Wu and Ma (2015)

8.7 The Role of Microbial Inoculants in Fertilizer Efficiency

High-yield and high-quality timothy production continues to be hampered by limitations associated with the maintenance of soil physicochemical properties, decline of enzyme activity in the soil, and soil microbial community change, which are detrimental to productivity and quality of forage. In order to overcome these drawbacks, there have been several management options put forward, one of which is the use of alternative types of fertilizers, the integration of organic additions with metal-chelated fertilizers, and the use of farmyard manure, in one way or another. Among them, combined phosphorus and nitrogen fertilizers have become rather famous owing to their convenience, effectiveness, and efficiency of costs (He et al., 2024).

Nevertheless, overuse of chemical fertilizers has caused decreases in soil fertility in the soils, environmental pollution, and decreased soil microbial diversity. Ahsan et al. (Ahsan et al., 2024) researchers found that compound fertilizers combined with microbial inoculants tremendously brought improvements in the soil enzyme activity of the rhizosphere and the abundance of bacteria. Given that microbial inoculants are the source of plant growth-promoting rhizobacteria (PGPR), their major functions are numerous in terms of growth-inducing impacts on plants. They are soil structure, microbiome diversity, plant defensive induction, and synthesis of plant growth hormones, including indole-3-acetic acid (IAA), gibberellins (GA), and cytokinins (CTK), among others, that directly stimulate plant growth. Nanoparticles can also play a direct role in enhancing plant growth under various conditions (He et al., 2024; Assad et al., 2025).

More current uses of the microbial inoculants have been to increase crop yield in cash crops, to control disease, and for other narrow-focus agricultural purposes as well. As an illustration, the inoculation by PGPR is claimed to enhance the parameters of plant growth, soil urease (SURE), invertase (SIN), alkaline phosphatase (SAKP), and peroxidase (SCAT) activities and readily available levels of soil nutrients in oats, alfalfa, and cucumber (He et al., 2024).

Similarly, it has been emphasized that the role of PGPR is an environmentally friendly alternative to synthetic pesticides and conventional practices, contributing to sustainable agriculture through enhanced plant growth and stress tolerance. In the context of crop protection (Lv et al., 2023). Demonstrated that indigenous microbial inoculants effectively biocontrolled major potato diseases such as Fusarium dry rot and wilt.

Irrespective of these developments, the establishment of the broad-spectrum and adaptable PGPR agents has not yet been successfully undertaken, partially as a result of the variability and complexity of the environments where they are held to be used. PGPR survival and efficacy are very sensitive to factors like climatic conditions, soil moisture and temperature, physicochemical properties, and cultivation practices. Moreover, the studies conducted on interactive influences of microbial inoculants on chemical fertilizers have been small-scale, hindering their practical usage to be used as a foundation on which synthetic fertilizer consumption needs cutbacks.

Combining the use of microbial inoculants and chemical fertilizers would bring a new wind to the contemporary forage production system. This integrated practice provides both soluble nutrients and long-term biological advantages, compared to protein applications used with other methods of applying microbial inoculants only. Reported that these combinations increase the nutrient content of soils, enzyme activities, and microbial diversity, which are the key indicators of fertility. Enzymes formed in the soil serve as catalysts in the cycles of nutrients and energy flow and appear to regulate the health of soil ecosystems (Liu et al., 2019).

Although the agronomic significance of timothy makes it a valuable crop, potential impacts on soil enzyme activity and microbial diversity by microbial inoculant-fertilizer combinations are largely unexamined. This paper will fill the above knowledge gap by conducting a field experiment using a specially designed forage microbial inoculant invented by our research group and used in combination with different amounts of chemical fertilizers. The effects of seven fertilization treatments on timothy growth, bacterial and fungal diversity in the rhizosphere, and soil enzyme activities were tested. The research is expected to determine the best fertilization practices to reduce the yield and quality, and, at the same time, conserve soil using these factors evaluated in field conditions. The results will offer a scientific foundation for incorporating microbial inoculants in sustainable fertilization plans that make timothy production stable and intermittent (He et al., 2024).

8.8 Strategies for Enhancing Soil Health and Fertilizer Use Efficiency

To improve fertilizer efficiency, there is a need to not only be more innovative technologically but also to have improved policy frameworks as well as economic incentives. The best practices of nutrient management can be implemented as a result of efficient governance, and the economic factor of subsidizing, carbon credits, and market-based instruments can motivate a farmer to utilize fertilizer sustainably. The flowchart (Figure 8.2) illustrated summarizes major policy and economic approaches that could spearhead environmental protection and also the productivity of agricultural activities.

8.9 Traditional Agronomic Practices

Sustainable farming systems have benefited from particularly durable soil management traditions, all of which provide easy but extremely efficient methods of bettering the health of the soil and optimum use of nutrients. Organic amendments, like compost, manure, and crop residues, may be added to add carbon and a broad array of nutrients, plus enliven

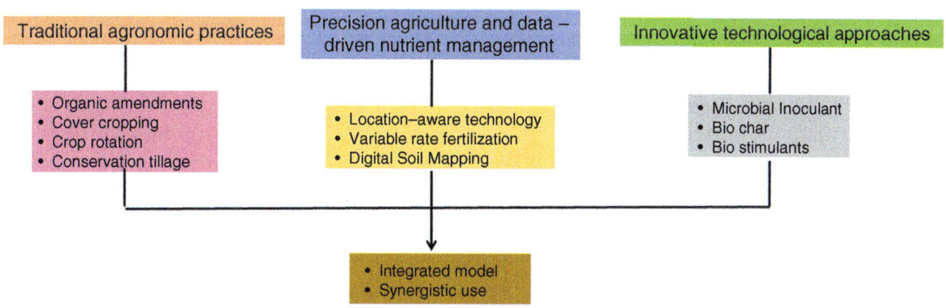

Figure 8.2 Major policy and economic approaches for enhancing agriculture productivity.

useful microbe action. Not only does this enhance the structure, aggregation, and water retention of the soil, but this also acts as a source of slow-release nutrients, making it less dependent on applying synthetic fertilizers and increasing nutrient retention in the soil (Valiyev et al., 2024).

Another widely used technique is cover cropping, which ensures no soil erosion at the soil surface and increases organic biomass building by the growing roots and shoots. Legume cover crops are especially useful in nitrogen fixation through atmospheric nitrogen and thus cut the need for using synthetic nitrogen fertilizers. Moreover, the practice elevates biodiversity in the soil, organic content, and solid stability to facilitate better uptake of nutrients (Mamatha et al., 2024).

These advantages are supplemented with crop rotation for different nutrient requirements, depth of rooting, and plant-soil organism associations, and to interrupt pest and disease cycles. Such pattern variation in cropping supports the avoidance of loss of nutrients, the further enhancement of soil structure depth, and the development as well as stimulation of more diverse soil microbiomes, which can lead to these systems eventually becoming less dependent on specific fertilizer applications (Mamatha et al., 2024). Conservation tillage, such as reduced or no-tillage systems, also contributes to soil health by reducing physical disturbance and retaining organic matter, moisture, and habitats of beneficial soil organisms. Better infiltration of water during conservation tillage promotes the availability of fertilizers in solution to reach the roots of the plants, as losses of the nutrients in runoff are reduced (Angers and Eriksen-Hamel, 2008).

8.10 Precision Agriculture and Data-Driven Nutrient Management

Precision agriculture is a paradigm-shifting concept of generally treating fields the same to using data and site-specific practices. Its fundamental idea is to use the right nutrients, at the right amount, in the right place, and at the right time (the 4Rs). This is done through leveraging a suite of cutting-edge technologies that enable farmers to go from simply displaying in-field variability in their fields to modeling and being able to manage it and maximize FUE while reducing environmental damage due to over-fertilization and excess fertilizer loss.

Central to such an approach is location-aware technology (the combination of geographic information systems (GIS), remote sensing, and various terrestrial sensors). Satellite technologies and those captured by aerial drones are excellent sources of remote sensing and invaluable spatial data of crop health, biomass, and nutrient status. This information is then interpreted into fine-scale maps using GIS platforms, which provide area boundaries according to soil properties and crop requirements. An example is in mapping salinity progression in the soil, where multi-temporal satellite imagery has shown great results and can be better addressed through remediation to boost the productivity of crops. Equally, soil sensing technologies that determine the electrical conductivity can detect the spatial variations in soil texture and organic matter, which directly affect the availability of nutrients (Nakachew et al., 2024).

Representing the real potential of such data-driven insights, there is variable rate fertilization (VRF). Rather than applying the equivalent uniform fertilizer rate over an entire field, VRF uses bearing-based, GPS-guided equipment to change application rates in real-time to adjust to varying rates over time within preestablished management zone maps. This guarantees the delivery of the appropriate amount of nutrients to the less fertile areas and the surplus in those already fertile areas, where uptake is maximized.

A compelling case study from the West Nile Delta in Egypt demonstrates the practical application of these technologies. Using Digital Soil Mapping (DSM), researchers identified widespread deficiencies in N, P, K, S, B, and Cu, along with specific areas where Mg-induced K shortages existed. This detailed assessment allowed for highly tailored recommendations—including conservation tillage, specific N-P-S-B-Cu fertilizer blends, and targeted lime application—obviating the need for generalized, inefficient interventions and maximizing crop potential (Nakachew et al., 2024).

These principles are also highly effective for managing specific nutrients in staple crops. In rice farming, precision techniques such as the deep placement of urea briquettes, drip fertigation, and variable-rate nitrogen application have consistently outperformed traditional broadcast methods. These strategies have been shown to significantly increase both crop yields and NUE, demonstrating how technology can close the gap between agricultural productivity and environmental sustainability (Nakachew et al., 2024). By integrating these advanced tools, modern agriculture can move toward systems that are both highly productive and ecologically sound.

8.11 Innovative Technological Approaches

Equally significant is soil microbiome management through the use of microbial inoculants and biostimulants, which target the biological drivers of soil fertility. These approaches enhance nutrient cycling, particularly phosphorus solubilization, suppress soilborne pathogens, and promote robust root development. By unlocking nutrients already present in the soil and improving plant uptake of applied fertilizers, microbiome management can substantially reduce the need for synthetic nutrient inputs.

The development of agricultural science and technology has presented high-specificity practices that augment the optimal efficiency of fertilizers and preserve the integrity of the environment. With GPS mapping and remote sensing, precision agriculture uses data analytics to allow site-selected nutrient management and variable-rate management. This will make fertilizers be used where and when needed with minimal wastage areas, fewer environmental influences, and maximum crop response. The other promising innovation is the use of biochar. This immobile, porous, carbon-rich, high cation exchange capacity (CEC) material enhances air, water, and nutrient retaining capacity and capacity of the soil, as well as the ability to sequester carbon on a long-term basis. Biochar enhances the availability of the fertilizers used, as it prevents the leaching of major nutrients like potassium and calcium to maintain the growth of plants over a long period of time.

Another challenge is equally important, which is the management of the soil microbiome using microbial inoculants and biostimulants, which the intervener breaks down to biological determinants of soil fertility. The methods can increase the recirculation of nutrients, especially the solubilization of phosphorus that inhibits soil diseases, and stimulate strong root growth. It is considered that microbiome management can greatly decrease the necessity of synthetic inputs in nutrients by making the ones already present in the soil available to the plants and enhancing their ability to absorb the ones added through fertilizers (Mamatha et al., 2024).

8.12 Toward an Integrated Framework

The interaction of traditional practices and modern innovations gives a powerful and integrated model of sustainable intensification when they come together. This type of hybrid treatment is key in facilitating fertilizer application efficacy and soil wellness to occur hand-in-hand with productivity objectives and the long-term ecological stability that is always the focus. The harmonization of proven agronomic practices and data-based, precision-dependent interventions will see agricultural systems sustain high yields and reduce loss of nutrients and damaging impacts on the environment, heralding the path of truly resilient food systems.

8.13 Policy and Economic Considerations for Improving Fertilizer Efficiency

The efficiency of fertilizers is determined not only by the agronomic manner of fertilizer use but also by the system of government policies, economic incentives, and institutional capacities that regulate the agricultural systems. The Food and Agriculture Organization (FAO) has a code of conduct dedicated to better use and management of the fertilizer, which identified that the sustainable use of nutrients is achieved by balancing productivity objectives and maintaining long-term soil health and environmental sustainability. Varying levels of the use of fertilizers across regions related to the variance in the development of regulatory environments, economic means, and the availability of fertilizer to farmers underscore the importance of context-dependent policymaking and economies.

In developed countries, INM systems coupled with a sound regulatory framework and economic incentives to practice sustainability are used to increase the efficiency of fertilizer. As an illustration, the substitution of 3040% of chemical nitrogen by composted manure in Japan caused a rise in microbial evenness of 20% or an improvement in rice production of 1012% (Patra et al., 2021). This is done through measures that are strengthened by policies that support precision farming and soil testing to maximize the use of nutrients. At the European Union level, the Nitrates Directive has prescribed tight regulatory limits on the use of nitrogen-based fertilizer in a bid to safeguard groundwater. Nations such as Denmark and Germany utilize economic and regulatory processes that promote the use of 50% biogas slurry in substitution of synthetic nitrogen with a Shannon diversity index exceeding 5.0 (Meegoda et al., 2025). The further results of the long-term experiments in France indicate that mixed organic and synthetic fertilization has the potential to boost arbuscular mycorrhizal fungi abundance by 35%, promoting phosphorus uptake levels (Campos et al., 2018). Such examples show that policy requirements, when coupled with economic incentives, can induce the take-up of practices that do not compromise yields but maintain soil biodiversity.

Farmers in developing nations tend to use substantial amounts of organic fertilizers like manure, compost, and crop residues, as policy and economic restrictions seemingly limit the use of chemical-based fertilizers. Although the practices enhance soil structure and microbial resistance, they are reduced in nutrient content, which would pose serious yield gaps. To give a specific example, a level of maize yields is still 30–40% lower in Ethiopia under pure organic fertilizing in comparison with the aggregated provisions integrating the organic and mineral fertilizers (Abebe et al., 2022). Programs to mitigate these problems are underway, such as the FAO-supported Africa Fertilizer Summit, which aims to expand access to mineral fertilizers, especially through market reforms, subsidy schemes, and investments in infrastructure, and at the same time, encourages the adoption of integrated soil fertility management practices that would help to avoid long-term degradation.

The international difference in fertilizers saw the necessity of adaptive policies on the one hand, thinking about the economy without considering the environment, depicting the requirement of adaptive policies with environment-friendly thinking. National agricultural policies can be aligned with those of FAO, farmer education programs can be expanded, and rural infrastructure can be added to increase the efficiency of fertilizer application in various situations. Economically, the integrated options minimize the need to rely on expensive inputs of chemical elements and binary inputs and preserve the long-run soil yield capacity, as well as enhance stability on price fluctuations across markets. No more innovative hints should be seen as how to combine the regulatory measures, economic incentives, and locally adapted agronomic changes in order to avoid the existence of yield gaps but not to harm the ecological basis of agricultural systems (Xing et al., 2025).

8.14 Conclusion

Global food security will crucially rely on addressing a major inefficiency in modern food production: the low amount of applied nitrogen that is reported to be absorbed by crops in the range of 30–50%, with the remaining nitrogen lost to processes that worsen the quality of the air, water, and soil. There is no individual intervention that holds the key to the solution, and this chapter demonstrates that the solution exists in a collective model that brings together the wisdom of agrarian systems, the innovation of the biological systems, and precision technology. An example of such a venture would be INM. Substituting a part of the mineral fertilizers with organic materials has been found to raise microbial richness 12–15% and key soil enzymes, including 3-glucosidase, by more than 122%. Such biological benefits are immediately realized through productivity and, in measurements, increases in lettuce biomass of 75% and rice yields by 15–20%, with soil organic carbon doubling (100%), a key indicator of long-term fertility and climate resilience. Such outcomes are further increased through technological advances. With CRFs and nitrification inhibitors, it is possible to minimize nutrient losses significantly: by 84% applying inhibitors, in the case of ammonia volatilization. Precision agriculture is the application of the 4R Nutrient Stewardship concept, namely right nutrient, right rate, right place, and right time, by applying data-based technologies, like DSM and VRF, to fine-tune inputs with surgical precision. The way ahead requires the move to knowing-intensive management rather than the input-intensive agricultural practice. When the restorative potential of INM and microbial inoculants is introduced with the affordability advantages of new fertilizer technology and the supportive policy infrastructure, yield gaps that could not be closed to date can indeed be brought under control, increasing farmer profits and preserving the ecological base of food production. Obtaining this balance is not only a strategic decision but also an ecological and moral duty to feed people and the earth to the upcoming generations.

References

Abebe, T. G., Tamtam, M. R., Abebe, A. A., Abtemariam, K. A., Shigut, T. G., Dejen, Y. A., & Haile, E. G. (2022). Growing use and impacts of chemical fertilizers and assessing alternative organic fertilizer sources in Ethiopia. *Applied and Environmental Soil Science*, 2022, 4738416.

Ahsan, T., Tian, P.-C., Gao, J., Wang, C., Liu, C., & Huang, Y.-Q. (2024). Effects of microbial agent and microbial fertilizer input on soil microbial community structure and diversity in a peanut continuous cropping system. *Journal of Advanced Research*, 64, 1–13.

Amin, F., Al-Huqail, A. A., Ullah, S., Khan, M. N., Kaplan, A., Ali, B., Iqbal, M., Elsaid, F. G., Ercisli, S., & Malik, T. (2024). Mitigation effect of alpha-tocopherol and thermo-priming in brassica Napus L. under induced mercuric chloride stress. *BMC Plant Biology*, 24, 108.

Angers, D., & Eriksen-Hamel, N. J. S. S. S. O. A. J. (2008). Full-inversion tillage and organic carbon distribution in soil profiles: A meta-analysis. *Soil Science Society of America Journal*, 72, 1370–1374.

Assad, N., Abbas, A., Ur Rehman, M. F., & Naeem-Ul-Hassan, M. (2024). Photo-catalytic and biological applications of phyto-functionalized zinc oxide nanoparticles synthesized using a polar extract of equisetum diffusum D. *RSC Advances*, 14, 22344–22358.

Assad, N., Ahmad, H., Abbas, A., Rehman, M. F. U., Naeem-Ul-Hassan, M., Sher, M., Riaz, T., Laila, M. B., Zahra, I., Ullah, F., & Assad, Y. (2025). Green synthesis of selenium nanoparticles using equisetum diffusum: Characterization, antibacterial potential, effects on plant growth. *IEEE Transactions on Nanobioscience*, 24(4), 443–453.

Barłóg, P. (2023a). Improving fertilizer use efficiency-methods and strategies for the future. *Plants (Basel)*, 12, 3658.

Barłóg, P. J. P. (2023b). *Improving fertilizer use efficiency—methods and strategies for the future*. Mdpi.

Campos, P., Borie, F., Cornejo, P., López-Ráez, J. A., López-García, Á., & Seguel, A. (2018). Phosphorus acquisition efficiency related to root traits: Is mycorrhizal symbiosis a key factor to wheat and barley cropping? *Frontiers in Plant Science*, 9, 752.

Cassim, B. M. A. R., Besen, M. R., Kachinski, W. D., Macon, C. R., De Almeida Junior, J. H. V., Sakurada, R., Inoue, T. T., & Batista, M. A. (2022). Nitrogen fertilizers technologies for corn in two yield environments in South Brazil. *Plants*, 11, 1890.

Congreves, K. A., Otchere, O., Ferland, D., Farzadfar, S., Williams, S., & Arcand, M. M. (2021). Nitrogen use efficiency definitions of today and tomorrow. *Frontiers in Plant Science*, 12, 637108.

Ghaffar, S., Abbas, A., Naeem-Ul-Hassan, M., Assad, N., Sher, M., Ullah, S., Alhazmi, H. A., Najmi, A., Zoghebi, K., Al Bratty, M., Hanbashi, A., Makeen, H. A., & Amin, H. M. A. (2023). Improved photocatalytic and antioxidant activity of olive fruit extract-mediated Zno nanoparticles. *Antioxidants*, 12, 1201.

Grzebisz, W., Diatta, J., Barłóg, P., Biber, M., Potarzycki, J., Łukowiak, R., Przygocka-Cyna, K., & Szczepaniak, W. J. P. (2022). Soil fertility clock—crop rotation as a paradigm in nitrogen fertilizer productivity control. *Plants*, 11, 2841.

Gul, H., Khan, H. A., Nasreen, Z., Assad, N., Turab, S. A., & Hanif, M. (2025). Exploring the antibacterial, anticoagulant, and hemolytic potential of green-synthesized Fe_2O_3 nanoparticles by cucurbita pepo pulp. *IEEE Transactions on Nanobioscience*, 24(4), 411–420.

Hafeez, A., Batool, R., Arshad, A., Khan, M. N., Ali, S., Singh, N., Garhwal, V., Javed, M. A., Fatima, E., & Suleman, F. (2025). *Soil and water conservation under changing climate. Challenges and solutions of climate impact on agriculture*. Elsevier.

He, S., Zhang, Y., Yang, X., Li, Q., Li, C., & Yao, T. J. A. (2024). Effects of microbial inoculants combined with chemical fertilizer on growth and soil nutrient dynamics of timothy (Phleum Pratense L.). *Agronomy*, 14, 1016.

Hussain, S. B., Zahra, S., Hussain, S., Mahmood, H., Haq, A. U., Shafique, M., Gill, A. R., Ali, A., Javed, M. A., Khan, M. N., Kaplan, A., Assad, N., Aljowaie, R. M., & Elshikh, M. S. (2025). Characterization and comparative antibacterial activities of gold nanoparticles synthesized by Tagetes Patula L. (Asteraceae) flower extract. *Microscopy Research and Technique*, 88, 1869–1880.

Krasilnikov, P., Taboada, M. A., & Agriculture, A. J. (2022). *Fertilizer use, soil health and agricultural sustainability*. Mdpi.

Liu, Z., Gao, F., Yang, J., Zhen, X., Li, Y., Zhao, J., Li, J., Qian, B., Yang, D., & Li, X. J. F. I. P. S. (2019). Photosynthetic characteristics and uptake and translocation of nitrogen in peanut in a wheat–peanut rotation system under different fertilizer management regimes. *Frontiers in Plant Science*, 10, 86.

Liu, J., Shu, A., Song, W., Shi, W., Li, M., Zhang, W., Li, Z., Liu, G., Yuan, F., & Zhang, S. (2021). Long-term organic fertilizer substitution increases rice yield by improving soil properties and regulating soil bacteria. *Geoderma*, 404, 115287.

Liu, H., Li, S., Qiang, R., Lu, E., Li, C., Zhang, J., & Gao, Q. J. F. I. E. S. (2022a). Response of soil microbial community structure to phosphate fertilizer reduction and combinations of microbial fertilizer. *Frontiers in Environmental Science*, 10, 899727.

Liu, X., Zhang, J., Wang, Q., Chang, T., Shaghaleh, H., & Hamoud, Y. A. (2022b). Improvement of photosynthesis by biochar and vermicompost to enhance tomato (Solanum Lycopersicum L.) yield under greenhouse conditions. *Plants*, 11, 3214.

Lu, Y., Xu, J., Liu, Z., Chen, Y., Zhang, X., & Chen, D. J. A. (2024). Management strategy of slow-release nitrogen fertilizers for direct-sown cotton after wheat harvest. *Agronomy*, 14, 536.

Lv, Q., Chi, B., He, N., Zhang, D., Dai, J., Zhang, Y., & Dong, H. (2023). Cotton-based rotation, intercropping, and alternate intercropping increase yields by improving root–shoot relations. *Agronomy*, 13, 413.

Mamatha, B., Mudigiri, C., Ramesh, G., Saidulu, P., Meenakshi, N., & Prasanna, C. L. J. A. J. S. S. P. N. (2024). Enhancing soil health and fertility management for sustainable agriculture: A review. *Asian Journal of Soil Science and Plant Nutrition*, 10, 182–190.

Meegoda, J. N., Chande, C., & Bakshi, I. (2025). Biodigesters for sustainable food waste management. *International Journal of Environmental Research and Public Health*, 22, 382.

Nakachew, K., Yigermal, H., Assefa, F., Gelaye, Y., & Ali, S. (2024). Review on enhancing the efficiency of fertilizer utilization: Strategies for optimal nutrient management. *Open Agriculture*, 9, 20220356.

Palansooriya, K. N., Ok, Y. S., Awad, Y. M., Lee, S. S., Sung, J.-K., Koutsospyros, A., & Moon, D. H. (2019). Impacts of biochar application on upland agriculture: A review. *Journal of Environmental Management*, 234, 52–64.

Panta, S., & Parajulee, D. (2021). Integrated nutrient management (Inm) in soil and sustainable agriculture. *International Journal of Applied Sciences and Biotechnology*, 9, 160–165.

Patra, A., Sharma, V. K., Nath, D. J., Ghosh, A., Purakayastha, T. J., Barman, M., Kumar, S., Chobhe, K. A., Anil, A. S., & Rekwar, R. K. (2021). Impact of soil acidity influenced by long-term integrated use of enriched compost, biofertilizers, and fertilizer on soil microbial activity and biomass in rice under acidic soil. *Journal of Soil Science and Plant Nutrition*, 21, 756–767.

Saeed, S., Ullah, A., Ullah, S., Noor, J., Ali, B., Khan, M. N., Hashem, M., Mostafa, Y. S., & Alamri, S. J. L. (2022). Validating the impact of water potential and temperature on seed germination of wheat (Triticum Aestivum L.) via hydrothermal time model. *Life*, 12, 983.

Selim, M. J. E. J. O. A. (2018). Potential role of cropping system and integrated nutrient management on nutrients uptake and utilization by maize grown in calcareous soil. *Egyptian Journal of Agronomy*, 40, 297–312.

Selim, M. M. (2020). Introduction to the integrated nutrient management strategies and their contribution to yield and soil properties. *International Journal of Agronomy*, 2020, 2821678.

Valiyev, S., Rajabov, T., Kabulova, F., Khujanov, A., & Urokov, S. J. J. O. P. B. (2024). Changes in the amount of photosynthetic pigments in the native artemisia diffusa in the semi-desert rangelands of Uzbekistan under the influence of different sheep grazing intensities and different seasons. *Journal of Plant Biota*, 3(1), 24–27.

Wei, X., Xie, B., Wan, C., Song, R., Zhong, W., Xin, S., & Song, K. J. A. (2024). Enhancing soil health and plant growth through microbial fertilizers: Mechanisms, benefits, and sustainable agricultural practices. *Agronomy*, 14, 609.

Wu, W., & Ma, B. (2015). Integrated nutrient management (Inm) for sustaining crop productivity and reducing environmental impact: A review. *Science of the Total Environment*, 512, 415–427.

Xing, Y., Xie, Y., & Wang, X. (2025). Enhancing soil health through balanced fertilization: A pathway to sustainable agriculture and food security. *Frontiers in Microbiology*, 16, 1536524.

Yang, Y., Li, T., Wang, Y., Cheng, H., Chang, S. X., Liang, C., An, S. J. S. B., & Biochemistry. (2021). Negative effects of multiple global change factors on soil microbial diversity. *Soil Biology and Biochemistry*, 156, 108229.

Yousaf, M., Li, J., Lu, J., Ren, T., Cong, R., Fahad, S., & Li, X. J. S. R. (2017). Effects of fertilization on crop production and nutrient-supplying capacity under rice-oilseed rape rotation system. *Scientific Reports*, 7, 1270.

Zhu, Z., Bai, Y., Lv, M., Tian, G., Zhang, X., Li, L., Jiang, Y., & Ge, S. J. H. P. J. (2020). Soil fertility, microbial biomass, and microbial functional diversity responses to four years fertilization in an apple orchard in North China. *Horticultural Plant Journal*, 6, 223–230.

9

Synergistic Effects of Diverse Fertilizer Bioactivities on Crop Growth and Soil Health

Tauqeer Ahmed Qadri[1], Amjid Khan[2], Rashid Abbas Khan[3], Shabbar Abbas[1], Abdul Wahab[1], Anam Fatima[4], Shah Fahad[5], and Muhammad Anas[4,6]

[1] Department of Biosciences, COMSATS University Islamabad, Islamabad, Pakistan
[2] UNESCO-UNISA Africa Chair in Nanoscience and Nanotechnology (U2ACN2), College of Graduate Studies, University of South Africa, Pretoria, South Africa
[3] Department of Biology, Faculty of Sciences, Prince Sonkla University, Songkhla, Thailand
[4] Department of Plant Sciences, Faculty of Biological Sciences, Quaid-i-Azam University, Islamabad, Pakistan
[5] Department of Agronomy, Abdul Wali Khan University Mardan, Mardan, Khyber-Pakhtunkhwa, Pakistan
[6] Department of Soil Science & Plant Nutrition, Selcuk University, Konya, Türkiye

9.1 Introduction

9.1.1 Fertilizers and Global Food Security

Fertilizers have become one of the most important technological advances in the history of agronomy as they have become the foundation of the current food-production processes in unprecedented efficiency. The utmost influence was perceived during the Green Revolution when the comprehensive application of nitrogen, phosphorus, and potassium (NPK) contributions generated oceans of yield rescue, and it prohibited extensive famines (Quitzow et al., 2025). They are now the sole solution to providing the increasing nutritional needs of a rapidly increasing world population that will exceed nine billion people by 2050. Without exogenous nutrient supplementation, the indigenous fertility of soils would be inadequate to support the levels of productivity that such an increase requires (Surya et al., 2021). Since the establishment of synthetic fertilizers, a continuum of organic amendments has been resorbed in both natural (through compost, animal manure, and crop residues) and biological (mycorrhizal fungi and plant growth-promoting rhizobacteria [PGPR]) forms that are appreciating growing attention to the ability to sustainably contribute to the cycling of nutrients, soil health, and crop resilience (de Andrade et al., 2023). Merging of these heterogeneous sources of fertilizer can be seen as evidence of the bricks facing many dimensions required to achieve food security on a global scale in light of climate change, land degradation, and depletion of natural resources (Maestre et al., 2025).

However, numerous disadvantages can be described as a complete package that was expressed by the extent of the dependence on conventional fertilizers. Excess use of chemical amendments causes nutrient imbalances, nutrient leaching, and runoff effects, which in turn trigger eutrophication of water bodies and contamination of groundwater stores (Akinnawo, 2023). Overfertilization of nutrients, especially nitrogen, is a huge contributor to greenhouse gas emissions in the form of nitrous oxide, which is elevated in regard to its global warming potential and which is much greater than that of carbon dioxide. Constant, intensive application of the mineral fertilizers frequently violates the balance of the key nutrients on the land border, drains the fertility of soil organic matter (SOM), and enlarges microbes (Ashitha et al., 2021). On the other hand, fertilizers are not ecologically friendly and have gained a lot of criticism in terms of the unpredictable release profile and not being able to meet the needs of a high-yield system (Priya et al., 2024). These synergistic benefits and failings imply that the conventional, exogenous-sole chemistry-based approach is empirically poor, as are synonymous biogenic deliveries, to meet the erasure dynamics between high productivity and ecological sustainability without limits (Raven & Wackernagel, 2020).

9.1.2 The Emergence of Fertilizer Synergy

Fertilizer synergy has become subject to growing attention in the literature. The term synergy is not about the simultaneous use of individual fertilizers, but it is about the very rational organization of positive relationships that could possibly offer the maximum help in terms of nutrition to the soil, in the way of securing its structural stability, and a health-related influence over the state of plants (Liu et al., 2024). For instance, co-inoculating plant-enhancing rhizobacteria populations with inorganic NPK fertilizers can reduce nutrient leaching, improve plant nutrient assimilation, and develop plant-root and microbiome-associated co-benefits (Phares et al., 2022). Also, fertilizers that are combined with polymer-based coverings or biochar have a gradual release that is favorable to the organic surroundings of microbes in the soil, which plays a role in assisting this nutrient cycling process (Firmanda et al., 2023). By using such synergies in combination with nanotechnological carrier systems, it is possible to deliver nutrients to certain target areas where cross-communicative reactions between the chemical and biological components have been identified to be most effectively carried out at the micro- and nanometer range (Phares et al., 2022).

The primary impact of such synergetic paradigms is that they can close the divide between environmental productivity and environmental care. Synergy does not end with specific assistances; it distinguishes interrelationships between plant physiology, soil matrix, fertilizer chemistry, microflora community, and subsequent formation of a vibrant system, wherever each of the constituents improves the action of the other (Khoso et al., 2024). This incorporated administration system preserves nutrient-use productivity and farm efficiency and helps to prevent the reduction of enduring soil fertility and the production of resistance against the harmful effects of ecological strains such as hypoxia, drought, and salinity (Muhammad et al., 2024). The fertilization interaction is submissive in the climate-transparent reformative agriculture (CTRA) revelation as a route for viable escalation that can preserve environmental service and ecological agriculture (Muhammad et al., 2024).

The scope of the present chapter is to conduct a multifunctional research study in identifying how the multifunctional bioaction as a result of using fertilizers can reorganize the future scholastic production and national position of soil health. It addresses the heterogeneity of bioactive components in fertilizers and reveals the mechanistic tools of synergism, as well as comments on the new technological solutions (coatings, nanocapsules, and biological consortia) that can be applied to exploit synergism. The long-term positive outcome of the fertilizer synergies on promoting soil fertility development and the process to overcome the resistance barriers, policy thresholds, and the economic constraints that govern the mainstream uptake are also criticized in this chapter. The chapter continues by saying that there is a paradigm change in the proper management of fertilizers: it cannot be done through only chemical/biological solutions but through synergies in constructing the food system in resilience and sustainability to address the challenges the next generation will face.

9.2 Understanding Bioactive Components in Fertilizers

Fertilizers are no longer projected as a mechanism to provide a practically required set of nutrients; increasingly, it is ambitious to grant them the degree of a complicated system that can incorporate bioactive constituents and which connects to soils and plant tissues to interact, influence crop productiveness, and ecosystem sustainability (Ammar et al., 2023). Such bioactive components could be chemical, biological, organic, or technologically engineered, with different characteristics and applications. Focusing on the contributions at an individual level creates the basis of the synergistic formulations of fertilizer strategies that are as agronomically beneficial as they are ecologically cost-effective (Mącik et al., 2020).

9.2.1 Chemical Fertilizers: NPK and Micronutrients

Chemical fertilizers are still the most used input in modern-day farming due to their direct and predictable effects on crop growth. The three most essential macronutrients are NPK, which are provided in concentrated forms of urea, ammonium nitrate, superphosphate, and potassium chloride (Kekeli et al., 2025). Their behavior is to form biofilms, providing highly easily accessible nutrients that cannot be dispensed with by the physiological processes of plants: nitrogen is used in chlorophyll synthesis and protein synthesis, phosphorus is involved in energy transmission and root growth, and potassium is involved in water regulation, enzyme activation, and denudement to stress (Chakraborty et al., 2022). Besides the macro elements, the essential bioactive elements are micronutrients such as zinc, iron, manganese, copper, and boron, which are provided via chemical formulations. Although needed in small amounts, shortages of these micronutrients deteriorate the yield and quality significantly, especially in intensive cropping systems where the soil crops run out (Darnton-Hill, 2018). Micronutrient fertilizers and foliar sprays are therefore becoming a necessity in allowing balanced nutrition of plants as well as correcting the nutrient imbalances. However, soils tend to cause losses of nutrients through leaching, volatilization, or fixation, which leads to reduced fertilizer efficiency due to high solubility and rapid release of chemical fertilizers (Priya et al., 2024).

Their bioactivity is, in most cases, intense, although it is often temporary and comes with environmental hazards. Such is a limiting factor that highlights the importance of using other types of fertilizers to complement the use of chemical inputs to stabilize or increase the utilization of nutrients (Nongbet et al., 2022).

9.2.2 Biological Inputs: PGPR, Mycorrhizae, and Biostimulants

Biological fertilizers add microorganisms that are living or biologically active in the soil–plant system. Some of the most prominent are the PGPR that colonize the rhizosphere and promote plant growth by nitrogen fixing, phosphate solubilization, and siderophore production as well as the synthesis of phytohormones (Wahab et al., 2025). The ability of representative species like the genera of Azotobacter, Pseudomonas, Bacillus, and Rhizobium to enhance nutrient uptake and boost the tolerance of plants to stress has been researched commonly (de Andrade et al., 2023).

The other important community is the arbuscular mycorrhizal fungi (AMF) that establish mutualistic interactions with plant roots and extend the effective root surface area and improve the uptake of phosphorus and micronutrients. The symbioses also bring in greater soil agglomeration and strength against biotic and abiotic challenges (Wahab et al., 2023).

Along with living microbial inoculants, biostimulants such as seaweed extracts, amino acids, and humic materials are becoming recognized as having bioactive efficacy in phytometabolic stimulation, stress-tolerance augmentation, and greater nutrient-use efficiency (Nowicka, 2022). Biostimulants are not directly nutritious, but rather they are able to promote or inhibit plant and soil mechanisms, thus giving a layering of activity that can be complementary to chemical or organic fertilizers (Asghar et al., 2022). Bioactivity of the biological inputs lies in their capacity to catalyze functions within the soil–plant–microbe continuum, to convert stagnant nutrient pools to accessible forms, and, in the process, enhance plant vitality and soil strength (Wang et al., 2024).

9.2.3 Organic Amendments: Compost, Manure, and Biochar

Organic fertilizers form an ancient but extremely beneficial category of bioactive inputs that enhance soil fertility by providing nutrients, enriching organic matter, and stimulating microorganisms (Gross et al., 2022). Compost will release macro- and microenvironmental nutrients in slow but constant amounts and increase the water-holding capacity and porosity of the soil. It also brings good microbes that promote nutrient cycling and suppression of the disease (Guo et al., 2020). Animal manures, rich in organic carbon and nutrients, function as both fertilizers and soil conditioners. Their deprivation can liberate biomolecules that comprise humic materials and organic acids that augment nutrient maintenance and initiate microbial growth. Composts also upsurge the buffering capacity of soils and cation exchange in case of foresighted usage (Seyedsadr et al., 2022).

Other more recent organic additions include the carbon-intensive product of the pyrolysis of plant biomass called biochar (Ma et al., 2024). Besides retaining nutrients, the outstanding porosity of biochar provides the microbiome with its sole habitat in addition to adsorbing toxic compounds and reducing the emission of greenhouse gases by soils (Qian et al., 2023). It has been thoroughly studied as a cosmic carrier of microbe inoculants and stabilizer of nutrient liberation and has become an honored ingredient of synergistic fertilizer preparations. The importance of unraveling these mechanisms is that the synergy is a trade-off not only of SO and SH but also of environmental footprints (Qian et al., 2023). Bioactivity of organic amendments can be more than just processes of supplying nutrients; the soil ecosystem is influenced by it in ways that improve fertility and strength in the long term. Their slow release of nutrients, as well as their carbon, makes them a perfect match for the fast-acting chemical fertilizers (Ebrahimi et al., 2024).

9.2.4 Nanofertilizers and Smart Fertilizers

Nanotechnology has revolutionized the field of food systems, diagnostics, therapeutics, pharmaceuticals, the agriculture sector, and nutraceuticals (Arshad et al., 2021). Nanotechnology in the agricultural sector has come with a new category of fertilizers to facilitate performance beyond the capability of standard fertilizers. Nanofertilizers use nanoparticles/nanoporous carriers to administer nutrients in controlled, effective, and targeted ways (Nongbet et al., 2022). An example is nano-encapsulated nitrogen fertilizers, which deliver nutrients synchronously as demanded by plants, resulting in the minimization of losses and the enhancement of nutrient use efficiency. To the same degree, nano-forms of micronutrients like zinc oxide or iron oxide are more bioavailable due to more surface area and reactivity (Nongbet et al., 2022).

Smart fertilizers, in addition to nanofertilizers, have been developed integrating with advanced delivery systems such as polymeric coating, hydrogels, and biodegradable film, where release of the nutrients is dependent on the moisture status, pH, and/or growth phases of the plant (El-Shal et al., 2022). Some smart fertilizers have also been fortified with biological ingredients in the form of encapsulated PGPR and mycorrhizae and are therefore multifunctional products with chemical, biological, and physical bioactivity (Kaur et al., 2023). The bioactivity of these up-to-date formulations is no longer based on the amount of nutrients themselves but on their precision, efficacy, and flexibility. Accordingly, they provide a major competitive advantage in the development of synergistic fertilizer systems meeting productivity and sustainability objectives through the mitigation of losses and the improvement of targeted delivery (Kekeli et al., 2025).

9.3 Mechanism of Fertilizers Synergy: Interactions Among Bioactivities

Synergistic fertilizer systems are not crude mesohistoric composites of chemical, organic, biological, or nano-technologically necessitated inputs but embodied and evolved interactive mechanistic coupling between the soil–plant interface and the entire continuum (Ayamba et al., 2023). Such interactions induce changes in nutrient cycles, microbiota, or plant physiology at a level that exceeds the true potential of single-fertilizer applications. It is significant to unravel these mechanisms since the synergy is a trade-off not only between SO and SH but also among environmental footprints (El-Sharkawy et al., 2025; Lu et al., 2023). The most common mechanisms reported included (i) increased nutrient uptake and use; (ii) change of soil enzymatic activity through nutrient cycling; (iii) stress tolerance; and (iv) faster rates of nutrient return and recycling through soil enzyme activity (Lu et al., 2023).

9.3.1 Nutrient Availability and Uptake Efficiency

Improvement of nutrient levels and increased effectiveness of uptake is one of the most significant characteristics of fertilizing synergy. Standard chemical fertilizers are usually supplied in easy-to-solve mixtures, but significant proportions in these formulations are lost by means of drainage, vaporization, or entrapment. Through a combination of organic amendments or biological inputs, losses of nutrients are minimized and long-term availability is increased (Gowda et al., 2024). For instance, co-application of NPK fertilizers with compost or manure enhances nutrient retention in soils. The presence of organic matter augments ion interchangeability and the value of water retention, consequently, allowing nutrients to stay available over an extensive length of time (Nieves-Cordones et al., 2020). Similarly, a porous store is provided by biochar, which locks the soluble nutrients of the mineral fertilizers and emits them gradually as much and as frequently as the demand by plants. The interactions not only increase nutrient-use efficiency but also lower the number of fertilizer applications needed (Nieves-Cordones et al., 2020). Biological inputs further increase the efficiency of uptake. Phosphate-solubilizing bacteria and mycorrhizal fungi apply organic acids and enzymes that liberate, as a result, previously unavailable phosphorus and micronutrients (Majed et al., 2025). These biological agents increase the reservoir of nutrients that plants have to use, making the crops exploit fertilizers when the supply sources are chemical fertilizers. Siderophores, which are produced by PGPR, chelate iron and make it plant available, supplementing the level of nutrients supplied by synthetic fertilizers (Tang, 2024). Nanofertilizers boost synergies by providing nutrients with high specificity, providing sustained and controlled release that is much more soluble and reactive. Encapsulation and nanosizing enable this controlled release and provide a slow release of nutrients with reduced leaching and volatilization. For example, nano-urea in combination with PGPR is able to not only deliver nitrogen selectively but also create a rhizospheric environment suitable for microbial colonization and thus increase nitrogen uptake efficiency. All these interactions suggest that the ineffectiveness of the traditional single-input fertilization regimes could be surpassed through the combined use of different fertilizer formulations (Khambalkar et al., 2025).

9.3.2 Soil–Plant–Microbe Signaling

The major impacts of fertilizer synergy are not merely linked to nutrient regulations but rather relate to signaling networks governing mutual interactions between plants and microbes or between plants and soil. This signaling network includes chemical and biochemical cues, which regulate the growth of plants, defense, and stress responses (Ku et al., 2025). The mycorrhizal fungi use plant root signaling molecules, including strigolactones, to stimulate colonization of root tissue by fungal hyphae. Mineral fertilizers enhance mycorrhiza association, nutrient uptake, and root architecture signaling pathways (Gilbert & Johnson, 2017). Also, the phytohormones produced by PGPR, such as indole-3-acetic acid (IAA), gibberellins, and cytokinins, improve the elongation of the root, leading to proliferation of the root surface area for effective uptake of chemically fertilized nutrients (Park et al., 2017).

Signaling molecules are also found as organic stimulants. Humic substances induced by compost or radiocarbonous decomposition trigger the expression of stress responsive and nutrient-transport genes in plants. These are the bioactive signals, which bridge the soil–plant–microbe continuum and bind these components, which are in good interaction with one another (Zhuang et al., 2013). Nanofertilizers go a step further to deliver the signals specifically, like the engineered carriers that deliver the signaling molecules or micronutrients based on the pH and water content of the soil, which has a direct impact on plant–microbe signaling. It has been found that this accuracy is required to produce signals at the correct spatiotemporal position, potentially useful to cooperative processes (Nongbet et al., 2022). Simple nutrient donors are transformed into the signaling action of fertilizers through the synthesis of components of the earth system control. Synaptic fertilizers can therefore more efficiently enhance interactions among crops by enhancing positive communication pathways in growth, defense, and resource utilization (Jakhar et al., 2022).

9.3.3 Stress Tolerance Mechanisms

The other significant feature of fertilizer synergy is that it promotes tolerance of plants to abiotic and biotic stress. Contemporary agriculture increasingly confronts challenges such as drought, salinity, heavy-metal contamination, and pathogen outbreaks. Although the fertilizers can partially reduce these stresses, synergistic blends develop a stronger protective network (Abdelrahman et al., 2022). As an example, mineral fertilizers, when combined with plant-growing-stimulating rhizobacteria, increase the ability to endure drought conditions via the production of osmolytes like proline, which is used to sustain cell water equilibrium. PGPR also produces 1-aminacyclopropane-1-carboxylate (ACC) deaminase, an enzyme that reduces ethylene levels in plants in stress, thus decreasing the growth inhibition. These biological processes, coupled with sufficient nutritional inputs by way of chemical fertilizers, mean that plants can remain productive even in the suboptimal environment (Yuan et al., 2022).

Compost and biochar are also examples of organic amendments that only takes bedrock to be stress-tolerant and enhance water retention and soil structure. Adsorption of toxic compounds and heavy metals also lowers the bioavailability of these elements, preventing plant toxicity by biochar (Abd-Elhalim et al., 2025). Nanofertilizers promote stress resistance by releasing nutrients and delivering the same in a controlled manner. Just to give an example, nanofertilizers, such as nano iron and nano zinc, enhance the activity of antioxidant enzymes in plants and thus are able to help plants fight oxidative stress caused by drought or heat. Whenever such nanofertilizers are combined with organic carriers or microbial inoculants, healthier and more productive plants are produced in the stressed setting (Bashir et al., 2025). Fertilizer synergy establishes a web of defensive responses that protect crops against a wide stress spectrum and maintain stability in an unstable atmosphere via the interplay of multiple sources of bioactivity (Kekeli et al., 2025).

9.3.4 Synergy in Nutrient Cycling and Soil Enzyme Activity

Fertilizer synergy also plays a transformational role in the cycling of nutrients, which are largely brought about by soil enzymes. Some of the soil enzymes that are important in converting organic and inorganic materials to forms that plants can use include urease, phosphatases, dehydrogenase, and cellulase. Their activity reflects the activity of such microbes and soil health (Ding et al., 2024). Biological fertilizers, for example, PGPR and arbuscular mycorrhizal fungi, may stimulate enzymes that mobilize nutrient tracers locked up in organic matter or mineral complexes. In an example, microbially released phosphatases are converting insoluble phosphates into the accessible forms, and nitrogen-fixing bacteria are increasing urease-please, which causes nitrogen mineralization (Daunoras et al., 2024). The availability of substrates and energy sources is raised when the application of chemical fertilizers is released at the same time, thus increasing enzyme activity and the efficiency of nutrient cycles (Ding et al., 2024).

Microbial metabolism uses organic amendments as inputs, and enzyme activity uses organic amendments as stabilizing environments. Compost and manure enrich the soils with organic carbon, thus supplying the growth and production of microbes and enzymes (Ho et al., 2022). The porous structure of biochar acts as a physical medium, providing enzyme stabilization by shielding enzymes against degradation and lengthening their working life. The stabilization effect of the nutrient release potential is increased by enzymes, especially in association with mineral fertilizers (Mehdaoui et al., 2024).

The release of nutrients has been enhanced by nanofertilizers through close interaction within an active microbial community. This nutrient–microbe coordination maximizes enzyme utilization, reduces nutrient loss, and enhances soil fertility (Neemisha & Rani, 2022). Through these interactions, simulated fertilizers are dynamic bioreactors where chemical–biological–organic reactions combine to sustain the cycling of nutrients. Subsequently, the accessibility of additional nutrients to crops is accompanied by increased soil structure, fertility, and resilience (Omokaro et al., 2025).

Figure 9.1 illustrates the conceptual framework of fertilizer synergies at the soil–plant–microbe interface, showing how organic, chemical, biological, and nanocarrier-based inputs collectively enhance nutrient availability, microbial activity, and plant uptake efficiency.

9.4 The Role of Fertilizer Coatings and Carriers in Enhancing Synergies

The efficiency of fertilizer is not only dictated by the content of a given fertilizer but also by how the fertilizer is applied to a soil–plant system. Due to the volatilization or leaching of the traditional linear fertilizers, productivity is not satisfactory, and the following environmental problems are observed (Lawrencia et al., 2021). An even more recent innovation in fertilization has been the coating of fertilizers with or without the use of carrier materials to improve nutrient dynamics, to preserve bioactive compounds, and to further optimize the microbial crop inoculants, to add a further layer of facilitated synergistic activity of fertilization. Resource use efficiency and sustainable agriculture development can result in a higher nutrient-use efficiency and increased compatibility with the soil microbial communities to satisfy the long-term needs of the crops (Riseh et al., 2023).

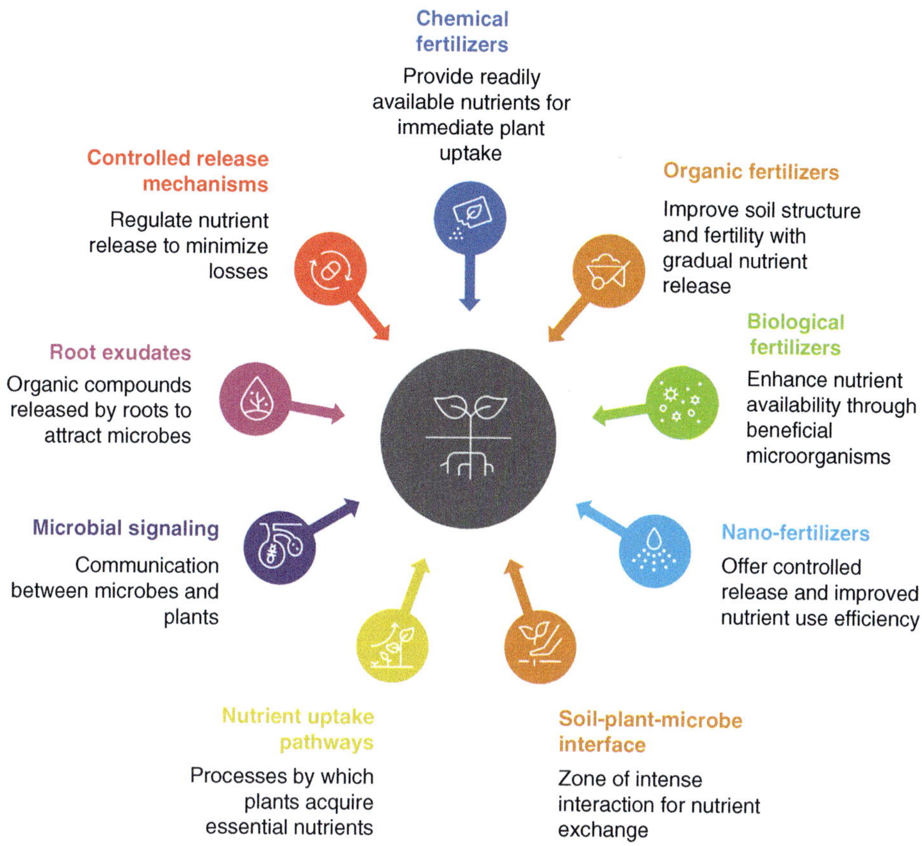

Figure 9.1 Conceptual framework of fertilizer synergies at the soil–plant–microbe interface. Organic amendments, chemical fertilizers, biological inoculants, and nanocarrier-based inputs all boost both nutrient availability and microbial activity in addition to improving the uptake efficiency of plants. A set of synergistic mechanisms, including controlled release of nutrients, stimulation of root exudates, and promotion of beneficial microbial communities, ensures the crop remains highly productive, the soil remains healthy, and the environment remains sustainable.

9.4.1 Slow-Release and Controlled-Release Coatings

The primary limitation of conventional fertilizer is that it is a supply uptake. For example, nitrogenous fertilizers (urea) are highly hydrolyzed and leached, and thus a large amount of nutrients are in the environment, as well as produced greenhouse gases (Al-Rawajfeh et al., 2021). This problem can be ameliorated by the application of slow- and controlled-release fertilizers (SRFs and CRFs) in which granules are encapsulated with polymeric sulfur- or resin-based coatings to help control the release of nutrients over time. In these coatings, fertilizer release is timed to match crop growth stages, and required application rates are reduced (El-Sharkawy et al., 2024; Firmanda et al., 2024). This controlled release spares waste and ensures synergies between the nutrients: a limited nitrogen release at the same time as a high phosphorus and potassium uptake, enabling balanced nutrition (Lawrencia et al., 2021). Furthermore, what it is doing is allowing the release of nutrients to slow down the decomposition process, which is why the microflora of the soil matter is able to be kept steady since no nutrients are being shocked with such high spikes. This leads to an enhancement of microbial activity, which is essential in the turnover of organic matter and enzyme production to deliver complementary effects of fertilizer bioactivity (Kumar et al., 2025).

9.4.2 Nanocarriers and Polymer-Based Fertilizers

Nanotechnology has also introduced another twist to the fertilizer innovation role in the form of nanocarriers and polymer-based systems. When structured into nanocarriers, nanoclays, chitosan, or even carbon-based nanomaterials, they are very surface area and porosity tunable and thus are novel in the sense that they can be used to encapsulate nutrients, growth stimulants, or even as many as microbial metabolites. Their fine size is such that they can deliver APU nutrients to plant roots or leaves and enhance their uptake efficiency, thus minimizing the wastage (Machado et al., 2022). Polymer fertilizers based on the proper fertilizer polymer may contain biodegradable or smart polymers and may react to the varying soil pH, moisture, or temperature, thus emitting nutrients under favorable conditions. A so-called smart release mechanism offers the supply of nutrients at the right time and in the right amount, which prefers synergistic interactions of plants and microorganisms (Dutta et al., 2022). An example of this, the encapsulation of micronutrients (including zinc or iron) within nanocarriers can enhance their bioavailability, yet at the same time reduce antagonism between them and macronutrients. In the same way, multifunctional fertilizers that improve the stress tolerance and resiliency of crops may be developed by the combination of nanocarriers with biostimulants or by the combination of nanocarriers with hormones (Machado et al., 2022). They are also conducted with such nanocarrier and polymer technologies to protect nonbioactive molecules options in registry-operation, which are subjected to biodegradability within a soil environment, and thus ensure the longevity of their functional capacity. This is particularly critical to sensitive materials such as plant hormones or signaling molecules that might play a role in mediating interactions among soil plants and microbes (Garg et al., 2023).

9.4.3 Carriers for Microbial Inoculants

Fertilizers contain inoculations of PGPR, mycorrhizal fungi, and nitrogen-fixing bacteria (Ghorui et al., 2025; Sivaram et al., 2023). One of the very greatest problems, however, in microbial fertilization is the preservation of these microorganisms during storage and in the interval after application. Carriers are needed to maintain the activity of inoculated microbes and their colonization in the rhizosphere (Sivaram et al., 2023). Conventionally used peat, clay, and compost provide the microbes with a conducive environment and nutrients (Younis et al., 2022). In addition to enhancing the synergies between the microbes and chemical fertilization, advanced carriers, biochar, alginate beads, and nanostructured gels not only enable the survival of the microbes but also increase the interplay between these microbes and chemical fertilization (Sivaram et al., 2023). For instance, the carbonaceous structure of biochar harbors minerals and microbe cells, and their juxtaposition in the composition of the carbonaceous material triggers the potassium cycling (Singh & Kumar, 2024). They may also use carriers to co-serve divergent strains of nitrogen-fixing, phosphate-solubilizing, and stress-alleviating microorganisms. The microbial fertilizers could form a synergy effect that is further increased through the symbiotic character of the microbiomes in general, since pooled microbiomes have a cumulative effect on enzyme activity in the soil, availability of nutrients, and plant health (Khan et al., 2024). Recent approaches to fertilization are moving towards the use of complex bioactive systems involving both microbial inoculants and nutrient carriers to enhance crop yield and maintain soil fertility (Singh & Kumar, 2024).

9.5 Biological Fertilizer Innovations: Exploring Synergies with Advanced Technologies

The active cultivation of the lands of these areas makes the creation of more intelligent and ecologically oriented agronomical methods of fertilization, which involves not only the production of nutrients, urgent. Agro-fertilizers, which have also been the subject of extensive scientific research because of their ability to mitigate adverse ecological impacts, are not different from beneficial microbes and organic fertilizers (Shanmugavel et al., 2023). These farm inputs have the potential to be synergistic with new technologies, such as synthetic biology (and genetically modified microorganisms more broadly) and precision farming. This integration of AMF offers a nutritional advantage for crop growth, improves soil health where the crop is grown, confers resistance towards stress, and assures sustainability in general terms in agro-ecosystems (Singh et al., 2024).

9.5.1 Microbial Consortia and Engineered Inoculants

Biofertilizers are traditionally based on single microbial taxa (such as Azotobacter or Rhizobium, or phosphate-solubilizing bacteria). Despite being functional to a level extent, inoculants containing a single strain can be associated with limited flexibility in heterogeneous soil and crop systems (Khan, 2022). This has led to the formation of microbial consortia and the development of microbe samples, namely, selectively assembled colonies of bacteria, fungi, and actinomycetes. There is synergism in the work of such consortia in the sense that they perform complementary biochemical activities such as nitrogen fixation, phosphate solubilization, and potassium mobilization, and use phytohormones as one of the synergies (Khan, 2022).

Well-designed inoculants move the paradigm. Advances in molecular biology enable genome editing of microorganisms in order to increase bioactivity. As an example, strains of *Bacillus subtilis* gene edited in order to secrete higher levels of antimicrobial peptides can inhibit soilborne pathogens, and engineered species of *P. aeruginosa* can be modified to increase root colonization and nutrient solubilization capacity. These inoculants not only increase crop yields but also reduce the use of synthetic pesticides and fertilizers, thus catering to the sustainability distortions of modern agriculture (Liu et al., 2023).

9.5.2 Synthetic Biology and Gene-Edited Microbes

Synthetic biology has newly opened the doors of possibilities regarding designing a microbe with a designed feature. With new genetic circuits, researchers might program microorganisms to know when soil nutrients are in a specific state and to release nutrients of their own accord whenever they know that farming needs are the object of cultivation. For instance, the synthetic nitrogen-fixing microbe might be engineered to induce nitrogen expression in nitrogen-depleted soil to make the most of consumed resources (Li et al., 2022). Cas9 endonuclease gene-editing tools are used to produce microbial fertilizers. Additionally, these technologies can be used to cause genomic alterations that increase microbes' resistance to abiotic stressors like acidity, salinity, and drought (Wei & Li, 2023). To illustrate, mycorrhizal fungi that are more resistant to soil contaminants can be introduced to much more degraded (read: longer-term polluted) substrates. These additions increase the functionality of microbes, as per the type of soil and crop, by interacting with microbial associations on a community level to give positive effects (Kang et al., 2020).

9.5.3 Integration with Digital Agriculture and Precision Farming

Therefore, when digital farming is applied together with biologically produced fertilizers, nutrient management is transformed by enhancing nutrient cycling. Precision-agriculture technologies monitor the nutritional status and health of crops in real time, using GPS-based tools, soil sensors, and unmanned aerial vehicles. These data-driven solutions help farmers to use biofertilizers during the most opportune time, in the right quantities, and in the right places (Klerkx et al., 2019). A spatial map of the diversity of soil microbes can also be developed using digital facilities to determine the location of the various types of sites where a specific type of inoculant would be most effective. As an example, phosphate-sensing sensors can be used to apply phosphate-solubilizing microbial consortia in a very specific way, which, in addition to reducing waste, allows phosphate-solubilizing microbial consortia to be applied only in the soil that does not require phosphate (Canfora et al., 2021). In addition, biofertilizer-chemical input interactivity can be predicted using AI-driven models to enable agronomists to integrate scientific-based nutrient management responses in an intended agroecosystem (Klerkx et al., 2019). Digital farming is associated with the efficiency of applying microbial fertilization, and grower viability is also improved. The ensuing conjoined reduced input prices, increased output, and maintained healthy soils, which in turn also contribute to sustainable production and thus food security in the long term (Shang et al., 2021).

Figure 9.2 illustrates how integrated fertilizer interactions regulate nutrient uptake, strengthen plant–microbe communication, improve stress resilience, and accelerate soil enzymatic cycling—ultimately supporting sustainable crop productivity and soil fertility.

9.6 Soil Health and Fertilizers Synergies: Long-Term Benefits for Soil Fertility

Sustainable agriculture is made possible by the health of soils and subsequently determines the capacity of the ecosystem to supply crops with resilience, water, and nutrients. It is a well-known fact that fertilizer affects productivity, but the rate of fertilizer

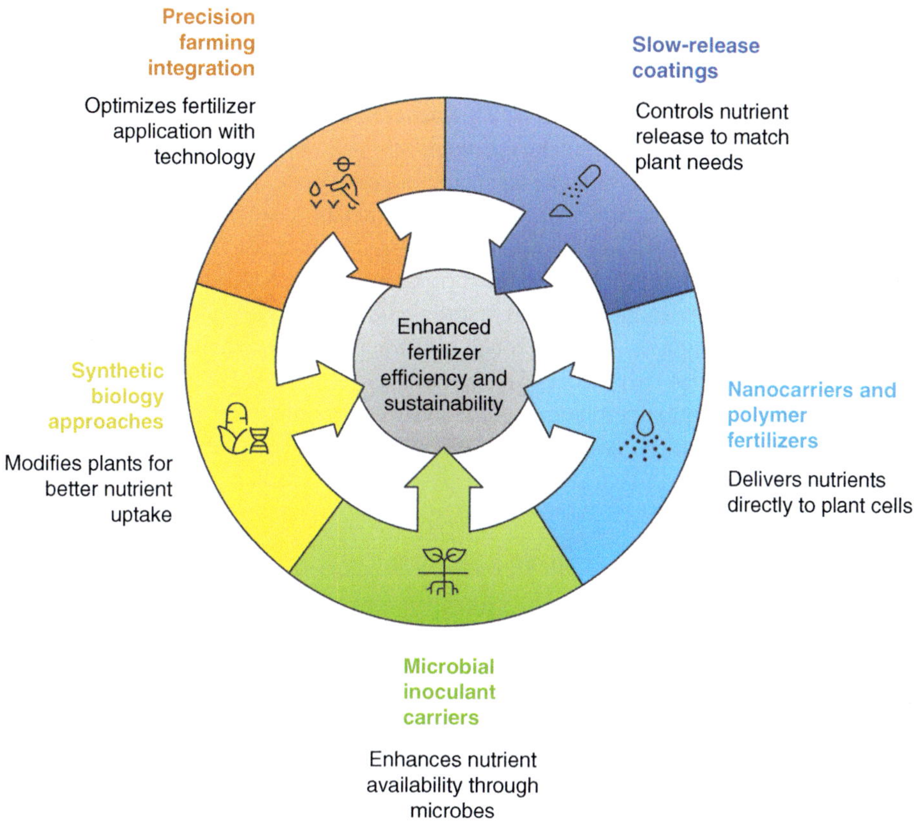

Figure 9.2 Systematic pathways of fertilizer interaction in plant–soil systems. Efforts of interaction between biology and physicochemistry promote nutrient uptake functionality, modulate soil–plant microbe signaling upon interacting, enhance stress tolerance, and accelerate soil–enzyme-mediated nutrient cycling. The combination of all these mechanisms ensures sustainable crop production, soil fertility, and climate stability.

application needs to balance the persistence of fertilizer in a community where ecological processes and microbial activity occur (Xing et al., 2025). When the application of chemical, organic, and biological fertilizers is combined, the combined impact on SOM, microbial diversity, and ecological functioning can be better than the combined impact of individual effects (Jin et al., 2022). In this way, enhancement of nutrient cycling, the soil composition, and retention of carbon can ensure soil conservation and short-term crop production (Xing et al., 2025).

9.6.1 Soil Organic Matter (SOM) Buildup

Soil fertility includes organic matter in soil, which may affect water retention, nutrient availability, and soil stability. While nutrients can be provided through the use of traditional synthetic fertilizers, they frequently do not add enough to increase SOM concentrations in a manner that is not detrimental, as overapplication could potentially facilitate the degradation of organic matter (Cotrufo & Lavallee, 2022). Conversely, organic amendments such as compost, manure, and biochar have been found to naturally stimulate SOM accretion (i.e., building up) through the addition of carbon-rich material, facilitating increased soil organic stocks (Cotrufo & Lavallee, 2022; Hasnat et al., 2022). Organic additives are supplementary to the inorganic fertilizers. Hazardous waste fertilizers (waste chemistry) offer a boon for plant tillage in the short term, and organic waste cultures enrich soil ecology and mitigate nutrient leaching in the long run (nutrient balances). For instance, the inclusion of nitrogen fertilizer and composted manure has increased crop maturity, and decomposition of crop residues has led to the formation of humus (Sollins & Gregg, 2017). Furthermore, SOM serves as the structural material and energy source of the soil microbials, and consequently, it practices the microbial functions such as mineralization and nutrient cycling. Elevating SOM also reduces nutrient loss and enhances soil buffering capacity to pH changes, leading to greater tolerance of the soil to environmental stress. Consequently, the SOM regeneration, which goes along with the associated integrated fertilization, leads to fertile, high-productive, and climate change-adapted soils (He et al., 2025).

9.6.2 Microbial Diversity and Ecosystem Services

Microflora in soil is the recording of the ecology of agricultural ecosystems. Fertilizers both directly and indirectly affect the microbial communities by providing nutrients and changing the physical and chemical properties of soils, respectively (Delgado-Baquerizo et al., 2016). Too much dependence on synthetic fertilizers may lead to a reduction in microbial diversity, thus promoting the growth of a few species loved by nutrients to dominate the ecological balance. In contrast, microorganisms thrive in number and in diversity in the case of chemical fertilizers coupled with organic inputs as well as biofertilizers (Nizamani et al., 2024).

PGPR, mycorrhizal fungi, and microbial consortia are important with respect to promoting nutrient availability, degradation of pathogens, and tolerance to stress. Through the administration of microbial inoculants and organic amendments, the soil will be enriched with useful organisms, which will also help with performing ecosystem services like biological nitrogen fixation and phosphate solubilization, and it is also associated with the production of phytohormones. They supplement the nutrient content provided by chemical fertilizers, accordingly forming a more efficient and balanced nutrient cycle (Gayan et al., 2023). Studies of the microbial diversity also provide resilience to soil. Microbiological complexes guarantee functional redundancy: that is, when one set of microbes is compromised due to changing environmental conditions, other sets can still keep performing a crucial role (nutrient cycling, pathogen control, and so on) (Zhai et al., 2024). This biological insurance normalizes the soil productivity and forms the basis of minimizing the use of outside chemical ones. Microbial diversity, therefore, leads to land fertility and ecosystem resilience in the long run by facilitating the interaction of fertilizers (Zhai et al., 2024).

9.6.3 Soil Structure and Aggregation

Water infiltration, aeration, and root penetration are directly controlled by soil structure, which is in turn controlled by the aggregation of soil particles into aggregates. Treatments in fertilization have a considerable effect on soil aggregates. Chemical fertilizers can stimulate plant and root biomass, thereby indirectly supplying the soil with organic residues, but they have a rather weak capacity to trigger structural change compared to organic or biological amendments (Schlüter et al., 2020). Organic amendments such as compost or manure increase the aggregate formation as a binding agent. Moreover, it is justified by the impact of microbial activity: the production of glomalin by mycorrhizal fungi and the addition of bacterial extracellular polymeric substances to soil aggregates (Basset et al., 2023). In combination with organic/microbial inputs, the presence of an enlarged root system that is sensitive to soil microbes results in a positive feedback loop that promotes an increase in aggregate stability (Ali et al., 2024). The quality of soil at that time increases to increase the porosity, water-holding capacity, and resistance to erosion. The long-term impacts of fertility are far more extensive in scale, with healthy soils preventing nutrient loss and encouraging the presence of useful microbial communities. Moreover, fertilizer synergies are not only able to enhance crop yield but also provide physical stability to soil structure, which guarantees the continuous production of products over generations (Ali et al., 2024).

9.6.4 Nutrient Cycling and Enzyme Activity

In the optimization of nutrient cycling, fertilizer synergies are important. Phosphorus, potassium, and nitrogen are the macronutrients that can be attained relatively easily through inorganic fertilized materials, which are, however, prone to loss through leaching, volatilization, or fixation and are not readily taken up by the crop. These losses can be slowed by organic fertilizers and microbial inoculants, which enhance nutrient retention of the soil–plant system as well as nutrient transformations. By improving nitrogen fixation of the atmosphere to plant-available forms and phosphorus solubilization of insoluble fractions by phosphate-solubilizing microorganisms (Song et al., 2025). The reaction of microbial communities in soil breaks down organic matter and enhances the rate of nutrient mineralization through soil enzymes, including urease, phosphatase, and dehydrogenase. Fertilizer synergies and Senge box—these processes are augmented through the synergistic stimulation of microbial activity (through the supply of organic matter) and nutrient triggers (through chemical fertilizers) (Feng et al., 2025). An example is that when phosphate fertilizers are added together with phosphate-solubilizing bacteria, the uptake of approximately 25% more phosphate is increased, and biochar addition improves microbial enzyme activity by creating stable surfaces on which the microbes can grow. The reason behind such interactions is that the nutrients are recycled effectively in the soil apparatus such that external inputs are minimized and the long-run soil fertility is maintained (Song et al., 2025).

9.6.5 Carbon Sequestration and Climate Benefits

Other than improved yield, there is also the implication of fertilization synergies on soil processes related to carbon cycling at the global level. Due to the formation of organic amendments, microbial activities enhance the structure of the soil and, in turn, carbon sequestration as well as the stability of SOM. An example of such is the biochar, which is sequestered in the soil over a period of decades, keeping carbon demineralized, not to mention increasing soil fertility. Combined with inorganic fertilizers, biochar has a greater nutrient use efficiency with less excretion of greenhouse gas emissions like N_2O (Dhingra & Kumar, 2025). The inoculants also modify the interaction with carbon, with net photosynthetically fixed carbon being released into the soil by rhizodeposition, promoting the growth of macrobiotic biomass and positive feedback, which promotes further fertility and sequestration. In addition, the advantages of carbon sequestration fertilization can be increased soil fertility and a decreased concentration of CO_2 in the atmosphere (Hu et al., 2022).

9.6.6 Long-Term Sustainability Perspective

Greater yields, resistance, ecological soundness, and sustainability are also the possible benefits of systems of synergistic fertilization to soil welfare. Integrated fertilization through incorporation of fertilizer can preserve the surface soil structure, the biodiversity, and the integrity of soil architecture with organisms compared to pure chemical fertilization. Also, they facilitate the recycling of nutrients, minimize the environmental losses, and make the agricultural system sustainable and without depleted natural resources (Wu et al., 2017). Combining the concepts of fertility and soil health can enable farmers and scientists to reach their short-term production goals, as well as safeguard the long-term ecological foundations of agriculture (i.e., soil health) (Xing et al., 2025). Thus, the synergies can be a vital channel to sustainable intensification that will guarantee increased food production, effective recycling of resources, and erosion reduction (Wu et al., 2017).

Figure 9.3 highlights the long-term benefits of fertilizer synergies, showing how combined chemical, biological, organic, and nanofertilizer inputs enhance soil organic matter, microbial diversity, soil structure, and carbon sequestration for climate-smart, regenerative agriculture.

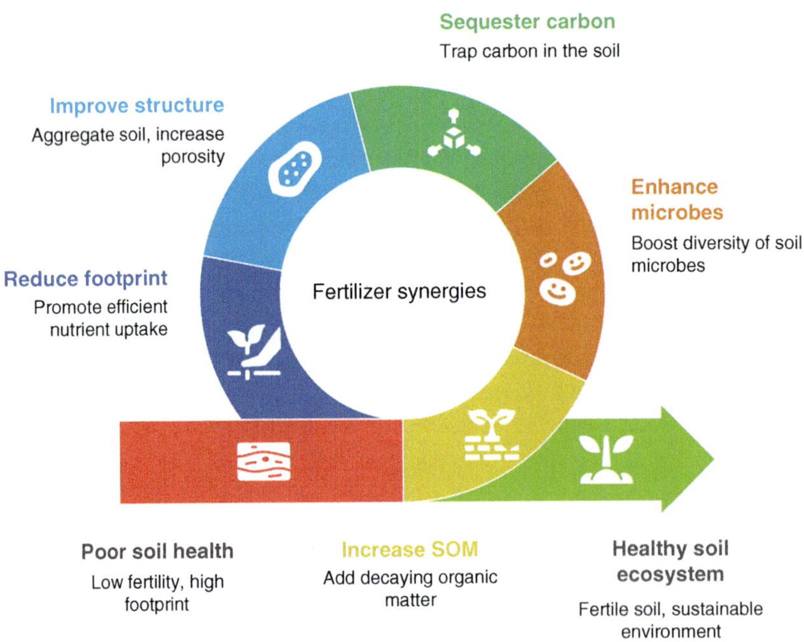

Figure 9.3 Long-term benefits of fertilizer synergies on soil health and sustainability. Chemical, biological, organic, and nanofertilizer integration facilitates the accumulation of soil organic matter, increases microbial diversity, improves soil structure, and helps in carbon sequestration, which leads to regenerative and climate-smart agriculture (CSA).

9.7 Challenges in Fertilizer Synergy

Despite the high expectations of synergistic fertilizer bioactivities, there are a number of challenges in translating this potential into success, which can be replicated in field implementations. These barriers are usually caused by the inherent complexity of agricultural systems as chemical, biological, and environmental factors dynamically interact (Pandian et al., 2024). An understanding and solution to these inadequacies are of paramount importance to stakeholders across the board to ensure that synergistic approaches do not just stay within experimental settings but spread to the mainstream of sustainable agriculture (Ejedegba, 2024).

9.7.1 Compatibility Issues Between Inputs

One of the biggest challenges may be the compatibility of chemical inputs and biological inputs. The majority of synthetic pesticides, fertilizers, or soil amendments may suppress or increase the diversity of the microbes or degrade inoculant efficiencies, thereby threatening the anticipated synergies. Excess nitrogen fertilizers, as shown in one example, may inhibit colonization of mycorrhiza fungi or nitrogen-fixing bacteria (Townsend, 2021). Likewise, not all coatings or funerals can live with biological inoculants, and they reduce their survival. These compatibility problems can be overcome by making a wise formulation strategy, timing application, and creating a blend of fertilizers that are specifically developed with co-functionality (Hannan et al., 2020).

9.7.2 Soil–Climate Variability

The other difficulty is that the use of fertilizers is heavily affected by regional soil and climatic conditions. The manifestations of the fertilizers' synergies are influenced by the soil pH and its texture, the existence of organic matter, which varies widely across the regions, and the communities of microorganisms. Similarly, climatic conditions, including rainfall change, drought conditions, or temperature fluctuations, may modify the dynamics of microbes and the pattern of nutrient discharge (Martey & Kuwornu, 2021). Such site-specificities yield the problem that one may not be able to provide a general set of formulations that works under any of the conditions with equal success. Therefore, precautionary field testing, responsive management, and accuracy cultivation are required to adjust the tactic for such a fertilizer combination with the uniqueness of various agroecosystems (Liu & Basso, 2020).

9.7.3 Environmental Trade-Offs and Unintended Impacts

Even though the overall benefit of using synergizes-type fertilizer is to ease environmental burdens, chances are present that it can lead to unintended consequences when not controlled. Inappropriate release of nutrients by CRFs may also contribute to water eutrophication, whereas new nanomaterials may present ecotoxicity problems in soil and water systems (Chen et al., 2025). Also, engineered microbial inoculants or gene-edited strains can create biosafety and ecological issues of long-term persistence and interactions with native microbiota. As such, stringent regulations on environmental risk, biosafety, and regulations should be achieved by providing strict environmental risk tests, biosafety, and guidelines that help clarify the cost and benefits of fertilizer synergies as far as trade-offs are aforementioned (Van Lanen et al., 2025).

9.7.4 Knowledge and Adoption Gaps

There is still a gap in knowledge amongst farmers, extension workers, and even policymakers, which is a leading stumbling block. Still, a number of smallholder farmers are not conversant with the tenets of fertilizer synergy, and the extra price of the improved formulations might deter their use in resource-poor contexts (Prajapati et al., 2025). To cover this gap requires effective knowledge promotion, a set of farmer training, and financial incentives to promote early adoption. The policy support regarding sustainable inputs will be in the form of public–private partnerships, and these relations will be significant in mainstreaming synergistic fertilizers (Mkenda et al., 2020). Table 9.1 provides an integrative overview of fertilizer synergies, summarizing key components, underlying mechanisms, associated benefits, and potential challenges within soil–plant–microbe systems.

9.8 Policy and Economic Considerations for Fertilizer Synergy Adoption

The widespread use of synergistic fertilizer strategies not only demands their scientific feasibility but also demands their economic feasibility and policy protective provisions. Cost, risk, and initial returns are the main considerations of farmers and, more specifically, the smallholders when making their decisions. As such, economic incentives, clear regulations, and institutional assistance are necessary to make sure that the specific benefits of fertilizer synergies are sustained on a scale (Ejedegba, 2024).

9.8.1 Cost–Benefit Analysis for Farmers

For the majority of agricultural producers, the transition from conventional fertilizers to synergistic inputs hinges on the financial justification of the investment. Synergistic fertilizers can be effective in enhancing nutrient-use efficiency and crop productivity, but inexpensive synergistic fertilizers might cost more up-front investments, in particular when studying nanofertilizers, microbial consortia, or polymer-coated products (Akinyi et al., 2022). Adoption may thus be deterred by short-term cost pressures when long-term benefits with respect to soil fertility and ecological benefits are large. Costs and benefits: When used over a series of growing seasons, costs are always outweighed by benefits in the form of reduced rates of application of fertilizers, an increase in yield, and environmental penalties. These long-term economic benefits should be communicated in the form of case studies, case extension services, and demonstration plots to convince farmers about their profitability (Wang & Zhang, n.d.).

Table 9.1 Integrative overview of fertilizer synergies: components, mechanisms, benefits, and challenges.

Fertilizer input/approach	Bioactive mechanisms	Synergistic effects on plants and soil	Applications in sustainable agriculture	Challenges/limitations	Future opportunities	References
Chemical fertilizers (NPK, micronutrients)	Direct nutrient supply, improved root uptake	Rapid growth, higher yields	Baseline crop nutrition	Nutrient leaching, low efficiency	Smart formulations, nano-coatings	Singaravel et al. (2025)
Organic amendments (compost, manure, biochar)	SOM buildup, improved soil structure, CEC enhancement	Long-term soil fertility, carbon sequestration	Organic and regenerative agriculture	Variable nutrient release, bulk handling	Integration with nanofertilizers, precision composting	Elshony et al. (2025)
Biological inputs (PGPR, mycorrhizae, biostimulants)	Hormone production, nutrient solubilization, and ISR	Stress tolerance, enhanced nutrient uptake	Biofertilizers, climate resilience	Survival under harsh soils, inconsistent results	Engineered microbial consortia, synthetic biology	Sible et al. (2025)
Nanofertilizers and smart fertilizers	Controlled release, targeted delivery	Higher NUE, reduced losses, and precision farming	Climate-smart agriculture	Cost, regulatory hurdles, nanoparticle risks	Safe nanocarriers, digital integration	Bouhadi et al. (2025), Kaur et al. (2023)
Fertilizer coatings and carriers (polymers, nanomaterials, bio-based films)	Slow/controlled release, microbial protection	Prolonged availability, enhanced microbial activity	Next-gen formulations	Environmental persistence, cost	Biodegradable polymers, eco-safe nanocarriers	Mendonca Cidreira et al. (2025)
Synergistic mechanisms	Soil–plant–microbe signaling, enzyme activation, nutrient cycling	Improved nutrient use efficiency, yield stability	Holistic soil health management	Context-specific variability	AI-driven precision agriculture, big-data optimization	Shi et al. (2022)

(Continued)

Table 9.1 (Continued)

Fertilizer input/ approach	Bioactive mechanisms	Synergistic effects on plants and soil	Applications in sustainable agriculture	Challenges/ limitations	Future opportunities	References
Soil health impacts	Microbial diversity, SOM buildup, carbon sequestration	Ecosystem resilience, sustainable productivity	Regenerative systems	Time lag in benefits	Long-term soil restoration frameworks	Fu et al. (2021)
Policy and economic considerations	Subsidies, regulatory standards, and farmer incentives	Adoption at scale, reduced environmental footprint	Sustainable intensification	Policy gaps, affordability	Circular bioeconomy, carbon credits	Jie et al. (2023)
Future vision	Integration of bio, nano, and digital tools	Carbon neutrality, climate-smart farming	Global food security	Equity in adoption	Next-generation regenerative agriculture	Tucker (2023)

9.8.2 Incentives and Subsidies for Sustainable Inputs

Subsidies, tax breaks, or incentives accorded by the government to reduce barriers to entry are very essential. In several states, chemical fertilizers are already heavily subsidized, thus providing a lopsided level playing ground for biofertilizers, organic fertilizers, and advanced formulations. Reflecting or carrying over the subsidies to accommodate the sustainable inputs would support the implementation of synergistic strategies (Wang & Zhang, n.d.). Also, performance-oriented strategies (such as farmers who prove to have minimal nutrient run-offs or better soil conditions) may be used to promote their adoption. Another significant role is played by public–private partnerships: they should be able to present the innovative technologies created in the laboratories, making them able and affordable to end users (Jia & Guang, 2024).

9.8.3 Regulatory Challenges and Standards

There is another layer of complexity related to the regulatory environment of fertilizer innovation. Although the quality and safety of regular fertilizers are well defined, there are numerous cases of biofertilizers, microbial inoculants, and nanomaterials where there is a senseless setup regarding approval, labeling, and monitoring of the products (Freyer et al., 2024). Such a belief discourages commercialization and creates fear in farmers and distributors because of this regulatory uncertainty. In addition, the risk accountability issues topical in the biosafety of electronics of engineered microbes or nanocarriers require well-developed frameworks of risk appraisal (Xi, 2024). This requires a harmonized international standard with a transparent approval procedure and traceability capabilities to build a degree of confidence in the minds of the farmers, as well as consumers' safety. The policies must also be fine-tuned to the dynamics of technology to enable research, field validation, and exchange of data (Abimanyu & Faiz, 2023).

9.9 Conclusion and Future Perspectives

In recent times, developments in sustainable agriculture rely on the synergic application of fertilizers. Also, despite the fact that this solution is likely to achieve immense progress in global food security, the use of conventional fertilizers is linked to pollution of the environment and soil erosion and is not using the nutrients effectively. Over the past several decades, collaborative application of chemical fertilizers with bioactive substances like microbe inoculants, organic additions, and nanomaterials has been reported to give rise to synergistic effects resulting in enhancement of nutrient use efficiency, plant growth, and soil health in sustainable agriculture. These interactions have an effect beyond immediate productivity and increase SOM increment, soil microbial diversity, improved soil structure, and rapid carbon uptake. Whereas intrinsic fertilization synergies are actionable to overcome the adverse effect of (climatic) hazards by fostering stress resistance and ensuring production in the face of drought, salinity, and high temperature, the latter benefits of the same effects, stimulation responses, are still more significant. Although there are obstacles such as input prompts, mixed results among soil-climate gradients, and restraints to domestication, the indications of the contribution of the fertilizer synergies to bridge the division between production and environmental prodigal son appear quite evident through the correlative success. New generations of crop fertilizers based on the integration of emergent biological technologies (such as nanomaterials and precision delivery systems) will offer the foundation on which food production will be brought to sustainable food security.

This is a new path in fertilizer synergies that is increasingly being connected to world goals as to the sustainability and production systems of the future in agriculture. Circular economy models based on collaborating with the bioeconomy are likely to drive the valorization of agricultural and food waste into bioactive fertilizer, which will play a role in larger nutrient recycling and reduced dependence on finite resources. At the same time, CSA is expected to be positively affected by fertilization because of synergistic relationships that will reduce greenhouse gas emissions and increase crop resistance to unpredictable climatic conditions. These fertilization methods can become even more complex and flexible in the future as the advancements in artificial biology, gene-engineered microorganisms, and intelligent nano-carriers become available. Integrated fertilizers are also destined to become an important element of regenerative agriculture in the future at the global and local scales, as they will enable reversing the process of soil degradation, enhancing the ecosystem services, and achieving carbon neutrality. The endogenous frontier is more than the absorption of technology; it involves the creation of favorable policy, development of the agricultural capacity, and international collaboration that facilitates extensive distribution. The key to making modern agriculture a resource-efficient business resistant to climate change, which could be reached in the event of successful synergy between science, policy, and practice, may lie in such synergistic fertilizers.

References

Abd-Elhalim, H. M., Hamed, R. R., El Sangry, D. A. M., & Radwan, K. H. (2025). Characterization and isolation of 1-aminocyclopropane1-carboxylate (ACC) deaminase-producing, plant growth-promoting rhizobacteria from the rhizosphere of Egyptian wheat cultivars for improved drought resilience. *Brazilian Journal of Biology*, 84, e285457. https://doi.org/10.1590/1519-6984.285457

Abdelrahman, M., Tran, L. P., & Shigyo, M. (2022). Editorial: Physiological and molecular perspectives of stress tolerance in vegetables. *Frontiers in Plant Science*, 13, 1004093. https://doi.org/10.3389/fpls.2022.1004093

Abimanyu, A., & Faiz, I. A. (2023). ASEAN halal markets analysis: Regulatory and harmonization challenges. *Global Review of Islamic Economics and Business*, 11(1), 25–44.

Akinnawo, S. O. (2023). Eutrophication: Causes, consequences, physical, chemical and biological techniques for mitigation strategies. *Environmental Challenges*, 12, 100733. https://doi.org/10.1016/j.envc.2023.100733

Akinyi, D. P., Karanja Ng'ang'a, S., Ngigi, M., Mathenge, M., & Girvetz, E. (2022). Cost-benefit analysis of prioritized climate-smart agricultural practices among smallholder farmers: Evidence from selected value chains across sub-Saharan Africa. *Heliyon*, 8(4).

Ali, N., Abbas, S. A. A. A., Sharif, L., Shafiq, M., Kamran, Z., Haseeb, M., & Shahid, M. A. (2024). Chapter 13—microbial extracellular polymeric substance and impacts on soil aggregation. In K. A. Abd-Elsalam & H. I. Mohamed (Eds.), *Bacterial secondary metabolites* (pp. 221–237). Elsevier. https://doi.org/10.1016/B978-0-323-95251-4.00021-1

Al-Rawajfeh, A. E., Alrbaihat, M. R., & AlShamaileh, E. M. (2021). Chapter 4—characteristics and types of slow- and controlled-release fertilizers. In F. B. Lewu, T. Volova, S. Thomas, & K. R. Rakhimol (Eds.), *Controlled release fertilizers for sustainable agriculture* (pp. 57–78). Academic Press. https://doi.org/10.1016/B978-0-12-819555-0.00.04-2

Ammar, E. E., Rady, H. A., Khattab, A. M., Amer, M. H., Mohamed, S. A., Elodamy, N. I., Al-Farga, A., & Aioub, A. A. A. (2023). A comprehensive overview of eco-friendly bio-fertilizers extracted from living organisms. *Environmental Science and Pollution Research International*, 30(53), 113119–113137. https://doi.org/10.1007/s11356-023-30260-x

de Andrade, L. A., Santos, C. H. B., Frezarin, E. T., Sales, L. R., & Rigobelo, E. C. (2023). Plant growth-promoting rhizobacteria for sustainable agricultural production. *Microorganisms*, 11(4). https://doi.org/10.3390/microorganisms11041088

Arshad, R., Gulshad, L., Haq, I. U., Farooq, M. A., Al-Farga, A., Siddique, R., Manzoor, M. F., & Karrar, E. (2021). Nanotechnology: A novel tool to enhance the bioavailability of micronutrients. *Food Science & Nutrition*, 9(6), 3354–3361.

Asghar, W., Akça, M. O., Akça, H., Tarf, O. J., Kataoka, R., & Turgay, O. C. (2022). Chapter 1—alternative strategies to synthetic chemical fertilizers: Revitalization of soil quality for sustainable agriculture using organic-based approaches. In H. B. Singh & A. Vaishnav (Eds.), *New and future developments in microbial biotechnology and bioengineering* (pp. 1–30). Elsevier. https://doi.org/10.1016/B978-0-323-85581-5.00003-3

Ashitha, A., Rakhimol, K. R., & Mathew, J. (2021). Chapter 2—fate of the conventional fertilizers in environment. In F. B. Lewu, T. Volova, S. Thomas, & K. R. Rakhimol (Eds.), *Controlled release fertilizers for sustainable agriculture* (pp. 25–39). Academic Press. https://doi.org/10.1016/B978-0-12-819555-0.00.02-9

Ayamba, B. E., Abaidoo, R. C., Opoku, A., & Ewusi-Mensah, N. (2023). Mechanisms for nutrient interactions from organic amendments and mineral fertilizer inputs under cropping systems: A review. *PeerJ*, 11, e15135. https://doi.org/10.7717/peerj.15135

Bashir, M. A., Raza, Q.-U.-A., Ahmad, I., Ali, M. A., Raza, H. M. A., & Rehim, A. (2025). Chapter 18–control release of nanofertilizer against stress in plants: Solutions, threats and opportunities. In H. Tombuloglu, G. Tombuloglu, K. R. Hakeem, F. S. Baloch, & M. A. Ansari (Eds.), *Nanomaterials for enhanced plant-based food production* (pp. 263–271). Academic Press. https://doi.org/10.1016/B978-0-443-23688-4.00011-7

Basset, C., Abou Najm, M., Ghezzehei, T., Hao, X., & Daccache, A. (2023). How does soil structure affect water infiltration? A meta-data systematic review. *Soil and Tillage Research*, 226, 105577. https://doi.org/10.1016/j.still.2022.105577

Bouhadi, M., Javed, Q., Jakubus, M., Elkouali, M. H., Fougrach, H., Ansar, A., Ban, S. G., Ban, D., Heath, D., & Černe, M. (2025). Nanoparticles for sustainable agriculture: Assessment of benefits and risks. *Agronomy*, 15(5), 1131.

Canfora, L., Costa, C., Pallottino, F., & Mocali, S. (2021). Trends in soil microbial inoculants research: A science mapping approach to unravel strengths and weaknesses of their application. *Agriculture*, 11(2), 158.

Chakraborty, S., Mandal, M., Chakraborty, A. P., & Majumdar, S. (2022). Chapter 12–zinc solubilizing rhizobacteria as soil health engineer managing zinc deficiency in plants. In R. C. Dubey & P. Kumar (Eds.), *Rhizosphere engineering* (pp. 215–238). Academic Press. https://doi.org/10.1016/B978-0-323-89973-4.00018-1

Chen, T., Feng, S., Xie, Y., Wang, K., Wang, X., Zhao, J., & Zhang, J. (2025). Ecological trade-offs in shallow lake restoration: Assessing ecosystem responses to aquatic vegetation restoration and biomanipulation strategies against the background of fishing ban. *Ecological Indicators*, 178, 113932.

Cotrufo, M. F., & Lavallee, J. M. (2022). Chapter one—soil organic matter formation, persistence, and functioning: A synthesis of current understanding to inform its conservation and regeneration. In D. L. Sparks (Ed.), *Advances in agronomy* (Vol. 172, pp. 1–66). Academic Press. https://doi.org/10.1016/bs.agron.2021.11.002

Darnton-Hill, I. (2018). Chapter 2—prevalence, causes, and consequences of micronutrient deficiencies. the gap between need and action. In M. G. V. Mannar & R. F. Hurrell (Eds.), *Food fortification in a globalized world* (pp. 13–28). Academic Press. https://doi.org/10.1016/B978-0-12-802861-2.00002-X

Daunoras, J., Kačergius, A., & Gudiukaitė, R. (2024). Role of soil microbiota enzymes in soil health and activity changes depending on climate change and the type of soil ecosystem. *Biology (Basel)*, 13(2). https://doi.org/10.3390/biology13020085

Delgado-Baquerizo, M., Maestre, F. T., Reich, P. B., Jeffries, T. C., Gaitan, J. J., Encinar, D., Berdugo, M., Campbell, C. D., & Singh, B. K. (2016). Microbial diversity drives multifunctionality in terrestrial ecosystems. *Nature Communications*, 7, 10541. https://doi.org/10.1038/ncomms10541

Dhingra, G., & Kumar, A. (2025). Carbon capture and sequestration: Cutting-edge technologies to combat climate change. *Sustainable Energy Technologies and Assessments*, 75, 104226. https://doi.org/10.1016/j.seta.2025.104226

Ding, Y., Gao, X., Shu, D., Siddique, K. H. M., Song, X., Wu, P., Li, C., & Zhao, X. (2024). Enhancing soil health and nutrient cycling through soil amendments: Improving the synergy of bacteria and fungi. *Science of the Total Environment*, 923, 171332. https://doi.org/10.1016/j.scitotenv.2024.171332

Dutta, S., Pal, S., Panwar, P., Sharma, R. K., & Bhutia, P. L. (2022). Biopolymeric nanocarriers for nutrient delivery and crop biofortification. *ACS Omega*, 7(30), 25909–25920. https://doi.org/10.1021/acsomega.2c02494

Ebrahimi, M., Gholipour, S., Mostafaii, G., & Yousefian, F. (2024). Biochar-amended food waste compost: A review of properties. *Results in Engineering*, 24, 103118. https://doi.org/10.1016/j.rineng.2024.10.118

Ejedegba, E. O. (2024). Synergizing fertilizer innovation and renewable energy for improved food security and climate resilience. *International Journal of Research Publication and Reviews*, 5(12), 3073–3088.

El-Shal, R. M., El-Naggar, A. H., El-Beshbeshy, T. R., Mahmoud, E. K., El-Kader, N. I. A., Missaui, A. M., Du, D., Ghoneim, A. M., & El-Sharkawy, M. S. (2022). Effect of nano-fertilizers on alfalfa plants grown under different salt stresses in hydroponic system. *Agriculture*, 12(8), 1113.

El-Sharkawy, M., Li, J., AL-Huqail, A. A., Hamed, M. A., Du, D., & EL-Khamisy, R. R. (2024). Slow-released fertilizers optimization and experimental impacts on soil fertility and wheat-maize cropping system. *Scientia Agricola*, 81, e20230234.

El-Sharkawy, M., Li, J., AL-Huqail, A. A., Du, D., EL-Khamisy, R. R., & El-Gamal, B. A. (2025). Sustainable microbial strategies for enhancing soil fertility and wheat (Triticum aestivum L.) production. *Journal of Soil Science and Plant Nutrition*, 25(1), 496–513.

Elshony, M., Farid, I., Elkamar, F., Abbas, H., Ali, I., & Abbas, M. H. (2025). Effects of organic amendments on soil productivity, its physical and chemical properties: A review of compost and biochar applications. *Biochar and Compost Technology*, 2(1), 72–91.

Feng, D., Meng, L., Wen, Y., Uwiragiye, Y., AbuQamar, S. F., Okoth, N., Zhu, Q., Wu, Z., Wu, Y., Müller, C., Zhang, J., & Elrys, A. S. (2025). Edaphic and climatic factors control the response of nutrient-cycling enzyme activity to common heavy metals in soils. *Journal of Hazardous Materials*, 494, 138475. https://doi.org/10.1016/j.jhazmat.2025.138475

Firmanda, A., Fahma, F., Syamsu, K., Sari, Y. W., Suryanegara, L., Wood, K., & Saito, Y. (2023). Factors influencing the biodegradability of agro-biopolymer based slow or controlled release fertilizer. *Journal of Polymers and the Environment*, 31(5), 1706–1724.

Firmanda, A., Fahma, F., Syamsu, K., Mahardika, M., Suryanegara, L., Munif, A., Gozan, M., Wood, K., Hidayat, R., & Yulia, D. (2024). Biopolymer-based slow/controlled-release fertilizer (SRF/CRF): Nutrient release mechanism and agricultural sustainability. *Journal of Environmental Chemical Engineering*, 12(2), 112177. https://doi.org/10.1016/j.jece.2024.112177

Freyer, O., Wiest, I. C., Kather, J. N., & Gilbert, S. (2024). A future role for health applications of large language models depends on regulators enforcing safety standards. *The Lancet Digital Health*, 6(9), e662–e672.

Fu, B., Chen, L., Huang, H., Qu, P., & Wei, Z. (2021). Impacts of crop residues on soil health: A review. *Environmental Pollutants and Bioavailability*, 33(1), 164–173.

Garg, D., Sridhar, K., Stephen Inbaraj, B., Chawla, P., Tripathi, M., & Sharma, M. (2023). Nano-biofertilizer formulations for agriculture: A systematic review on recent advances and prospective applications. *Bioengineering (Basel)*, 10(9). https://doi.org/10.3390/bioengineering10091010

Gayan, A., Borah, P., Nath, D., & Kataki, R. (2023). Chapter 4—soil microbial diversity, soil health and agricultural sustainability. In M. Farooq, N. Gogoi, & M. Pisante (Eds.), *Sustainable agriculture and the environment* (pp. 107–126). Academic Press. https://doi.org/10.1016/B978.0-323-90500-8.00006-3

Ghorui, M., Chowdhury, S., & Burla, S. (2025). Recent advances in the commercial formulation of arbuscular mycorrhizal inoculants. *Frontiers in Industrial Microbiology*, 3, 1553472.

Gilbert, L., & Johnson, D. (2017). Chapter four—plant–plant communication through common mycorrhizal networks. In G. Becard (Ed.), *Advances in botanical research* (Vol. 82, pp. 83–97). Academic Press. https://doi.org/10.1016/bs.abr.2016.09.001

Gowda, R. S., Kaur, M., Kaushal, B., Kaur, H., Kumar, V., Sharma, R., Singh, T., Choudhary, A., Mehta, S., & Husen, A. (2024). Chapter 4—behavior, sources, uptake, interaction, and nutrient use efficiency in plant system under changing environment. In A. Husen (Ed.), *Essential minerals in plant-soil systems* (pp. 93–127). Elsevier. https://doi.org/10.1016/B978-0-443-16082-0.00017-5

Gross, C. D., Bork, E. W., Carlyle, C. N., & Chang, S. X. (2022). Biochar and its manure-based feedstock have divergent effects on soil organic carbon and greenhouse gas emissions in croplands. *Science of the Total Environment*, 806, 151337. https://doi.org/10.1016/j.scitotenv.2021.151337

Guo, X.-X., Liu, H.-T., & Zhang, J. (2020). The role of biochar in organic waste composting and soil improvement: A review. *Waste Management*, 102, 884–899. https://doi.org/10.1016/j.wasman.2019.12.003

Hannan, M. A., Ker, P. J., Lipu, M. S. H., Choi, Z. H., Rahman, M. S. A., Muttaqi, K. M., & Blaabjerg, F. (2020). State of the art of solid-state transformers: Advanced topologies, implementation issues, recent progress and improvements. *IEEE Access*, 8, 19113–19132.

Hasnat, M., Alam, M. A., Khanam, M., Binte, B. I., Kabir, M. H., Alam, M. S., Kamal, M. Z. U., Rahman, G. K. M. M., Haque, M. M., & Rahman, M. M. (2022). Effect of nitrogen fertilizer and biochar on organic matter mineralization and carbon accretion in soil. *Sustainability*, 14(6), 3684.

He, Z., Zhu, X., Zhou, F., Zhu, M., Bao, X., Deng, F., He, H., & Zhang, X. (2025). How bacterial traits scale to soil organic matter pool buildup revealed by long-term maize straw mulching experiment. *Agriculture, Ecosystems & Environment*, 393, 109816. https://doi.org/10.1016/j.agee.2025.109816

Ho, T. T. K., Tra, V. T., Le, T. H., Nguyen, N.-K.-Q., Tran, C.-S., Nguyen, P.-T., Vo, T.-D.-H., Thai, V.-N., & Bui, X.-T. (2022). Compost to improve sustainable soil cultivation and crop productivity. *Case Studies in Chemical and Environmental Engineering*, 6, 100211. https://doi.org/10.1016/j.cscee.2022.100211

Hu, Y., Zhang, Q., Hu, S., Xiao, G., Chen, X., Wang, J., Qi, Y., Zhang, L., & Han, L. (2022). Research progress and prospects of ecosystem carbon sequestration under climate change (1992–2022). *Ecological Indicators*, 145, 109656. https://doi.org/10.1016/j.ecolind.2022.109656

Jakhar, A. M., Aziz, I., Kaleri, A. R., Hasnain, M., Haider, G., Ma, J., & Abideen, Z. (2022). Nano-fertilizers: A sustainable technology for improving crop nutrition and food security. *NanoImpact*, 27, 100411. https://doi.org/10.1016/j.impact.2022.100411

Jia, X., & Guang, W. (2024). Can innovation improve corporate ESG performance? The moderating effect of internal and external incentives. *Sustainability*, 16(15), 6582.

Jie, H., Khan, I., Alharthi, M., Zafar, M. W., & Saeed, A. (2023). Sustainable energy policy, socio-economic development, and ecological footprint: The economic significance of natural resources, population growth, and industrial development. *Utilities Policy*, 81, 101490.

Jin, N., Jin, L., Wang, S., Li, J., Liu, F., Liu, Z., Luo, S., Wu, Y., Lyu, J., & Yu, J. (2022). Reduced chemical fertilizer combined with bio-organic fertilizer affects the soil microbial community and yield and quality of lettuce. *Frontiers in Microbiology*, 13, 863325.

Kang, M., Choe, D., Kim, K., Cho, B. K., & Cho, S. (2020). Synthetic biology approaches in the development of engineered therapeutic microbes. *International Journal of Molecular Sciences*, 21(22). https://doi.org/10.3390/ijms21228744

Kaur, H., Hussain, S. J., Mir, R. A., Chandra Verma, V., Naik, B., Kumar, P., & Dubey, R. C. (2023). Nanofertilizers—emerging smart fertilizers for modern and sustainable agriculture. *Biocatalysis and Agricultural Biotechnology*, 54, 102921. https://doi.org/10.1016/j.bcab.2023.102921

Kekeli, M. A., Wang, Q., & Rui, Y. (2025). The role of nano-fertilizers in sustainable agriculture: Boosting crop yields and enhancing quality. *Plants (Basel)*, 14(4). https://doi.org/10.3390/plants14040554

Khambalkar, P. A., Agrawal, S., Dhaliwal, S. S., Yadav, S. S., Sadawarti, M. J., Singh, A., Yadav, I. R., Yadav, K., Prasad, D., Singh, A., & Afreen, N. (2025). Sustainable nutrient management balancing soil health and food security for future generations. *Applied Food Research*, 5(2), 101087. https://doi.org/10.1016/j.afres.2025.101087

Khan, S. T. (2022). Consortia-based microbial inoculants for sustaining agricultural activities. *Applied Soil Ecology*, 176, 104503. https://doi.org/10.1016/j.apsoil.2022.104503

Khan, M., Khan, T., Tabassum, B., & Hashim, M. (2024). Microbiome-driven soil fertility: Understanding symbiotic relationships. In *Progress in soil microbiome research* (pp. 77–115). Springer.

Khoso, M. A., Wagan, S., Alam, I., Hussain, A., Ali, Q., Saha, S., Poudel, T. R., Manghwar, H., & Liu, F. (2024). Impact of plant growth-promoting rhizobacteria (PGPR) on plant nutrition and root characteristics: Current perspective. *Plant Stress*, 11, 100341.

Klerkx, L., Jakku, E., & Labarthe, P. (2019). A review of social science on digital agriculture, smart farming and agriculture 4.0: New contributions and a future research agenda. *NJAS—Wageningen Journal of Life Sciences*, 90-91, 100315. https://doi.org/10.1016/j.njas.2019.100315

Ku, Y. S., Cheng, S. S., Luk, C. Y., Leung, H. S., Chan, T. Y., & Lam, H. M. (2025). Deciphering metabolite signalling between plant roots and soil pathogens to design resistance. *BMC Plant Biology*, 25(1), 308. https://doi.org/10.1186/s12870-025-06321-3

Kumar, A., Singh, H. S., Kumar, A., Singh, M. P., Singh, A., Maurya, N. K., Darbha, S., & Kumar, R. (2025). A comprehensive review on plant-soil interactions: Microbial dynamics, nutrient cycling and sustainable crop production. *Asian Journal of Soil Science and Plant Nutrition*, 11(2), 44–62.

Lawrencia, D., Wong, S. K., Low, D. Y. S., Goh, B. H., Goh, J. K., Ruktanonchai, U. R., Soottitantawat, A., Lee, L. H., & Tang, S. Y. (2021). Controlled release fertilizers: A review on coating materials and mechanism of release. *Plants (Basel)*, 10(2). https://doi.org/10.3390/plants10020238

Li, W., Huang, C., & Chen, J. (2022). The application of CRISPR/Cas mediated gene editing in synthetic biology: Challenges and optimizations. *Frontiers in Bioengineering and Biotechnology*, 10, 890155. https://doi.org/10.3389/fbioe.2022.890155

Liu, L., & Basso, B. (2020). Impacts of climate variability and adaptation strategies on crop yields and soil organic carbon in the US Midwest. *PLoS One*, 15(1), e0225433.

Liu, X., Mei, S., & Salles, J. F. (2023). Inoculated microbial consortia perform better than single strains in living soil: A meta-analysis. *Applied Soil Ecology*, 190, 105011. https://doi.org/10.1016/j.apsoil.2023.105011

Liu, Q., Pang, Z., Sun, H., Zeng, X., Kong, X., Li, S., & Shen, Y. (2024). Unveiling the maize-benefit: Synergistic impacts of organic-inorganic fertilizer cooperation on rhizosphere microorganisms and metabolites. *Applied Soil Ecology*, 193, 105171. https://doi.org/10.1016/j.apsoil.2023.10.171

Lu, C., Zhang, Z., Guo, P., Wang, R., Liu, T., Luo, J., Hao, B., Wang, Y., & Guo, W. (2023). Synergistic mechanisms of bioorganic fertilizer and AMF driving rhizosphere bacterial community to improve phytoremediation efficiency of multiple HMs-contaminated saline soil. *Science of the Total Environment*, 883, 163708. https://doi.org/10.1016/j.scitotenv.2023.163708

Ma, G., Shi, Q., Wu, Y., Liu, Y., Han, L., Hu, J., Mao, H., & Zuo, Z. (2024). Effects of biochar on the growth and physiological and mechanical properties of cucumber plug seedlings before and after transplanting. *Agriculture; Basel*, 14(11), 2012.

Machado, T. O., Grabow, J., Sayer, C., de Araújo, P. H. H., Ehrenhard, M. L., & Wurm, F. R. (2022). Biopolymer-based nanocarriers for sustained release of agrochemicals: A review on materials and social science perspectives for a sustainable future of agri- and horticulture. *Advances in Colloid and Interface Science*, 303, 102645. https://doi.org/10.1016/j.cis.2022.102645

Mącik, M., Gryta, A., & Frąc, M. (2020). Chapter two—biofertilizers in agriculture: An overview on concepts, strategies and effects on soil microorganisms. In D. L. Sparks (Ed.), *Advances in agronomy* (Vol. 162, pp. 31–87). Academic Press. https://doi.org/10.1016/bs.agron.2020.02.001

Maestre, F. T., Guirado, E., Armenteras, D., Beck, H. E., AlShalan, M. S., Al-Saud, N. T., Chami, R., Fu, B., Gichenje, H., Huber-Sannwald, E., Speranza, C. I., Martínez-Valderrama, J., McCabe, M. F., Orr, B. J., Tang, T., Metternicht, G., Miess, M., Reynolds, J. F., Stringer, L. C., & Duarte, C. M. (2025). Bending the curve of land degradation to achieve global environmental goals. *Nature*, 644(8076), 347–355. https://doi.org/10.1038/s41586-025-09365-5

Majed, N., Wang, Z., Baldwin, M. T., Uddin, F., Saif, M., Dash, P., & Gu, A. Z. (2025). Advancements in phosphorus species profiling and bioavailability assessment with implications for phosphorus sustainability. *Current Opinion in Biotechnology*, 93, 103295. https://doi.org/10.1016/j.copbio.2025.103295

Martey, E., & Kuwornu, J. K. (2021). Perceptions of climate variability and soil fertility management choices among smallholder farmers in northern Ghana. *Ecological Economics*, 180, 106870.

Mehdaoui, I., Mahmoud, R., Majbar, Z., Berrada, S., Ben Abbou, M., Elshikh, M. S., Ajmal Ali, M., Chen, T.-W., Taleb, M., & Rais, Z. (2024). Comparing how compost and manure affect soil organic matter using a complete factorial design. *Journal of King Saud University, Science*, 36(10), 103471. https://doi.org/10.1016/j.jksus.2024.103471

Mendonca Cidreira, A. C., Wei, L., Aldekhail, A., & Islam Rubel, R. (2025). Controlled-release nitrogen fertilizers: A review on bio-based and smart coating materials. *Journal of Applied Polymer Science*, 142(3), e56390.

Mkenda, P. A., Ndakidemi, P. A., Stevenson, P. C., Arnold, S. E., Darbyshire, I., Belmain, S. R., Priebe, J., Johnson, A. C., Tumbo, J., & Gurr, G. M. (2020). Knowledge gaps among smallholder farmers hinder adoption of conservation biological control. *Biocontrol Science and Technology*, 30(3), 256–277.

Muhammad, M., Waheed, A., Wahab, A., Majeed, M., Nazim, M., Liu, Y.-H., Li, L., & Li, W.-J. (2024). Soil salinity and drought tolerance: An evaluation of plant growth, productivity, microbial diversity, and amelioration strategies. *Plant Stress*, 11, 100319. https://doi.org/10.1016/j.stress.2023.100319

Neemisha, & Rani, N. (2022). Chapter 9—rejuvenating soil health using organic manures for sustainable agriculture. In H. B. Singh & A. Vaishnav (Eds.), *New and future developments in microbial biotechnology and bioengineering* (pp. 181–198). Elsevier. https://doi.org/10.1016/B978-0-323-85579-2.00014-9

Nieves-Cordones, M., Rubio, F., & Santa-María, G. E. (2020). Editorial: Nutrient use-efficiency in plants: An integrative approach. *Frontiers in Plant Science*, 11, 623976. https://doi.org/10.3389/fpls.2020.623976

Nizamani, M. M., Hughes, A. C., Qureshi, S., Zhang, Q., Tarafder, E., Das, D., Acharya, K., Wang, Y., & Zhang, Z.-G. (2024). Microbial biodiversity and plant functional trait interactions in multifunctional ecosystems. *Applied Soil Ecology*, 201, 105515. https://doi.org/10.1016/j.apsoil.2024.105515

Nongbet, A., Mishra, A. K., Mohanta, Y. K., Mahanta, S., Ray, M. K., Khan, M., Baek, K. H., & Chakrabartty, I. (2022). Nanofertilizers: A smart and sustainable attribute to modern agriculture. *Plants (Basel)*, 11(19). https://doi.org/10.3390/plants11192587

Nowicka, B. (2022). Modifications of phytohormone metabolism aimed at stimulation of plant growth, improving their productivity and tolerance to abiotic and biotic stress factors. *Plants (Basel)*, 11(24). https://doi.org/10.3390/plants11243430

Omokaro, G. O., Kornev, K. P., Nafula, Z. S., Chikukula, A. A., Osayogie, O. G., & Efeni, O. S. (2025). Biochar for sustainable soil management: Enhancing soil fertility, plant growth and climate resilience. *Farming System*, 3(4), 100167. https://doi.org/10.1016/j.farsys.2025.10.167

Pandian, K., Mustaffa, M. R. A. F., Mahalingam, G., Paramasivam, A., John Prince, A., Gajendiren, M., Rafiqi Mohammad, A. R., & Varanasi, S. T. (2024). Synergistic conservation approaches for nurturing soil, food security and human health towards sustainable development goals. *Journal of Hazardous Materials Advances*, 16, 100479. https://doi.org/10.1016/j.hazadv.2024.100479

Park, C. H., Yeo, H. J., Park, Y. J., Morgan, A. M., Valan Arasu, M., Al-Dhabi, N. A., & Park, S. U. (2017). Influence of indole-3-acetic acid and gibberellic acid on phenylpropanoid accumulation in common buckwheat (fagopyrum esculentum moench) sprouts. *Molecules*, 22(3). https://doi.org/10.3390/molecules22030374

Phares, C. A., Amoakwah, E., Danquah, A., Afrifa, A., Beyaw, L. R., & Frimpong, K. A. (2022). Biochar and NPK fertilizer co-applied with plant growth promoting bacteria (PGPB) enhanced maize grain yield and nutrient use efficiency of inorganic fertilizer. *Journal of Agriculture and Food Research*, 10, 100434. https://doi.org/10.1016/j.jafr.2022.100434

Prajapati, C. S., Priya, N. K., Bishnoi, S., Vishwakarma, S. K., Buvaneswari, K., Shastri, S., Tripathi, S., & Jadhav, A. (2025). The role of participatory approaches in modern agricultural extension: Bridging knowledge gaps for sustainable farming practices. *Journal of Experimental Agriculture International*, 47(2), 204–222.

Priya, E., Sarkar, S., & Maji, P. K. (2024). A review on slow-release fertilizer: Nutrient release mechanism and agricultural sustainability. *Journal of Environmental Chemical Engineering*, 12(4), 113211. https://doi.org/10.1016/j.jece.2024.113211

Qian, S., Zhou, X., Fu, Y., Song, B., Yan, H., Chen, Z., Sun, Q., Ye, H., Qin, L., & Lai, C. (2023). Biochar-compost as a new option for soil improvement: Application in various problem soils. *Science of the Total Environment*, 870, 162024. https://doi.org/10.1016/j.scitotenv.2023.162024

Quitzow, R., Balmaceda, M., & Goldthau, A. (2025). The nexus of geopolitics, decarbonization, and food security gives rise to distinct challenges across fertilizer supply chains. *One Earth*, 8(1), 101173. https://doi.org/10.1016/j.oneear.2024.12.009

Raven, P., & Wackernagel, M. (2020). Maintaining biodiversity will define our long-term success. *Plant Diversity*, 42(4), 211–220. https://doi.org/10.1016/j.pld.2020.06.002

Riseh, R. S., Vazvani, M. G., & Kennedy, J. F. (2023). The application of chitosan as a carrier for fertilizer: A review. *International Journal of Biological Macromolecules*, 252, 126483. https://doi.org/10.1016/j.ij.iomac.2023.126483

Schlüter, S., Sammartino, S., & Koestel, J. (2020). Exploring the relationship between soil structure and soil functions via pore-scale imaging. *Geoderma*, 370, 114370. https://doi.org/10.1016/j.geoderma.2020.114370

Seyedsadr, S., Šípek, V., Jačka, L., Sněhota, M., Beesley, L., Pohořelý, M., Kovář, M., & Trakal, L. (2022). Biochar considerably increases the easily available water and nutrient content in low-organic soils amended with compost and manure. *Chemosphere*, 293, 133586. https://doi.org/10.1016/j.chemosphere.2022.133586

Shang, L., Heckelei, T., Gerullis, M. K., Börner, J., & Rasch, S. (2021). Adoption and diffusion of digital farming technologies—integrating farm-level evidence and system interaction. *Agricultural Systems*, 190, 103074. https://doi.org/10.1016/j.agsy.2021.103074

Shanmugavel, D., Rusyn, I., Solorza-Feria, O., & Kamaraj, S.-K. (2023). Sustainable SMART fertilizers in agriculture systems: A review on fundamentals to in-field applications. *Science of the Total Environment*, 904, 166729. https://doi.org/10.1016/j.scitotenv.2023.166729

Shi, H., Liang, D., Deng, H., Xie, F., Chen, Z., Chen, Y., Lu, Q., Liu, X., & Zhang, C. (2022). Bio-based superhydrophobic polymer coatings for slow-release fertilizers via a UV-curing encapsulation method. *Industrial Crops and Products*, 188, 115580.

Sible, C. N., Seebauer, J. R., & Below, F. E. (2025). Biostimulant or biological? The complexity of defining, categorizing, and regulating microbial inoculants. *Agricultural & Environmental Letters*, 10(2), e70027.

Singaravel, R., Arthi, M., Kamaraj, S., & Seenivasan, M. (2025). Micronutrient management in sandy soils through organic materials enriched with multi-micronutrients. *E3S Web of Conferences*. Second International Conference on Green Energy, Environmental Engineering and Sustainable Technologies 2024 (ICGEST 2024) Nov 29th, 2024 - Nov 30th, 2024, Bhiali India

Singh, V., & Kumar, B. (2024). A review of agricultural microbial inoculants and their carriers in bioformulation. *Rhizosphere*, 29, 100843. https://doi.org/10.1016/j.rhisph.2023.100843

Singh, S., Singh, R., Singh, K., Katoch, K., Zaeen, A. A., Birhan, D. A., Singh, A., Sandhu, H. S., Singh, H., & Sahrma, L. K. (2024). Smart fertilizer technologies: An environmental impact assessment for sustainable agriculture. *Smart Agricultural Technology*, 8, 100504. https://doi.org/10.1016/j.atech.2024.100504

Sivaram, A. K., Abinandan, S., Chen, C., Venkateswartlu, K., & Megharaj, M. (2023). Chapter two—microbial inoculant carriers: Soil health improvement and moisture retention in sustainable agriculture. In D. L. Sparks (Ed.), *Advances in agronomy* (Vol. 180, pp. 35–91). Academic Press. https://doi.org/10.1016/bs.agron.2023.03.001

Sollins, P., & Gregg, J. W. (2017). Soil organic matter accumulation in relation to changing soil volume, mass, and structure: Concepts and calculations. *Geoderma*, 301, 60–71. https://doi.org/10.1016/j.geoderma.2017.04.013

Song, D., Jin, G., Su, Z., Ge, C., Fan, H., & Yao, H. (2025). Influence of biodegradable microplastics on soil carbon cycling: Insights from soil respiration, enzyme activity, carbon use efficiency and microbial community. *Environmental Research*, 266, 120558. https://doi.org/10.1016/j.envres.2024.120558

Surya, B., Menne, F., Sabhan, H., Suriani, S., Abubakar, H., & Idris, M. (2021). Economic growth, increasing productivity of SMEs, and open innovation. *Journal of Open Innovation: Technology, Market, and Complexity*, 7(1), 20. https://doi.org/10.3390/joitmc7010020

Tang, L. (2024). Soil fertility, plant nutrition and nutrient management. *Plants (Basel)*, 14(1). https://doi.org/10.3390/plants14010.34

Townsend, J. T. (2021). Issues and models concerning the processing of a finite number of inputs 1. In *Human information processing* (pp. 133–185). Routledge, Taylor & Francis Group.

Tucker, J. (2023). The future vision (s) of AI health in the Nordics: Comparing the national AI strategies. *Futures*, 149, 103154.

Van Lanen, N. J., Duchardt, C. J., Pejchar, L., Shyvers, J. E., & Aldridge, C. L. (2025). Considering multiecosystem trade-offs is critical when leveraging systematic conservation planning for restoration. *Global Change Biology*, 31(1), e70020.

Wahab, A., Muhammad, M., Munir, A., Abdi, G., Zaman, W., Ayaz, A., Khizar, C., & Reddy, S. P. P. (2023). Role of arbuscular mycorrhizal fungi in regulating growth, enhancing productivity, and potentially influencing ecosystems under abiotic and biotic stresses. *Plants (Basel)*, 12(17). https://doi.org/10.3390/plants12173102

Wahab, A., Batool, F., Abdi, G., Muhammad, M., Ullah, S., & Zaman, W. (2025). Role of plant growth-promoting rhizobacteria in sustainable agriculture: Addressing environmental and biological challenges. *Journal of Plant Physiology*, 307, 154455. https://doi.org/10.1016/j.jplph.2025.154455

Wang, Y., & Zhang, S. (2025). Input-oriented or output-oriented? optimizing subsidy incentives for multi-tier sustainable supply chains.

Wang, L., Lu, P., Feng, S., Hamel, C., Sun, D., Siddique, K. H. M., & Gan, G. Y. (2024). Strategies to improve soil health by optimizing the plant–soil–microbe–anthropogenic activity nexus. *Agriculture, Ecosystems & Environment*, 359, 108750. https://doi.org/10.1016/j.agee.2023.108750

Wei, J., & Li, Y. (2023). CRISPR-based gene editing technology and its application in microbial engineering. *Engineering Microbiology*, 3(4), 100101. https://doi.org/10.1016/j.engmic.2023.10.101

Wu, L., Subramanian, N., Abdulrahman, M. D., Liu, C., & Pawar, K. S. (2017). Short-term versus long-term benefits: Balanced sustainability framework and research propositions. *Sustainable Production and Consumption*, 11, 18–30. https://doi.org/10.1016/j.spc.2016.09.003

Xi, W. (2024). Regulatory changes and compliance challenges. In Emerald insight (Ed.), *Strategic financial management: A managerial approach* (pp. 119–134). Emerald Publishing Limited.

Xing, Y., Xie, Y., & Wang, X. (2025). Enhancing soil health through balanced fertilization: A pathway to sustainable agriculture and food security. *Frontiers in Microbiology*, 16, 1536524. https://doi.org/10.3389/fmicb.2025.1536524

Younis, A., Ahsan, M., Akram, A., Lim, K.-B., Zulfiqar, F., & Tariq, U. (2022). Use of organic substrates in sustainable horticulture. In M. Hasanuzzaman, B. Hawrylak Nowak, T. Islam, & M. Fujita (Eds.), *Biostimulants for crop production and sustainable agriculture* (pp. 122–138).

Yuan, Y., Zu, M., Sun, L., Zuo, J., & Tao, J. (2022). Isolation and screening of 1-aminocyclopropane-1-carboxylic acid (ACC) deaminase producing PGPR from Paeonia lactiflora rhizosphere and enhancement of plant growth. *Scientia Horticulturae*, 297, 110956. https://doi.org/10.1016/j.scienta.2022.110956

Zhai, C., Han, L., Xiong, C., Ge, A., Yue, X., Li, Y., Zhou, Z., Feng, J., Ru, J., Song, J., Jiang, L., Yang, Y., Zhang, L., & Wan, S. (2024). Soil microbial diversity and network complexity drive the ecosystem multifunctionality of temperate grasslands under changing precipitation. *Science of the Total Environment*, 906, 167217. https://doi.org/10.1016/j.scitotenv.2023.167217

Zhuang, X., Gao, J., Ma, A., Fu, S., & Zhuang, G. (2013). Bioactive molecules in soil ecosystems: Masters of the underground. *International Journal of Molecular Sciences*, 14(5), 8841–8868. https://doi.org/10.3390/ijms14058841

10

Economic Impacts: How Next-Gen Fertilizers Can Transform Global Agriculture

Tauqeer Ahmed Qadri[1], Amjid Khan[3], Rashid Abbas Khan[4], Shabbar Abbas[1], Abdul Wahab[1], Muhammad Rizwan Khan[2], and Muhammad Anas[2,5]

[1] *Department of Biosciences, COMSATS University Islamabad, Islamabad, Pakistan*
[2] *Department of Plant Sciences, Faculty of Biological Sciences, Quaid-i-Azam University, Islamabad, Pakistan*
[3] *UNESCO-UNISA Africa Chair in Nanoscience and Nanotechnology (U2ACN2), College of Graduate Studies, University of South Africa, Muckleneuk Ridge, Pretoria, South Africa*
[4] *Department of Biology, Faculty of Sciences, Prince Sonkla University, Songkhla, Thailand*
[5] *Department of Soil Science & Plant Nutrition, Selcuk University, Konya, Türkiye*

10.1 Introduction

10.1.1 From Conventional to Cutting-Edge: Why Fertilizers Matter

The fertilizers have always been perceived as the backbone of the new era in agriculture. According to World Bank statistics, the total fertilizer consumption increased by 12 times in the 10 years from 2006 to 2016, mainly due to the increase in fertilizer application in developing countries (Xiao et al., 2021). The Green Revolution firmly depends on soil fertilizers or synthetic nitrogen- phosphorus- potassium (NPK) that has enabled growth of a larger agricultural output than ever before and has maintained global agricultural production at its peak levels, preventing the impending dread (in the form of massive famine) that was imminent in various parts of the planet since the middle of the 20th century (Carmo et al., 2017). Food and Agriculture Organization (FAO) estimates that use of fertilizers has contributed directly to about 50% of the modern global food supply. This fact serves to amplify the fact that fertilizers are essential to meet the nutritional requirements of a fast-growing global population that is set to reach nearly ten billion in 2050 (Sida et al., 2025). As a result, farm fertilizers become not only inputs of agricultural life but also the invaluable instruments of the economy that affect food prices, farmer earnings, national food security, and even the balance of international markets (Verma et al., 2019). However, fertilizers themselves will be required to change as farming remains actively developing in the conditions of climate change, the shortage of resources, and changing consumer tastes. The traditional more fertilizer, more yield paradigm becomes more unsustainable. The ability to balance productivity goals with economies, environmental sustainability, and resiliency of farm systems in the long term underpins the economic potential of next-generation fertilizers (bin Radzali et al., 2025).

10.1.2 Economic Challenges of Traditional Inputs

In spite of their demonstrated government contributions to food security in the world, conventional fertilizers pose serious economic problems. They are known to be efficiency-challenged, firstly with a low efficiency of a mere 30–50% of the applied nitrogen and a precise toughness of 20% of phosphorus being absorbed by crops, and the remaining is washed off by leaching, volatilization, or runoff (Mekonnen, 2021). This has been leading to farmers spending too much on inputs but not achieving the best possible returns, hence a vicious economic effect of high input costs combined with diminished marginal returns (Reader et al., 2018).

Second, the international fertilizer market is very volatile, as it depends on the price of constantly changing energy prices, geopolitical issues, and trade barriers. As an example, the 2022 world fertilizer crisis led to importation prices increasing threefold across most of the nations, with much of it affecting smallholder agriculturalists within developing economies. This instability not only increases the costs of production but also increases food prices, hence instigating inflationary pressure, consequently undermining food security (Ejedegba, 2024).

Third, the hidden or external costs of traditional fertilizers, which include soil erosion and eutrophication of water and greenhouse-gasses globalize an extra economic cost on society (Gallardo, 2024). Billions of dollars are being spent by governments in environmental cleanup, health care costs, complications related to nitrate-contaminated water, and climate-updating considerations. Such indirect costs show that, although the traditional fertilizers increase short-term production, they still have long-term liabilities on the economy that compromise agricultural sustainability (Pretty & Bharucha, 2014).

10.1.3 Next-Gen Fertilizers as Catalysts for Agricultural Transformation

In attempting to overcome as many of these problems as possible, the agricultural industry is progressively adopting next institute generation fertilizers that may include controlled-release formulations, nanofertilizers, biofertilizers, and intelligent delivery systems (El-Shal et al., 2022). These emerging technologies are designed to promote nutrient use efficiency (NUE), reduce wastage, and make decisions on the timing and quantities of the proportion of the diet that is released (Sonea et al., 2023). The next-gen fertilizers will increase crop yields due to their ability to enhance the delivery and effectiveness of nutrients, helping a farmer achieve greater yields using less input expenses of the fertilizer and which is a direct financial need (Thakur et al., 2024).

As an example, controlled-release fertilizers that have been modified with biodegradable polymers can extend the duration of nutrient supply to several weeks and months, thus reducing the rate of losses to leachates (Hou et al., 2021). Similarly, nanofertilizers use carriers at nanoscales to administer nutrients into plant tissues with maximum absorption and through low application across the board (Iqbal et al., 2019). Biofertilizers are microorganisms; microorganisms that are beneficial in some way, like nitrogen fixers or phosphate solubilizers, make up biofertilizers, giving cheap, environmentally-friendly replacements that leave the soil healthier in the long run (Soni et al., 2022).

The economic impact of such technologies is multidimensional. Individual farmers face decreased fertilizer inputs, increased and more reliable yields, and increased soil productivity in the long run (Ogbeide, 2015). On national grounds, economies enjoy the cheap importation of fertilizers, decreased cost of healthcare, and environmental remediation and enhancement of food security systems. Next-gen fertilizers are new markets that can offer significant growth potential to investors and agribusinesses driven by growing and evolving interest in sustainable and climate-smart agriculture (Wanat et al., 2025).

Figure 10.1 presents a timeline depicting the transition from traditional fertilizers to advanced solutions such as biofertilizers, nanofertilizers, controlled-release formulations, and AI-integrated smart fertilizers, emphasizing the movement toward more efficient and sustainable nutrient management.

10.1.4 Aims and Scope of the Chapter

In this chapter, the economic aspects of next-generation fertilizers are discussed, emphasizing their role as GDP game changers to agriculture within local, national, and global settings. It starts by giving an overview of the existing fertilizer economy that highlights the inefficiency and the unaccounted costs of traditional inputs. This point is followed by an explanation of how next-gen fertilizers are able to enhance more efficient use of nutrients, help curb the negative impact on the economy, and make the environment more sustainable. Specific attention is drawn to their consequences to achieve food security on an international level, particularly in dairy areas struggling to meet yield discrepancies and increasing input prices.

A later section evaluates ways in which technologies rich in precision agriculture increase the commercial value of already superior fertilizers, and comparative evaluations show their effects on smallholder farmers and large-scale agribusinesses. Another topic the chapter covers is the decisive role of policies, subsidies, and international cooperation in positively influencing adoption and consistency with the United Nations Sustainable Development Goals (SDGs). Lastly, it builds up to a forward-looking view of future economic trends of the fertilizer industry, highlighting innovation, models of circular-economy, and sustainable agricultural progress.

10.2 The Current Economic Landscape of Fertilizer Use

10.2.1 Global Market Dynamics

Global fertilizer market is a subset of the modern agriculture industry, and its yearly market prices have surpassed USD 200 billion. Nitrogen-based fertilizers dominate the market, with the next in order being phosphate and potash products, thus creating the very important NPK triad that forms the backbone of most farming systems (Boberski et al., 2025). The

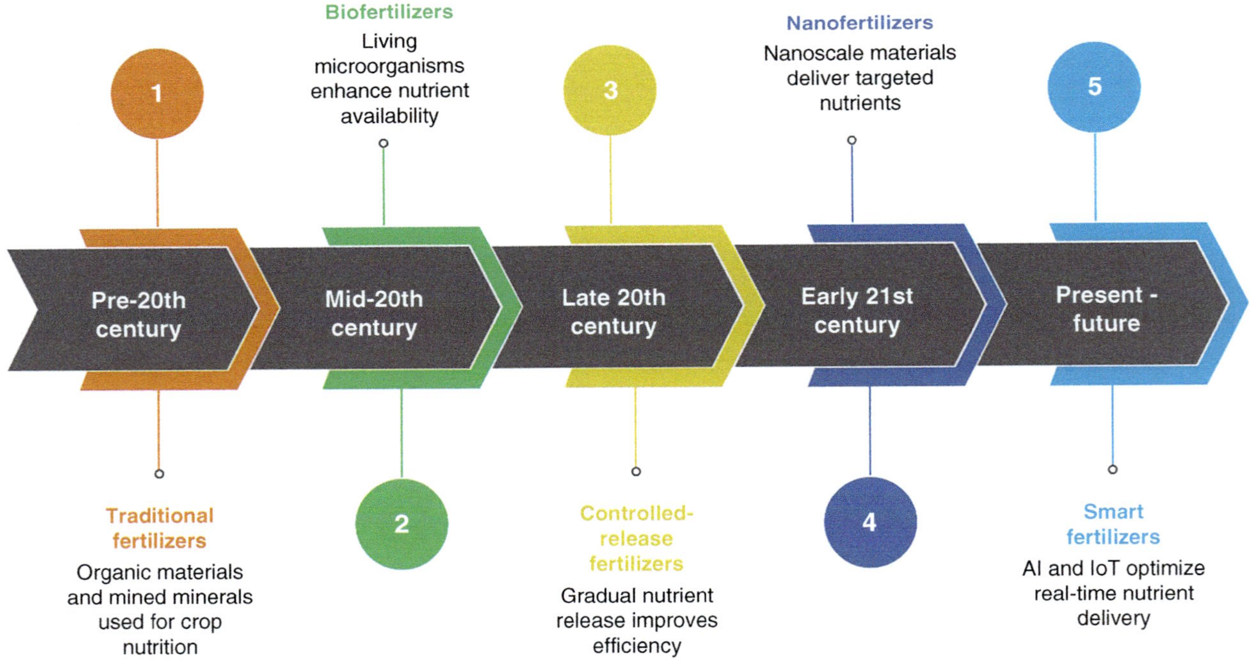

Figure 10.1 Evolution of fertilizer technologies: from conventional inputs to next-generation solutions. A timeline showing the transition from traditional chemical fertilizers to biofertilizers, nanofertilizers, controlled-release formulations, and AI-integrated smart fertilizers, emphasizing the shift toward sustainability and efficiency.

small group of countries controls a disproportionate production and supply of: China is the leader in nitrogen production, Morocco in large phosphate reserves, leaving Canada and Russia in the top exporting countries in potash. Such centralization creates a geopolitical/geo-sensitive market where occurrences of supply failure, or changes in policy in a limited number of countries, are reflected across the globe (Tanchum, 2022; Wang et al., 2011).

The patterns of demand are extremely varied regionally. With nearly 60% of the world's consumption of fertilizers, Asia leads to emergence of the world's biggest consumer countries, including China and India (Randive et al., 2021). Although Africa possesses a large agricultural potential, the clothing continent consumes less than 5° at the age of affordability aspects. Such uneven distribution can be described as an indication not only of varying agricultural intensification but also varying economic accessibility and infrastructure (Binswanger-Mkhize & Savastano, 2017). That, in turn, increases disparities within the open world market: the better off economies with the financial resources to spend on this source out their resources, leaving the cash-strapped economies with lack of fertilizers and facing constant yield threats (Ehui & Pender, 2005).

10.2.2 Economic Pressure on Farmers

Fertilizers are among the greatest recurrent costs of production for farmers. The cost of fertilizers in intensive systems like rice and maize production may represent 30–40% of total farm inputs (Gregory et al., 2010). Though the idea of fertilizers increasing yields unquestionably is true, the fact that the modern conventional inputs impose low NUE suggests significant percentages of the mentioned expenses are wasted (Mukherjee, 2013). Pragmatic research shows that agrarians will often count exactly a fraction of their monetary investment in the form of harvest when it comes to fulfilling their investment in applying the fertilizer directly or when the earth is overly porous (Rhodes, 2012).

Especially, smallholder farmers are susceptible. Higher fertilizer prices in sub-Saharan Africa, most of which have subsistence handicaps of many of their phenomena into lower inputs of input exploitation, further reproduce the poor performance (Chianu et al., 2012). On the other hand, farmers in high-input areas like East Asia face diminishing returns because fertilizer overapplication triggers disequilibrium of soil nutrients, diminishing returns, and expenditure on expenditures on correcting low soil nutrients (Goud et al., 2022). There is, therefore, an inevitable trade-off between the economic necessities of the application of fertilizers on the farm and profitability. There is simply no true profit without the reliance on fertilizers, leaving farmers perpetually in debt and insecure circles (Pannell, 2017).

10.2.3 Volatility and Trade Dependence

The fertilizer industry is closely connected with the world markets of energy and international trade. Such products as nitrogen fertilizers show great dependency on natural gas, which makes the prices of these products very sensitive to energy prices (Huan & Liu, 2023). The 2021–2022 global energy crisis, combined with geopolitical tension, spurred a price increase by over 150 in many areas, leading to a series of responses: higher farm input prices, higher food prices, and inflationary pressures across world economies (Agram & Rahal, 2024).

It is also vulnerable due to import dependency. Many nations, mainly of the African and South Asian countries, depend greatly on imported fertilizers because little has been produced domestically (Sandström et al., 2024). In cases when global supply chains are disrupted, such nations face acute shortages and price increases, thus putting national food security in jeopardy. Conversely, the fertilizer-exporting countries have the ability to leverage their status both financially and politically, even focus mainly on the home demands or limit the exports in case of a crisis as could be seen with the pyroxene embargo performed by China in 2021 (Snow, 2023).

10.2.4 The Hidden Bill: Environmental Liabilities

In addition to direct monetary expenses, the conventional fertilizers have hidden economic expenses in the form of environmental decay. Uncontrolled runoff of nitrogen and phosphorus also leads to water pollution, eutrophication, and the formation of dead zones in water bodies, thus reducing the productivity of fisheries systems and provoking expensive measures of water treatment (Hashimi & Hashimi, 2020). According to the US Environmental Protection Agency (EPA), costs of nitrogen pollution are estimated to cost the economy and human lives billions of dollars annually due to the destruction of ecosystems and health safety (Sobota et al., 2015).

The other serious liability is soil degradation. Unbalanced and uninterrupted use of fertilizers impurifies soil organic matter, disturbs the microbial community, and serves to hasten acidification (Yan et al., 2024). The interventions that are usually used to restore the degraded soils are costly, thus liming, organic amendments, or further-sophisticated soil management practices. Such costs, which can rarely be reflected in the economics of fertilizers, are a major cost of long-term liability to governments and the farmers (Finger et al., 2019).

Moreover, fertilizers increase the emission of greenhouse gases, especially nitrous oxide (N_2O), which has a global warming potential of close to 300 times that of carbon dioxide (Kudeyarov, 2020). Economic impacts are not limited to agriculture since nations are having a price of adapting to climate, preparing for carbon mitigation, and internationally entering into a climate agreement. Therefore, the externality of traditional fertilizers is not only a question of environmental issues, but a major economic issue that has a global bearing (Penuelas et al., 2023).

10.3 Fertilizer Efficiency as an Economic Game Changer

10.3.1 Nutrient Use Efficiency: Getting More from Less

One of the main concepts of fertilizer economics is NUE, which is defined as the percentage of added nutrients that are utilized and used by the plants. Traditional fertilizers are continuously characterized by an ineffectively low level of NUE when global values are approximately 30–50% of nitrogen and less than 20% of phosphorus. Therefore, these run inefficiencies lead to wastage of resources, an increase in the cost of inputs used by producers, and a harmful effect on the environment through losses of nutrients (Ebbisa, 2022).

10.3.2 Smart-Release and Nano-formulations: Precision in Practice

Next-generation fertilizers have advanced delivery systems that enable nutrients to be released in accordance with crop demand. Slow-release fertilizers, often coated with biodegradable polymers, release nutrients slowly over a period of weeks to months, assuaging the sudden surges of fertilizers that could exceed the ability of plants to absorb the nutrients and, by extension, reduce losses through leaching or volatilization (Almutari, 2023). Nanofertilizers take this a step further by using materials on the nanoscale such as clays, polymers, and biochar to trap nutrients. Small size of the particles enhances surface area, motility, and bioavailability, thus ensuring effective localization and time control of nutrient delivery. Nanocarriers can also be engineered to respond to soil moisture and pH or to root exudates in some cases, so that they can supply nutrients in a more dynamic and adaptive way (Yadav et al., 2023).

The financial consequences are high. Manufacturers are able to limit the use of repetitive applications and hence decrease labor costs and enhance the total achievement of nutrient absorption (Huang et al., 2013). Empirical investigations suggest that nanofertilizers have the capacity to minimize the application of nitrogen to reduce the nitrogen delivered by applications by 20–30% of application without product losses, resulting in benefits in monetary terms and elevation of profit margins. The technologies take fertilizers off generic bulk commodities into high-value precision inputs, redefining the financial cost-benefit in the agricultural production cycle (Avila-Quezada et al., 2022).

10.3.3 Economic Returns on Reduced Inputs

A combination of reduced input use, increased yields, and reduced external costs gives rise to the financial benefits of next-generation fertilizers. Improved nutrient absorption allows farmers to cut farmer use of fertilized by 15–30%, holding yields constant or boosting yields. This reduction in input expenditures is a very large saving, particularly in locations where fertilizers cost is a large percentage of total production expenditures (Arora et al., 2024). Labor savings further reinforce profitability. Controlled fertilizers alleviate labor requirements, machine applications, and the related power consumption through only one application of the controlled fertilizers per season, compared to notionally many releases using conventional fertilizers. This is especially pronounced in labor-harmed or high-wage economies, wherein the application costs constitute a conspicuous cost (Shah & Wu, 2019). In addition, efficiency achievements only make up environmental externalities that are increasingly becoming internalized in the cost regimes in the agricultural sector. Farmers that employ efficient fertilizers can receive governmental subsidies, carbon credits, or low compliance expenses relating to pollution control. These nonmonetary benefits are gained over time and enhance the ROI as compared to the traditional inputs (Hayo & Hasegawa, 2024).

10.3.4 Case Studies: Yield–Cost Ratios in Major Crops

Economic promise of effective fertilizers is supported by empirical crop-level quantified observations. In China, where there are field trials of polymer-coated urea in rice production, there has been a 15% reduction in nitrogen input with a 10% increase in yields, effectively increasing the net farmer income by 20–25% percentages (Chandra et al., 2019). This was also the case in the maize system of sub-Saharan Africa that uses biofertilizers with nitrogen-fixing bacteria; the use of biofertilizers reduced the amount of synthetic nitrogen needed in farming by 25% and increased yields by 12–15% thus making the farming system more lucrative (Aloo et al., 2021).

Nanofertilizers used in European wheat systems have automated yields comparable to or exceeding yields obtained using conventional fertilizers with 20–30% less inputs. This enhances the yield-cost ratio and cushions the producers against no price fertilization fluctuations by reducing the average reliance on external fertilizers (Goyal et al., 2025).

The various case studies demonstrate a common pattern: new fertilizers redefine the cost-benefit equation in favor of farmers due to the reduction of nonessential spending and maintenance of or improvement of output. By scaling these microscale benefits to a macroeconomic scale (over millions of hectares), one would expect to see the overall economic gains borne by the regions in the form of reduced costs of imports, better trade balances, and greater food security (Cochrane et al., 2024).

10.4 Environmental-Economic Nexus: Sustainability Pays

10.4.1 Environmental Damage as an Economic Loss

Although traditional fertilizers have been effective in helping increase crops, they have significant environmental costs that are long-term economic disadvantages. Uncontrolled nitrogen and phosphorus leaching sparks eutrophication in rivers, lakes, and ocean waters, and thus creates dead zones that destroy fisheries and economies. The dead zone in the Gulf of Mexico alone is estimated to cost fisheries, and related industries USD 2 billion or more annually. Similarly, water treatment infrastructures are expensive, and governments and end consumers have to bear the fiscal burden due to nitrate contamination of their groundwater (Acheampong & Opoku, 2023).

Impairment of soil health is another hidden cost. The use of too many synthetic fertilizers increases the acidity of the soil, reduces organic matter, and stabilizes the microbial biodiversity. Commodity costs of regeneration of degraded soils by liming, composting, or various techniques occasionally involve costs of several hundred dollars per hectare. Even though these expenditures cannot be directly attributed to fertilizer manufacturers, they are greatly imposed on farmers and the national economies, therefore, converting environmental harm by fertilizers into direct economic destruction at several levels (Acheampong & Opoku, 2023).

10.4.2 Green Fertilizers and Climate-Smart Solutions

The new generation fertilizers bring a twin benefit: maintaining crop output and lessening environmentally damaging effects. Biofertilizers, which use favorable microbes, including *Rhizobium, Azotobacter,* or phosphate-solubilizing bacteria, promote the availability of nutrients without the accompanying dangers of leaching or spray-loss. Likewise, formulations made with organic ingredients and the use of fertilizers enriched with biochar will enhance the health of soils, sequestration of carbon, and resilience in the long term (Shanmugavel et al., 2023).

The climate-smart fertilizer options, such as nitrification and urease inhibitors, lower the emission of nitrous oxide (N_2O)—the worst greenhouse gas. In reducing greenhouse gas production, fertilizers provide an additional advantage to environmental conservation and offer a pathway to farmers and nations to deliver on an international climate promise at a lower cost. With time, the use of such inputs reduces spending on controlling pollution, rehabilitation of soils, and adaptation to climate conditions, and as a result, the alternative is economically viable (Singh et al., 2024).

10.4.3 Valuing Ecosystem Services: Cleaner Water, Healthier Soil

Agricultural ecosystems offer diminishing services, which have high economic costs, namely clean water, fertility, pollination, and biodiversity. The incorrect usage of traditional fertilizers undermines such services, thus consuming natural capital. On the contrary, new fertilizers maintain and/or improve such services. As an example, controlled-release formulations minimize runoff, which, in turn, decreases water-treatment expenses of municipalities, and healthier soils reduce the amount of water required because they retain it better (Pereira et al., 2018).

In economic terms, the preservation of ecosystem services is equivalent to the procurement of sorts of natural insurance. Reducing the money that people spend on treating their health systems adds to the fact that cleaner water lowers expenses, so as to demonstrate that the highly productive soils in the long-term lead to better farm output, whereas balanced ecosystems decrease outbreaks of pests, thereby cutting down the expenses that are spent on purchasing paddocks sized insects. It has been estimated that billions are spent annually around the world due to the worsening of the ecosystem as agricultural losses, a cost that, by adopting greener types of fertilizer applications, could be significantly alleviated. The importance of implementing environmental sustainability into fertilizer policy and practice is determined by identifying these services as economic assets (Grizzetti et al., 2016).

10.4.4 Carbon Credits and the Bio-circular Economy

The emerging carbon economy promises a source of revenue to the farmers who implement sustainable fertilizers. Practices that reduce greenhouse gas emissions or increase the sequestration of soil carbon can be eligible to obtain carbon credits that can be traded in voluntary and compliance markets. As an example, the use of biofertilizers or fertilizers alongside biochar can be used to increase soil carbon stocks and therefore make farms net carbon sinks, instead of carbon sources. Such credits may help supplement the incomes of the farms and offer economic incentives to adopt (Wang et al., 2024). Moreover, bio-based fertilizers of next generation really fit into the use of the bio-railway economic circle. Turning waste in agriculture and industries into useful inputs, whether compost-led fertilizers, waste-derived biochar, or microbial inoculants, allows countries to reduce dependence on nonrenewable mineral reserves, but to nurture national green industries. The strategy not only empowers the economy when a crisis hits the country but also creates job opportunities in the rural setting, thus cementing the link between environmental stewardship and economic prosperity (Mesa et al., 2024).

10.4.5 Sustainability as a Payoff

A potent case can be made through the environmental-economic nexus: the financial gains gained as a result of environmental protection go straight to the benefit of society. The next-generation fertilizers that help to reduce the ecological harm and promote ecosystem services turn sustainability into an economic resource. They minimize intangible spending, introduce additional income sources into carbon markets, and enhance productivity in the future by preserving the soil and water resources. As a result, sustainability is not only an ethical decision; it is legally enforced (financial) and self-supporting (Xiao et al., 2018).

10.5 Fertilizers as Drivers of Food Security

10.5.1 Feeding Billions Through Efficiency

Fertilizers contribute a significant proportion to the worldwide food security. Other fertilizers, especially nitrogen, are used in the supply of food in the whole World; that is, about fifty percent of it is then lost this way, along with phosphorus and potassium. With their lack, crop production would plummet, subjecting the entire world to food insecurity for billions of people (Singh & Sapkota, 2025). The prevailing problem may not necessarily be the rise in the production of fertilizers, but rather making optimum use of said inputs. Increasing nutrient utilization efficiency (NUE) using next-generation fertilizers will allow farmers to produce larger yields using fewer inputs, also reducing the waste of farm inputs as well as resulting in reduced expenditures, which is a much-needed optimization given that arable land available per capita will decrease with phenomena such as urbanization, soil erosion, and climate change (Quitzow et al., 2025).

It has been shown in history that fertilizers have fueled the agricultural revolution. Increased farming practices led to food self-sufficiency in food-deficient Asia, with China and India having considerable initial food self-sufficiency due to the use of fertilizers (Pokhrel, 2020). However, large usage of standard fertilizers has created environmental strains, and the aspect of efficiency and creativity has been put on the priority agenda. On the other hand, African areas have high rates of yield differences, mostly due to the low average rates of fertilizer rotations at less than $20\,kg\,ha^{-1}$ in comparison to $150\,kg\,ha^{-1}$ in Asia (Lombe et al., 2024). Solving these inequalities would provide potential to multiply food production, to reduce hunger, and boost lower-income rural household incomes. Therefore, fertilizers are inevitable tools to increase the production of the arable lands and solve the existing and upcoming food security needs (Daniel et al., 2022).

10.5.2 Economic and Demographic Multipliers

Fertilizers produce significant socioeconomic benefits beyond increased crops. Efforts in achieving high proportions of agricultural efficiency boost agricultural earnings, maintain domestic market, and open up opportunities for jobs in fertilizer manufacturing, sale, and advisory services (Verma et al., 2019). The effective use of fertilizer on a national level minimizes reliance on food imports, enhances the balance of trade, and enhances international competitiveness. Agronically sustainable countries not only ensure their own food supply but also end up as major exporters, hence ensuring global food security (Cruz & Ahmed, 2018).

Demographic dynamics further underscore the importance of fertilizers. It is projected that these demands on food will increase by about 50–60% globally by 2050, that is, with population increase and urbanization. With the grey market experience of arable land growing in both fluctuations and fragmentation levels, production in left agricultural landscape will unavoidably increase (Hosan et al., 2022). During the development and application of fertilizers, the same increased demand can be assessed without a worsening of the deterioration of the environment, especially through adopting new and improved formulations of fertilizers, including slow-release, nano-engineered, and bio- and energy products (Goswami et al., 2024). Furthermore, changing diet patterns in the progress of urbanization generate more requirements for diverse and more competing foods. Biofortified and nutritionally enriched crops are supported by fertility fertilizers and thus reduce the problem of hidden hunger (Hosan et al., 2022).

Fertilizer factories go beyond simple agricultural inputs and emerge as drivers of the economy, social stabilizing forces, and lifeline nets to ensure the continuation of life as the population drastically increases. Global food security measures include attaining high-input systems efficiency and bridging yield gaps in low-performing areas and advanced fertilizers. They make sure that they achieve two things at once: feeding billions of people and maintaining the integrity of the environment (Guiné, 2024).

10.6 Next-Gen Fertilizers in Precision Agriculture

10.6.1 Smart Farming Meets Smart Fertilizers

The blending of accurate agriculture and high-tech fertilizers is reconstituting the use, control, and utilization of nutrients in arable land. The traditional blanket fertilization programs often lead to waste of nutrients, inefficiency in operations, and pollution. On the contrary, nanofertilizers like the slow-release formulations and nanofertilizers coated with slow-release chemicals can be designed to coordinate with the temporal application of nutrients by the crops to minimize waste and maximize the yield characteristics. Site- and crop-specified management, cited above, is supplemented by precision agriculture that optimally delivers fertilizers in the most suitable amount, at the necessary time, and to the right site. This synergy achieves high efficiency in nutrient use (NUE) and at the same time reduces the cost of the agricultural producers (Shanmugavel et al., 2023).

Smart fertilizers within the precision systems grow to greater levels than the inputs in attempts to supply nutrient contents; they act as data-driven growth regulators. They react dynamically to the soil environment are coordinated with the growth patterns of plants, and can be combined with biological inoculants to enhance plant resistance. A combination of these properties indicates a paradigm shift of reactive fertilization to proactive nutrition stewardship (Singh et al., 2024).

10.6.2 Digital Tools: Drones, IoT, and Sensor-Based Nutrient Delivery

The epicenter of this transformation is composed of digital technologies. Multispectral camera-operated remote-sensing drones are capable of creating fine-mapped maps of nutrient deficiencies in the agricultural fields, thus allowing the focused use of fertilizers. The Internet of Things (IoT), which includes equipment such as soil moisture and nutrient sensors, offers the real-time tracking of crop and soil conditions and introduces the data into farm management systems. This sentiment loop enables farmers to use fertilizers where they are needed, and overuse of fertilizers can be managed, and therefore the runoff is reduced (Islam et al., 2023).

Automated nutrient assumption systems as variable Nathalie rate applicators and fertigation units expand this feature by persisting in both static and flowing application rates relative to sensor data. As an example, nitrogen application can be customized in a way that supports stages in crop development and thus prevents under-application and overapplication. These inventions not only advance the operational efficiency level but also help reduce the environmental footprints by restricting NHG emissions and preventing eutrophication of water bodies (Bhatia et al., 2024).

10.6.3 Economic Efficiency of Real-Time Nutrient Management

Economic efficiency is the most crucial reason to incorporate the use of next-generation fertilizers with digital remedies. In contemporary farming, fertilizers' expenses are a large share of the input expenses. Precision systems yield large margins of cost reductions by avoiding infertile applications as well as boosting yield based on one unit of fertilizer. Nutrient management in real-time can save farmers 20–30% of their fertilizer costs and, at the same time, improve productivity, which only pleases profit margins (Ravikumar et al., 2024).

In addition, these systems make yield forecasting more precise, a feature that helps producers to maximize their choices when it comes to timing harvests, market prices, and supply-chain logistics. Lowering the input waste also cushions farmers against the fluctuations inherent in world fertilizer markets, whose price spikes may destabilize production budgets. In developing countries, the economic cushion strengthens the overall farmers and reduces reliance on expensive imports (Singh et al., 2023).

10.6.4 Synergistic Gains: AI, Data, and Smart Inputs

The concept of precision agriculture can be fulfilled using big data analytics and artificial intelligence (AI). This is done by using machine-learning algorithms to process satellite images, data gathered by drones, and light sensor readings on the ground, producing high-precision predictions of nutrient requirements. This makes a self-optimizing system when it is coupled with smart fertilizer that communally rereleases nutrients based on environmental conditions (Zafar et al., 2024).

These synergetic advantages are not only economical a short-term basis. The long-term benefits are that soil situation is improved, the carbon footprint also diminishes, and the ability to resist closer to climate variability (Lal, 2013). The farmers will be able to create

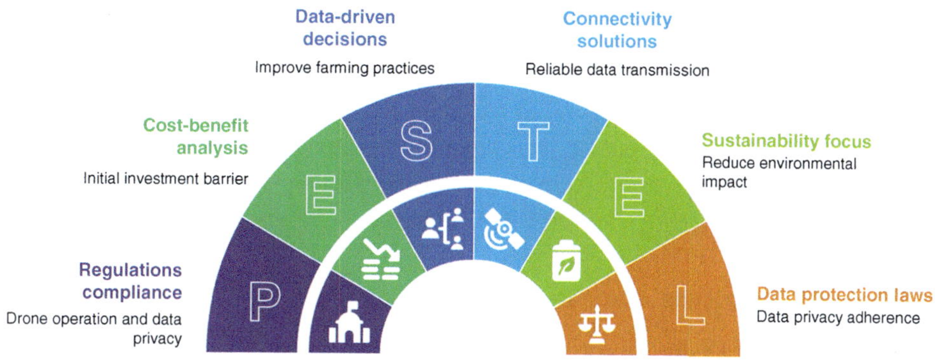

Figure 10.2 Integration of smart fertilizers with precision agriculture technologies. A schematic of how drones, IoT-based soil sensors, and AI-based platforms can be used to allow real-time monitoring and nutrient delivery in particular locations to minimize waste and increase productivity.

sustainable production mechanisms that combine the recommendations of AI-driven programs with nutrient-efficient inputs, thus receiving both profitability and stewardship. Additionally, predictive analytics can direct policymakers and agribusiness to allocate resources, thus ensuring food security at the macro level (Sun & Song, 2025).

Figure 10.2 presents a schematic demonstrating how drones, sensor networks, and AI-driven systems coordinate with smart fertilizers to provide targeted, real-time nutrient management for improved efficiency and crop performance.

10.7 Smallholder vs. Large-Scale Farming

10.7.1 Bridging Gaps for Smallholders: Access, Affordability, and Resilience

Access to the next stream of fertilizers and precision equipment is also one of the concerns that sustainably impact smallholder farmers all over the developing world. These producers usually work in a limited production space with limited access to credit, market, and inefficiency in linking to the market, and the use of traditional practices (Getahun et al., 2024). Fertilizers are very expensive, and therefore, the issue of affordability could be a decisive factor. Subsequently, an influx of subsidies, or schemes of cooperative purchasing and microfinance, catapults the smallholder with the power to integrate better quality fertilizers in their production process that, in turn, results in quantifiable yarn output and profitability levels (Zhong et al., 2023).

The gains are more than just monetary productivity gains. The improved accuracy of the fetus balances the loss of inputs as fertilizers, which minimizes the financial risks that lead to the loss of food supplies even during climatic stress (Penuelas et al., 2023). Smart farming promotes soil improvement, soil erosion, and agricultural resiliency in areas with low soil and common droughts. Expanding the yields intensifies available surpluses to the smallholders that promote sellable surpluses that enhance household food security and enable investment in farm diversification options, education, and access to health care (Zhong et al., 2023).

Moreover, there are high levels of socioeconomic spillovers. The elevated agricultural activity of the marginalized farmers triggers the rural economy in the region due to the enhanced demand for labor and the transportation and farming facilities. These ripple effects increase the strength of the resource-poor societies, break cycles of poverty, and enhance inclusive rural development. Thus, smallholder empowerment through affordable access to innovative fertilizers and technologies functions as both a food security indicator and a driver of socioeconomic transformation (Abraham & Pingali, 2020).

10.7.2 Scale Advantages: Efficiency, Profitability, and Global Trade

Economies of scale brought forth extensive value of large commercial farms as they have gained considerable advantage, especially through their application of next-generation fertilizers and precision agricultural technology, among others (Balafoutis et al., 2017). Such business ventures, usually having access to invest, will prefer using drones, IoT sensors, and automated nutrient delivery devices so that the optimism of smart fertilizers is achieved at full potential. Since the cost ratio reduces as the scale grows, commercial farms are capable of maximizing input usage and increasing the profit margins (Kim, 2021).

Use of fertilizers in large-scale farms also contributes to the large supply chain through to the global food and cooked food CE with regard to exporting. Their ability to adopt innovations fast and to handle large landholdings puts them at the forefront to stay up to the global food demand. Cereal farming in Asia and extensive maize and soybean farming in the Americas can serve as an example, where the surpluses created through precision fertility also promised to equalize global trade patterns (Zinn & Goldsby, 2020).

Nevertheless, it has been observed in studies that the benefits of scale do not exist only in financial terms. Broad-scale application of next-generation fertilizers promotes sustainability goals through a decreased number of greenhouse gases generated per unit of production and less nutrient leaching. On the one hand, commercial farms typically serve as the breeders of innovations; they are also the places where the technologies can be tried first and then implemented in relation to the smallholder. Knowledge developed via research collaborations and advanced nutrient-management modeling can be transferred downward, to the small-scale systems (Lau et al., 2014).

The social economic spillovers of extensive-scale operations, as compared to the large holder, are, in contrast, significantly different. Although increased efficiency increases supply competitiveness at the global level, it can be used to increase market concentration, which can discredit small players. Creating a balance between the efficiencies, extent of scale-just and rural equity-just considerations should thus be the central policy problem that practitioners will be facing (Baxter et al., 2024).

10.8 Policy, Governance, and Economic Incentives

10.8.1 Subsidies and Incentives: Making Innovation Affordable

The level of policy framework and economic provisions is definitive in the rate of absorption of the new technologies in agriculture. In the scenario of fertilizers, the subsidies are the most powerful tool for influencing both affordability and accessibility. However, it has often been pointed out that old-fashioned fertilizer subsidies have encouraged overuse, ineffectiveness, and erosion of soil. These dislocations are the societal objective of the modern move toward intelligent subsidies, in which targeted, intelligent subsidization aims to balance nutrient application and encourage the use of bio-based fertilizers and the use of sustainable practices by farmers (Baxter et al., 2024).

To take a case example, conditional subsidies on soil-testing outputs or on precision-farming systems ensure that fertilizers are applied at sensible rates to ensure that they do not waste economic value and also preserve the integrity of the environment. Similarly, manufacturers of eco-friendly fertilizers are tax-exempt and have access to credit facilities to reduce market-entry barriers, improving access to innovative products. These not only safeguard the livelihoods of smallholders but also encourage the inflow of investments, by private individuals, into research and development. By implication, subsidies and incentives act as gears in place to catalyze the migration to climate-smart fertilizers (Nyström et al., 2018).

10.8.2 Comparative Models: China, EU, USA, and Africa

Cross-regional analysis of fertilizer policy shows that different models are influenced by a range of differing economic interests and the availability of resources. China has led a priority of a no growth of fertilizer consumption policy, which has stabilized consumption and increased the efficiency of the landscape has been achieved with precision tools and with slow-release formulations. This measure has reduced wastage of nutrients, and high yields have remained. The European Union, on the other hand, focuses on environmental regulation in the form of the nitrates directive, accompanied by financial support to introduce the use of green fertilizers and organic amendments. Sustainability is put in the foreground of their model, and the interplay and emphasis are usually put on ecosystem services and productivity (Hu et al., 2024).

The US government is oriented toward supporting innovation through public–private research collaborations and the precision programs. Instead of heavy subsidies, the U.S has crop insurance schemes, conservation programs, and climate-modeled farming pioneer kisses, thus, developing a market-based ecosystem of fertilizer innovation. In the interim, Africa, which is faced by a combination of food insecurity and degraded soils, commonly uses direct subsidies of fertilizers to encourage smallholders on their use. Nevertheless, in the absence of comparable investments in the services of extension and distribution mechanisms, such subsidies occasionally do not convert to long-term productivity (Rádi, 2023). Together, these models show that, despite the lack of a single approach, effective policies usually work with affordability, sustainability, and incentives in innovation, which are balanced.

10.8.3 Public–Private Partnerships in Fertilizer Innovation

Governments, research institutions, and private enterprises are increasingly coming together in terms of innovation in regard to fertilizers. Public–private partnerships (PPP) close resource/expertise gaps, by giving governments a financial backdrop and regulatory clarity, and industry offering technical understanding and scalability. Projects like India partnerships to develop nano-fused fertilizers or African regional biofertilizer forums are a perfect example of efforts through which PPP can lead to innovation of products as well as penetration of those products in the market (Hermans et al., 2019).

Also, partnerships with digital agriculture companies improve the use of smart fertilizers through the integration of drones, IOT sensors, and nutrient-management systems. To farmers, it can be interpreted to mean not only the availability of fertilizers but also decision-support systems that maximize the use of the same. PPPs, therefore, can be referred to as the drivers of technological growth and spread of knowledge, hence making sure that innovations are no longer in the labs, but in the operational sphere as well (Hermans et al., 2019).

10.8.4 Risks of Policy Gaps: Falling Behind in Competitiveness

Although there are such developments, the presence of weak or inconsistent policy frameworks is also dangerous. Not employing incentives to incentivize sustainable fertilizer use can lead to countries facing reduced soil health, low levels of crop production, and increased exposure to the transforming climate. In addition, trade competitiveness may be skewed by uneven world policies, such as countries extensively following green fertilizers; they may be more efficient in plasticity of yield and in meeting international standards of sustainability (Martin et al., 2025).

Innovation is also hampered due to policy gaps. The lack of defined regulatory channels to certify bio-based fertilizers or nanofertilizers means that promising technologies are stifled in the process of commercializing them. Similarly, a general lack of investment in extension services also results in a situation where even in the presence of innovation, farmers do not have the necessary know-how to take advantage of it. These risks highlight the credibility of the need to match fertilizer policy with long-term sustainability and competitiveness goals.

Table 10.1 summarizes the economic, environmental, and policy dimensions of next-generation fertilizers, highlighting their cost implications, sustainability benefits, and regulatory considerations.

10.9 Fertilizers and the UN Sustainable Development Goals (SDGs)

10.9.1 SDG 2: Zero Hunger Through Better Yields

Sustainable Development Goal 2: Zero Hunger is central to food security across the globe, and fertilizers will be critical to attaining the set zero hunger goals. Fertilizers help farmers directly improve the quality and yield of a crop by restoring the nutrients in the soil and thus the adequate food supply to meet the growing population (Gil et al., 2019). The Green Revolution had shown this possibility, where the use of fertilizers (along with better varieties of crops) doubled productivity in most of the developing world. Nevertheless, agricultural practice in the future has to move beyond the yield-driven concept and become nutrient-effective and environmentally sound fertilization. Critical innovations are biofertilizers, nanofertilizers, and slow-release formulations that are difficult to create, which

Table 10.1 Next-generation fertilizers: economic, environmental, and policy dimensions.

Category	Traditional fertilizers	Next-gen fertilizers (bio, nano, smart, controlled-release)	Economic impact	Environmental impact	Policy/adoption needs	Reference
Nutrient use efficiency	Low to moderate (30–50%)	High (60–90%)	Reduces input costs; higher yield per unit of fertilizer	Less nutrient loss, reduced eutrophication	Promote R&D and farmer training	El-Sharkawy et al. (2024)
Yield gains	High but plateauing in many regions	Sustainable yield increases; site-specific	Boosts smallholder profitability and global food security	Balanced growth with soil health conservation	Subsidies for advanced inputs	Thierfelder and Mhlanga (2022)
Cost to farmers	Relatively low upfront, but high hidden costs (environmental loss, diminishing returns)	Higher upfront cost, but long-term savings via reduced usage	Improves economic resilience of farmers	Lower pollution cleanup costs	Incentives to offset initial investment	Jeanneaux et al. (2025)
Climate impact	Significant GHG emissions (N_2O release)	Lower emissions through precision release and bio-based options	Supports climate-smart agriculture	Reduced carbon footprint	Policies for carbon credits/green financing	Adnan et al. (2024)
Soil health	Degradation over time due to overuse	Enhanced microbial activity, better organic matter retention	Ensures long-term productivity	Restores ecological balance	Integration into sustainable farming schemes	Sprunger and Martin (2023)
Adoption barriers	Widely available but environmentally costly	Technology, awareness, and affordability gaps	Requires investment in digital tools	Transition phase may challenge resource-poor farmers	Public–private partnerships and extension programs	Bramlett and Radel (2017)

can be used to maintain productivity and reduce their negative impact on the environment. Fertilizers can play a role in eliminating hunger by sealing nutrient deficiencies in the exhausted soils, mainly in Africa and South Asia, and preserve the soil quality of future generations (Ladha et al., 2020).

10.9.2 SDG 8: Rural Livelihoods and Economic Growth

Fertilizers are also transformative to SDG 8: Decent Work and Economic Growth in the sense that they help to empower rural lives. High levels of crop output result in more farm revenues and thus smallholders can get out of poverty cycles. Supply chains of fertilizers also make people get jobs in the sphere of manufacturing, distribution, and extension services. Besides, new opportunities are presented by the introduction of value-added fertilizers, including fortified blends or bio-based products, which raise agribusiness and local economies (Rai et al., 2019).

However, to attain inclusive growth, the rural communities should have access to and affordability of these inventions. Policy, microcredit, and farmer cooperatives are necessary to prevent inequality and to enable the smallholders, who make up the majority of the agricultural labor pool, to have an opportunity to reap the fruits of productivity brought about by the use of fertilizers. This way, fertilizers can be regarded as not only a source of agricultural inputs but also as a form of rural economic empowerment (Zehri et al., 2024).

10.9.3 SDGs 12: Efficiency, Waste Reduction, and Smart Inputs

Such historical consumption of fertilizers has caused nutrient drainage, polluted water, and wasted energy, thus behaving contrary to SDG 12: Responsible Consumption and Production aims. To mitigate this response, a transformational shift toward resource-efficient, intelligent fertilizers to coordinate fertilizer delivery with plant demand is needed. Such technologies as precision farming technologies, coated urea, and nanofertilizers minimize the loss of nutrients and save farmers money spent on inputs. Use of organic waste to recycle into biofertilizers is a part of the circular economy practices, as the agriculture system should reduce waste and boost nutrient cycling (Castellano et al., 2024).

Fertilizers when used effectively do not only conserve resources but also increase the ability of the agroecosystems to endure in the long term. Fertilizers, hence, form focal instruments in attaining sustainable consumption and production in the sphere of agriculture.

10.9.4 SDGs 13 and 15: Climate Action and Healthy Ecosystems

Fertilizers are at cross crossroads with SDG 13 and SDG 15: Climate action and as life on land. Poor utilization of the fertilizer leads to emission of greenhouse gases and, most importantly, nitrous oxide, which is a great pollutant to the climate. On the other hand, agriculture can be able to mitigate carbon footprint development due to the existence of low-emission fertilizers, nitrification inhibitors, and microbial inoculants. Likewise, use of biofertilizers and carbon-absorbing soil amendments preserves the climate. Dissimilarity between the use of fertilizers and climate-smart farming (Yang et al., 2020).

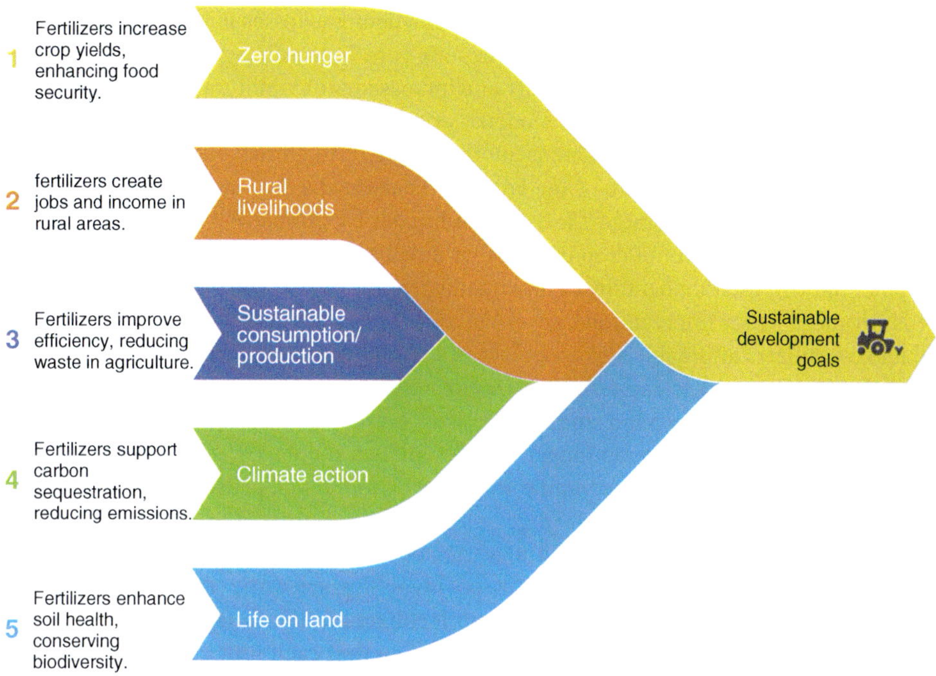

Figure 10.3 Linking fertilizer innovation with UN sustainable development goals (SDGs). A conceptual diagram connecting fertilizers' role to SDGs, including improved yields for zero hunger (SDG 2), rural economic growth (SDG 8), sustainable production and waste reduction (SDG 12), and climate action with ecosystem protection (SDGs 13 and 15).

Ecologically responsible fertilizer use conserves biodiversity by decreasing shortcomings of eutrophication, ground acidification, and habitat dilution. Nutrient-rich ways mitigate worn-out soils and enhance ecosystem services that require pollination, water purification, and carbon storage. When incorporated into the land management approach that is sustainable, fertilizers therefore lead to climate stabilization as well as ecosystem restoration (Campbell et al., 2018).

Figure 10.3 presents a conceptual diagram showing the connections between advanced fertilizer technologies and multiple UN SDGs, emphasizing benefits for food security, economic growth, sustainable resource management, and environmental conservation.

10.10 Conclusion and Future Perspectives

Fertilizers have traditionally formed one of the pillars of agricultural productivity. In the midst of the growing world challenges, such as climate change, soil erosion, food insecurity, and demographic activities, fertilizers have taken a more pivotal position. The availability of new generation fertilizers, including biofertilizers, nanoscale formulations, and controlled-release products, provides economic and ecological reasons to reconsider practices related to the management of nutrients. It is estimated that the innovation of this kind can increase the effectiveness of nutrient utilization, decrease losses in the environmental context, and lower the production costs in the long run. To the crop growers, they can be realized in terms of increased sustainability and increased profitability, as the environmental disadvantages that may undermine modern agricultural methodologies may be reduced or eliminated. At the same time, there is another strategic dilemma associated with fertilizer policy, weighing the welfare of the farmers versus food safety on a global scale. The fertilizers introspectively promote alleviation of hunger by stimulating an increased yield as well as the enhancement of crop quality that directly provides rural income. However, this potential will exist topography of an enabling environment. The globalized economy needs advanced policy interventions, governance reforms, and incentive mechanisms that will make fertilizers affordable and accessible to smallholders, who form the heart of food production. Additionally, a long-term partnership with businesses and governments should involve research encouragement and application of experimental solutions at the field level. Therefore, fertilizers are no longer traditional farm inputs, but they have been elevated into tools of strategic and economic security, environmental conservation, and social well-being.

The future should see the agricultural sector developing a robust competitive, and climate-sensitive attitude toward fertilizers, involving scientific innovation, policy agendas, and sustainable use. Initial work on fertilizers in the future will probably be shifted an efficiency versus sustainability balance, so that each unit used should maximize yield but should reduce its effects on the environment. A three-pronged approach to achieving this can be a trend to promote smart and bio-based fertilizers; increase site-specific actions, using digital and precision agriculture, as well as inbuilt circular economy aims, like organic waste setting aside, to utilize as inputs of nutrients. Global cooperation and local decentralization are also essential. A comparative analysis shows that the efficiency-based efforts of China, the environmentalization efforts of the European Union, and the access-based plans of Africa focus on different priorities; therefore, no magic bullet can fit the case. Concerted research, technology transfer, and fair policy frameworks are keys to the development of fertilizer future that can bring benefits to both the developed and developing world. In the end, fertilizers should develop into climate-friendly, resource-based, and farmer-focused instruments that bolster food systems against future shocks.

References

Abraham, M. and Pingali, P. (2020). Transforming smallholder agriculture to achieve the SDGs. In: *The role of smallholder farms in food and nutrition security*, 173–209. Cham: Springer International Publishing.

Acheampong, A.O. and Opoku, E.E.O. (2023). Environmental degradation and economic growth: Investigating linkages and potential pathways. *Energy Economics* 123: 106734. https://doi.org/10.1016/j.eneco.2023.106734.

Adnan, M., Xiao, B., Bibi, S. et al. (2024). Addressing current climate issues in Pakistan: An opportunity for a sustainable future. *Environmental Challenges* 15: 100887. https://doi.org/10.1016/j.envc.2024.100887.

Agram, I. and Rahal, F. (2024). The impact of crises on energy prices (2021-2023) European union case. *Revue des Economies financières bancaires et de management* 13 (2): 66–82.

Almutari, M.M. (2023). Synthesis and modification of slow-release fertilizers for sustainable agriculture and environment: A review. *Arabian Journal of Geosciences* 16 (9): 518.

Aloo, B., Mbega, E., Tumuhairwe, J., and Makumba, B. (2021). Advancement and practical applications of rhizobacterial biofertilizers for sustainable crop production in sub-Saharan Africa. *Agriculture & Food Security* 10 (1): 57.

Arora, P.K., Tripathi, S., Omar, R.A. et al. (2024). Next-generation fertilizers: The impact of bionanofertilizers on sustainable agriculture. *Microbial Cell Factories* 23 (1): 254.

Avila-Quezada, G.D., Ingle, A.P., Golińska, P., and Rai, M. (2022). Strategic applications of nano-fertilizers for sustainable agriculture: Benefits and bottlenecks. *Nanotechnology Reviews* 11 (1): 2123–2140.

Balafoutis, A., Beck, B., Fountas, S. et al. (2017). Precision agriculture technologies positively contributing to GHG emissions mitigation, farm productivity and economics. *Sustainability* 9 (8): 1339.

Baxter, H., Bearne, L., Stone, T. et al. (2024). The effectiveness of knowledge-sharing techniques and approaches in research funded by the national institute for health and care research (NIHR): A systematic review. *Health Research Policy and Systems* 22 (1): 41. https://doi.org/10.1186/s12961-024-01127-5.

Bhatia, M., Meenakshi, N., Kaur, P., and Dhir, A. (2024). Digital technologies and carbon neutrality goals: An in-depth investigation of drivers, barriers, and risk mitigation strategies. *Journal of Cleaner Production* 451: 141946. https://doi.org/10.1016/j.jclepro.2024.141946.

Bin Radzali, M.K., Hashim, A.M., Ali, M.H.M. et al. (2025). Exploring fertilizer-microbiome interactions through next-generation sequencing (NGS): Insights for sustainable agriculture. *Malaysian Journal of Science* 86–103.

Binswanger-Mkhize, H.P. and Savastano, S. (2017). Agricultural intensification: The status in six African countries. *Food Policy* 67: 26–40.

Boberski, P., Główka, M., Torchała, K. et al. (2025). Sustainable agriculture solutions: Biodegradable coatings for enhanced-efficiency fertilizers using cellulose and lignin. *Journal of Agricultural and Food Chemistry* 73 (22): 13105–13124.

Bramlett, M.D. and Radel, L.F. (2017). Factors associated with adoption and adoption intentions of nonparental caregivers. *Adoption Quarterly* 20 (1): 5–24. https://doi.org/10.1080/10926755.2016.1149534.

Campbell, B.M., Hansen, J., Rioux, J. et al. (2018). Urgent action to combat climate change and its impacts (SDG 13): Transforming agriculture and food systems. *Current Opinion in Environmental Sustainability* 34: 13–20. https://doi.org/10.1016/j.cosust.2018.06.005.

Carmo, M., García-Ruiz, R., Ferreira, M.I., and Domingos, T. (2017). The NPK soil nutrient balance of Portuguese cropland in the 1950s: The transition from organic to chemical fertilization. *Scientific Reports* 7 (1): 8111.

Castellano, R., De Bernardo, G., and Punzo, G. (2024). Sustainable well-being and sustainable consumption and production: An efficiency analysis of sustainable development goal 12. *Sustainability* 16 (17): 7535. Retrieved from https://www.mdpi.com/2071-1050/16/17/7535.

Chandra, M.S., Lal, M., Naresh, R. et al. (2019). Role of polymer coated fertilizers (PCFS) an advance technology for improving nutrient use efficiency and crop productivity: A review. *International Journal of Chemical Studies* 7 (6): 2667–2679.

Chianu, J.N., Chianu, J.N., and Mairura, F. (2012). Mineral fertilizers in the farming systems of sub-Saharan Africa. A review. *Agronomy for Sustainable Development* 32 (2): 545–566.

Cochrane, L., ElAlfy, A., Azadi, H., and Dejene, M. (2024). Defining "success" in large-scale agricultural investment: A typology based on different stakeholder perspectives. *Discover Sustainability* 5 (1): 173.

Cruz, M. and Ahmed, S.A. (2018). On the impact of demographic change on economic growth and poverty. *World Development* 105: 95–106. https://doi.org/10.1016/j.worlddev.2017.12.018.

Daniel, A.I., Fadaka, A.O., Gokul, A. et al. (2022). Biofertilizer: The future of food security and food safety. *Microorganisms* 10 (6): https://doi.org/10.3390/microorganisms10061220.

Ebbisa, A. (2022). *Nutrient use efficiency (NUE): Strategies to improve phosphorus. Sustainable crop production. Recent advances*, 3. Intech Open.

Ehui, S. and Pender, J. (2005). Resource degradation, low agricultural productivity, and poverty in sub-Saharan Africa: Pathways out of the spiral. *Agricultural Economics* 32: 225–242.

Ejedegba, N. (2024). Innovative solutions for food security and energy transition through sustainable fertilizer production techniques. *World Journal of Advanced Research and Reviews* 24 (3): 1679–1695.

El-Shal, R.M., El-Naggar, A.H., El-Beshbeshy, T.R. et al. (2022). Effect of nano-fertilizers on alfalfa plants grown under different salt stresses in hydroponic system. *Agriculture* 12 (8): 1113.

El-Sharkawy, M., Li, J., AL-Huqail, A.A. et al. (2024). Slow-released fertilizers optimization and experimental impacts on soil fertility and wheat-maize cropping system. *Scientia Agricola* 81: e20230234.

Finger, R., Swinton, S.M., El Benni, N., and Walter, A. (2019). Precision farming at the nexus of agricultural production and the environment. *Annual Review of Resource Economics* 11 (1): 313–335.

Gallardo, R.K. (2024). The environmental impacts of agriculture: A review. *International Review of Environmental and Resource Economics* 18 (1–2): 165–235.

Getahun, S., Kefale, H., and Gelaye, Y. (2024). Application of precision agriculture technologies for sustainable crop production and environmental sustainability: A systematic review. *The Scientific World Journal* 2024 (1): 2126734.

Gil, J.D.B., Reidsma, P., Giller, K. et al. (2019). Sustainable development goal 2: Improved targets and indicators for agriculture and food security. *Ambio* 48 (7): 685–698. https://doi.org/10.1007/s13280-018-1101-4.

Goswami, S., Singh, S.K., Mukherjee, S. et al. (2024). Nanofertilizers: A novel technology for enhancing nutrient use efficiency of crops and a relevance to agroforestry. In: *Agroforestry to combat global challenges: Current prospects and future challenges* (ed. Springer Nature), 293–322. Springer.

Goud, B.R., Raghavendra, M., Prasad, P. et al. (2022). Sustainable management and restoration of the fertility of damaged soils. *Agriculture Issues and Policies* 113: 113–131.

Goyal, A., Chavan, S.S., Mohite, R.A. et al. (2025). Emerging trends and perspectives on nano-fertilizers for sustainable agriculture. *Discover Nano* 20 (1): 97.

Gregory, D., Haefele, S., Buresh, R., and Singh, U. (2010). *Fertilizer use, markets, and management. Rice in the global economy. Strategic research and policy issues for food security*, 231–263. Los Banos, Philippines: International Rice Research Institute.

Grizzetti, B., Lanzanova, D., Liquete, C. et al. (2016). Assessing water ecosystem services for water resource management. *Environmental Science & Policy* 61: 194–203. https://doi.org/10.1016/j.envsci.2016.04.008.

Guiné, R.P.F. (2024). The challenges and strategies of food security under global change. *Foods* 13 (13): https://doi.org/10.3390/foods13132083.

Hashimi, R. and Hashimi, M.H. (2020). Effect of losing nitrogen fertilizers on living organism and ecosystem, and prevention approaches of their harmful effect. *Asian Soil Research Journal* 4 (2): 10–20.

Hayo, L. and Hasegawa, H. (2024). Enhancing emission reductions in South African agriculture: The crucial role of carbon credits in incentivizing climate-smart farming practices. *Sustainable Futures* 8: 100260.

Hermans, F., Geerling-Eiff, F., Potters, J., and Klerkx, L. (2019). Public-private partnerships as systemic agricultural innovation policy instruments—assessing their contribution to innovation system function dynamics. *NJAS—Wageningen Journal of Life Sciences* 88: 76–95. https://doi.org/10.1016/j.njas.2018.10.001.

Hosan, S., Karmaker, S.C., Rahman, M.M. et al. (2022). Dynamic links among the demographic dividend, digitalization, energy intensity and sustainable economic growth: Empirical evidence from emerging economies. *Journal of Cleaner Production* 330: 129858. https://doi.org/10.1016/j.jclepro.2021.129858.

Hou, P., Yuan, W., Li, G. et al. (2021). Deep fertilization with controlled-release fertilizer for higher cereal yield and N utilization in paddies: The optimal fertilization depth. *Agronomy Journal* 113 (6): 5027–5039.

Hu, W., Xie, D., Le, Y. et al. (2024). Evolution of food trade networks from a comparative perspective: An examination of China, the United States, Russia, the European Union, and African Countries. *Foods* 13 (18): https://doi.org/10.3390/foods13182897.

Huan, S. and Liu, X. (2023). Network modeling and stability improvement of the water-energy-fertilizer-food nexus flows based on global agricultural trade. *Sustainable Production and Consumption* 39: 480–494.

Huang, S.H., Liu, P., Mokasdar, A., and Hou, L. (2013). Additive manufacturing and its societal impact: A literature review. *The International Journal of Advanced Manufacturing Technology* 67 (5): 1191–1203.

Iqbal, M., Umar, S., and Mahmooduzzafar, f. (2019). Nano-fertilization to enhance nutrient use efficiency and productivity of crop plants. In: *Nanomaterials and plant potential* (ed. Springer Nature), 473–505. Springer.

Islam, M.R., Oliullah, K., Kabir, M.M. et al. (2023). Machine learning enabled IoT system for soil nutrients monitoring and crop recommendation. *Journal of Agriculture and Food Research* 14: 100880. https://doi.org/10.1016/j.jafr.2023.100880.

Jeanneaux, P., Wendling, E., Desjeux, Y. et al. (2025). Structure and determinants of the cost of setting up a farm: The case of young farmers in Central France. *Journal of Rural Studies* 114: 103583. https://doi.org/10.1016/j.jrurstud.2025.103583.

Kim, D. (2021). Economies of scale and international business cycles. *Journal of International Economics* 131: 103459. https://doi.org/10.1016/j.jinteco.2021.103459.

Kudeyarov, V. (2020). Nitrous oxide emission from fertilized soils: An analytical review. *Eurasian Soil Science* 53 (10): 1396–1407.

Ladha, J.K., Jat, M.L., Stirling, C.M. et al. (2020). Chapter two—achieving the sustainable development goals in agriculture: The crucial role of nitrogen in cereal-based systems. In: *Advances in agronomy*, vol. 163 (ed. D.L. Sparks), 39–116. Academic Press.

Lal, R. (2013). Soil carbon management and climate change. *Carbon Management* 4 (4): 439–462.

Lau, C.Y., Wang, C., Orsega, S. et al. (2014). International collaborative research partnerships: Blending science with management and diplomacy. *Journal of AIDS & Clinical Research* 5 (12): https://doi.org/10.4172/2155-6113.1000385.

Lombe, P., Carvalho, E., and Rosa-Santos, P. (2024). Drought dynamics in sub-saharan africa: Impacts and adaptation strategies. *Sustainability* 16 (22): 9902. Retrieved from https://www.mdpi.com/2071-1050/16/22/9902.

Martin, J., Ellström, P.-E., Wallo, A., and Elg, M. (2025). Bridging the policy-practice gap: A dual challenge of organizational learning. *The Learning Organization* 32 (7): 18–34. https://doi.org/10.1108/TLO-05-2023-0079.

Mekonnen, A. (2021). Recent advancement on the technology of enhancing fertilizer use efficiency and crop productivity: A review.

Mesa, J.A., Sierra-Fontalvo, L., Ortegon, K., and Gonzalez-Quiroga, A. (2024). Advancing circular bioeconomy: A critical review and assessment of indicators. *Sustainable Production and Consumption* 46: 324–342. https://doi.org/10.1016/j.spc.2024.03.006.

Mukherjee, D. (2013). Nutrient use efficiency for maximization of crop productivity. *Advances in Plant Physiology—An International Treatise Series. Scientific Publishers, Jodhpur, India* 14: 173–209.

Nyström, M.E., Karltun, J., Keller, C., and Andersson Gäre, B. (2018). Collaborative and partnership research for improvement of health and social services: Researcher's experiences from 20 projects. *Health Research Policy and Systems* 16 (1): 46. https://doi.org/10.1186/s12961-018-0322-0.

Ogbeide, O.A. (2015). Increasing agricultural productivity: A review of the multi-dimensional approach. *Mayfair Journal of Agribusiness Management* 1 (2): 1–24.

Pannell, D.J. (2017). Economic perspectives on nitrogen in farming systems: Managing trade-offs between production, risk and the environment. *Soil Research* 55 (6): 473–478.

Penuelas, J., Coello, F., and Sardans, J. (2023). A better use of fertilizers is needed for global food security and environmental sustainability. *Agriculture & Food Security* 12 (1): 1–9.

Pereira, P., Bogunovic, I., Muñoz-Rojas, M., and Brevik, E.C. (2018). Soil ecosystem services, sustainability, valuation and management. *Current Opinion in Environmental Science & Health* 5: 7–13. https://doi.org/10.1016/j.coesh.2017.12.003.

Pokhrel, S. (2020). Food availability and consumption in relation to developing strategies for sustained production and supply in Nepal. *Journal of Agriculture and Forestry University* 4 (1): 13–28.

Pretty, J. and Bharucha, Z.P. (2014). Sustainable intensification in agricultural systems. *Annals of Botany* 114 (8): 1571–1596.

Quitzow, R., Balmaceda, M., and Goldthau, A. (2025). The nexus of geopolitics, decarbonization, and food security gives rise to distinct challenges across fertilizer supply chains. *One Earth* 8 (1): 101173. https://doi.org/10.1016/j.oneear.2024.12.009.

Rádi, V. G. (22–26 May 2023). *Comparative analysis of the AI regulation of the EU, US and China from a privacy perspective*. Paper presented at the 2023 46th MIPRO ICT and Electronics Convention (MIPRO).

Rai, S.M., Brown, B.D., and Ruwanpura, K.N. (2019). SDG 8: Decent work and economic growth—a gendered analysis. *World Development* 113: 368–380. https://doi.org/10.1016/j.worlddev.2018.09.006.

Randive, K., Raut, T., and Jawadand, S. (2021). An overview of the global fertilizer trends and India's position in 2020. *Mineral Economics* 34 (3): 371–384.

Ravikumar, S., Vellingiri, G., Sellaperumal, P. et al. (2024). Real-time nitrogen monitoring and management to augment N use efficiency and ecosystem sustainability—a review. *Journal of Hazardous Materials Advances* 16: 100466. https://doi.org/10.1016/j.hazadv.2024.100466.

Reader, M.A., Revoredo-Giha, C., Lawrence, R.J. et al. (2018). Farmers' spending on variable inputs tends to maximise crop yields, not profit. *International Journal of Agricultural Management* 7 (1): 30–40.

Rhodes, C.J. (2012). Feeding and healing the world: Through regenerative agriculture and permaculture. *Science Progress* 95 (4): 345–446.

Sandström, V., Huan-Niemi, E., Niemi, J., and Kummu, M. (2024). Dependency on imported agricultural inputs—global trade patterns and recent trends. *Environmental Research: Food Systems* 1 (1): 015002.

Shah, F. and Wu, W. (2019). Soil and crop management strategies to ensure higher crop productivity within sustainable environments. *Sustainability* 11 (5): 1485.

Shanmugavel, D., Rusyn, I., Solorza-Feria, O., and Kamaraj, S.-K. (2023). Sustainable SMART fertilizers in agriculture systems: A review on fundamentals to in-field applications. *Science of the Total Environment* 904: 166729. https://doi.org/10.1016/j.scitotenv.2023.166729.

Sida, T. S., Desalegn, T., Abate, G. T., & Amede, T. (2025). Feeding a fast—growing population by 2050 through accelerated agricultural transformation: Potential avenues for Ethiopia *Pathways to African Food Security*, 1st edition. 35.

Singh, B. and Sapkota, T. (2025). Fertiliser policies in ensuring food security and emerging issues in India. *Indian Journal of Fertilisers* 21 (7): 616–636.

Singh, H., Halder, N., Singh, B. et al. (2023). Smart farming revolution: Portable and real-time soil nitrogen and phosphorus monitoring for sustainable agriculture. *Sensors (Basel)* 23 (13): https://doi.org/10.3390/s23135914.

Singh, S., Singh, R., Singh, K. et al. (2024). Smart fertilizer technologies: An environmental impact assessment for sustainable agriculture. *Smart Agricultural Technology* 8: 100504. https://doi.org/10.1016/j.atech.2024.100504.

Snow, M.S. (2023). The invasion of ukraine and the global food crisis of 2022: Responding to food supply shocks. In: *The great power competition volume 5: The russian invasion of ukraine and implications for the central region* (ed. Springer Nature), 351–387. Springer.

Sobota, D.J., Compton, J.E., McCrackin, M.L., and Singh, S. (2015). Cost of reactive nitrogen release from human activities to the environment in the United States. *Environmental Research Letters* 10 (2): 025006.

Sonea, C., Gheorghe-Irimia, R.A., Tapaloaga, D. et al. (2023). Optimizing animal nutrition and sustainability through precision feeding: A mini review of emerging strategies and technologies. *Annals of'Valahia'University of Târgovişte. Agriculture* 15 (2): 7.

Soni, S.K., Manhas, R., Jakhar, Y. et al. (2022). Biofertilizers for sustainable agriculture: Current trends and future perspective. In: *Genomic, proteomics, and biotechnology*, 331–356. Taylor & Francis Group.

Sprunger, C.D. and Martin, T.K. (2023). Chapter three—an integrated approach to assessing soil biological health. In: *Advances in agronomy*, vol. 182 (ed. D.L. Sparks), 131–168. Academic Press.

Sun, X. and Song, Y. (2025). Unlocking the Synergy: Increasing productivity through Human-AI collaboration in the industry 5.0 Era. *Computers & Industrial Engineering* 200: 110657. https://doi.org/10.1016/j.cie.2024.110657.

Tanchum, M. (2022). *Morocco's new challenges as a gatekeeper of the world's food supply: The geopolitics, economics, and sustainability of OCP's global fertilizer exports*. Middle East Institute.

Thakur, A.K., Mandal, K.G., Mohanty, R.K., and Sarangi, A. (2024). Next-gen rice farming: Ways to achieve food, nutritional and economic security under changing climatic conditions. *Current Science* 126 (4): 00113891.

Thierfelder, C. and Mhlanga, B. (2022). Short-term yield gains or long-term sustainability?—a synthesis of Conservation Agriculture long-term experiments in Southern Africa. *Agriculture, Ecosystems & Environment* 326: 107812. https://doi.org/10.1016/j.agee.2021.107812.

Verma, P., Singh, D., Pathania, I.P., and Aggarwal, K. (2019). Strategies to improve agriculture sustainability, soil fertility and enhancement of farmers income for the economic development. In: *Soil fertility management for sustainable development* (ed. Springer Nature), 43–70. Springer.

Wanat, L., Sikora, J., Majchrzak, L. et al. (2025). The economic and technological challenges of the agri-development implementation model in the case of the wielkopolska region in poland. *Agriculture* 15 (4): 412.

Wang, X., Cai, D., Hoogmoed, W.B., and Oenema, O. (2011). Regional distribution of nitrogen fertilizer use and N-saving potential for improvement of food production and nitrogen use efficiency in China. *Journal of the Science of Food and Agriculture* 91 (11): 2013–2023.

Wang, N., Bai, Y., Guo, Z. et al. (2024). Synergies between the circular economy and carbon emission reduction. *Science of the Total Environment* 951: 175603. https://doi.org/10.1016/j.scitotenv.2024.175603.

Xiao, C., Wang, Q., van der Vaart, T., and van Donk, D.P. (2018). When does corporate sustainability performance pay off? The impact of country-level sustainability performance. *Ecological Economics* 146: 325–333. https://doi.org/10.1016/j.ecolecon.2017.11.025.

Xiao, L., Liu, J., and Ge, J. (2021). Dynamic game in agriculture and industry cross-sectoral water pollution governance in developing countries. *Agricultural Water Management* 243: 106417.

Yadav, A., Yadav, K., and Abd-Elsalam, K.A. (2023). Nanofertilizers: Types, delivery and advantages in agricultural sustainability. *Agrochemicals* 2 (2): 296–336.

Yan, B., Deng, T., and Shi, L. (2024). Towards sustainable productivity of greenhouse vegetable soils: Limiting factors and mitigation strategies. *Plants* 13 (20): 2885.

Yang, S., Zhao, W., Liu, Y. et al. (2020). Prioritizing sustainable development goals and linking them to ecosystem services: A global expert's knowledge evaluation. *Geography and Sustainability* 1 (4): 321–330. https://doi.org/10.1016/j.geosus.2020.09.004.

Zafar, M.H., Langås, E.F., and Sanfilippo, F. (2024). Exploring the synergies between collaborative robotics, digital twins, augmentation, and industry 5.0 for smart manufacturing: A state-of-the-art review. *Robotics and Computer-Integrated Manufacturing* 89: 102769. https://doi.org/10.1016/j.rcim.2024.102769.

Zehri, C., Mohammed El Amin, B., Kadja, A. et al. (2024). Exploring the nexus of decent work, financial inclusion, and economic growth: A study aligned with SDG 8. *Sustainable Futures* 7: 100213. https://doi.org/10.1016/j.sftr.2024.100213.

Zhong, Z., Jiang, W., and Li, Y. (2023). Bridging the gap between smallholders and modern agriculture: Full insight into China's agricultural cooperatives. *Journal of Rural Studies* 101: 103037. https://doi.org/10.1016/j.jrurstud.2023.103037.

Zinn, W. and Goldsby, T.J. (2020). Global supply chains: Globalization research in a changing world. *Journal of Business Logistics* 41 (1): 4–5. https://doi.org/10.1111/jbl.12241. Epub 2020 Apr 6.

11

Enhancing Crop Nutritional Quality and Functionality Through Next-Generation Fertilizer Technologies

Salma Hameed[1], Muhammad Saqlain[1], Iqra Ghafoor[2], Sobia Riaz[1], Tehmeena Mukhtar[1], and Rahib Hussain[3]

[1] *Department of Environmental Sciences, University of Jhang, Jhang, Pakistan*
[2] *Institute of Soil and Environmental Science, University of Agriculture Faisalabad, Faisalabad, Pakistan*
[3] *Sustainable Development Study Centre, Government College University Lahore, Lahore, Pakistan*

11.1 Introduction—The Link Between Fertilizers and Crop Nutritional Quality

The cutting-edge nexus between fertilizer use and crop nutritional value in the current debate on agricultural production can be figured out as one of the most pertinent issues because they both encompass the dynamics of agricultural food production activities all over the world and the nutritional value of food to mankind (Barłóg et al., 2022). With the growing population, a serious effort on the part of the international community to respond to meet the new demand projected upwards to 9.7 billion population by 2050, simultaneous demands to increase both crop yields and nutritional density have begun to grow even more paramount. It is thus in the research presented herein that the complex interaction between fertilizer regimes and the nutritional make-up of food crops is explored, illustrating how wise nutrient control can accommodate the two competing goals of quality and quantity in agricultural practice.

11.1.1 The Contemporary Agricultural Paradigm: Yield vs. Nutritional Quality

11.1.1.1 The Dual Challenge in Modern Agriculture

Modern agriculture now faces what can be described as an unprecedented twofold challenge: the necessity to produce enough food and, at the same time, to maintain, and preferably, to improve the nutritional value of said crops. The recent extensive meta-analyses confirm that the use of fertilizer brings about a significant improvement in yields, an average of 30.9%, in addition to significant increases in crop nutritional health or 11.9% in the total carbohydrates, proteins, oil content, feed vitamin C, mineral nutrients, and total soluble solids (Chen et al., 2024) (Figure 11.1). The concurrence of these advantages supports the radical vagueness of strictly scientific efforts to fertilize in dealing with greater challenges of world nutrition.

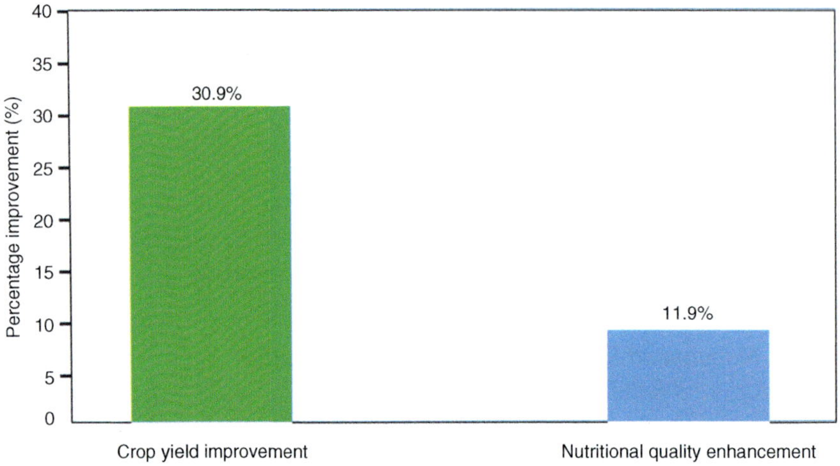

Figure 11.1 Effect of fertilizer application on crop yield and nutritional quality data. *Source:* Adapted from Chen et al. (2024).

The consequences of such a relationship go way beyond the cramped boundaries of agrarian output. The last 60 years of decades have seen a gradual decline in food quality supported by an acute loss of many nutritionally invaluable minerals, nutraceutical compounds in the major sown crops, vegetables, and fruits, which is widely thought to be as a result of the high concentration of CO_2 in the atmosphere and the high-intensive farming activities in the modern world (Bouis et al., 2024) (Figure 11.2). This nutritional dilution effect brings to the fore the decisive role that well-targeted fertilization regimes will have to play in undoing these negative trends.

11.1.2 Global Food Security Implications

Fellow workers, I believe that we should contend with the alarming trend, namely that the lack of balance in the application of fertilizers globally leaves us with a troubling trend indeed, namely that of areas that experience constant malnutrition of the land on the one hand, and, then, those which have seen blatant overuse leading to nutritional imbalances and environmental strain. The asymmetry of the present fertilizer regimes yields nutritional inequalities, both facilitating under- and over-nutrition, evidencing the urgent need for more equitable and eco-efficient dispensation and delivery approaches (Figure 11.2).

11.1.3 Mechanistic Foundations of Fertilizer–Nutrition Interactions

11.1.3.1 Macronutrient Pathways and Nutritional Enhancement

A complex repertoire of physiological and biochemical processes exists, thus coordinating the implications of fertilizer use and the nutritional value of crops. Since the basic macronutrient triad's nitrogen, phosphorus, and potassium form the essential skeleton of critical biomolecules, every single nutrient has a different, but complementary

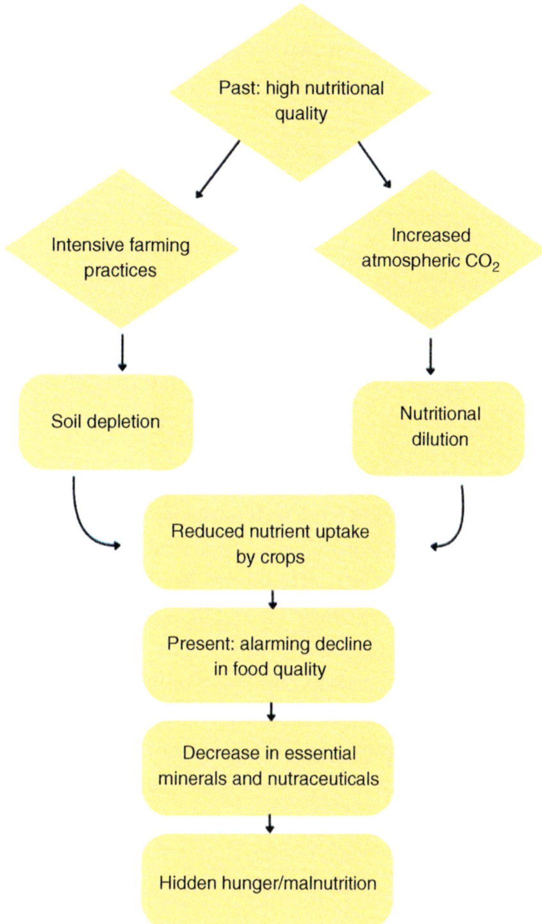

Figure 11.2 Pathways leading to the decline in crop nutritional quality and resulting in hidden hunger.

biochemical impact. An example is nitrogen, which plays an essential role in protein synthesis and specification of amino acids, almost the same case with phosphorus, which drives energy metabolism and nucleotide synthesis. Since nutrient acquisition in crops is closely related to existing soil nutrient conditions, sensible timing of irrigation and fertilizer additions significantly stimulates outright nutrient uptake performance by nitrogen, phosphorus, and potassium and contributes to faster growth and development (Xing et al., 2024).

11.1.4 Micronutrient Dynamics and Crop Quality

Looking at the nutrition of the plant, we can understand that micronutrients, although required in comparatively small amounts, have an inordinately large effect on the nutritional quality of crops (Kaushal, 2024). Good strategic use of both macro- and micronutrients is thus impossible to avoid, whether in terms of ensuring maximum yields or even attainment of premium nutritional value on the final products. Others like iron, zinc, selenium, and iodine not only provide the important minerals to the consumer but also play crucial roles as essential cofactors in enzymatic pathways involving vitamin biosynthesis of antioxidant production, and the generation of other healthy molecules (Abdoli, 2020).

New improvements in the technology of chelation and novel delivery devices have significantly increased the bioavailability and agronomic performance of micronutrients. Empirical evidence shows that compared to the traditional sources, chelated micronutrient formulations are highly effective in terms of nutrient absorption and, at the same time, reduce soil fixation, with the overall outcome being a significant increase in nutritional values of crops (Madhupriyaa et al., 2024). These advances form a dramatic paradigm breakthrough of modern agriculture that spells the birth of precision nutrition management as a plausible and effective methodological system.

11.1.5 Contemporary Fertilization Strategies and Innovations

11.1.5.1 Novel Fertilizer Technologies

The future of responsive plant nutrition is dependent on long-term innovation that can increase nutrient use efficiencies in heterogeneous agricultural environments. Recent studies in nanofertilizers, biofertilizers developed using biotechnology, and the new category of next-generation fertilizers that operate at high efficiency have provided evidence that must be taken seriously and weigh the nutrient-releasing function and level of agronomic performance and ecological consequence (Maaz et al., 2025). These modalities, owing to their improved accuracy, promise improvement of the level of achieving yields and nutritional standards, and at the same time, reducing ecological hoof-prints (Schaller et al., 2024)

11.1.6 Integrated Fertilization Approaches

As our current understanding of sustainable agricultural techniques with regard to fertilization, the combined scheme of both chemical and organic fertilization more often tends to be considered as one of the effective means of increasing soil fertility and maintaining agricultural production (Xing & Wang, 2024a). Nutritional quality optimized fertilization regimes combining chemicals and organic fertilizers are at the forefront in addressing this issue (Figure 11.3). The integration of the rapid delivery of nutrients made possible through synthetic fertilizers with the longevity-based advantages of organic amendments provides a synergetic balance between productivity and nutritional characteristics (Figure 11.3).

Experimental data suggest that organic fertilization techniques, especially, are very efficient ways of increasing the nutritional value of crops. The use of organic fertilizers improves the health of the soil and enhances the growth of plants by increasing the content of soil organic matter and regulating the availability of nutrients (Liu et al., 2024b). The nutrients released slowly with the use of organic sources enhance the soil condition, increase microbial diversity, and consequently lead to the maximum nutrient intake and nutritionally refined crops (Figure 11.3).

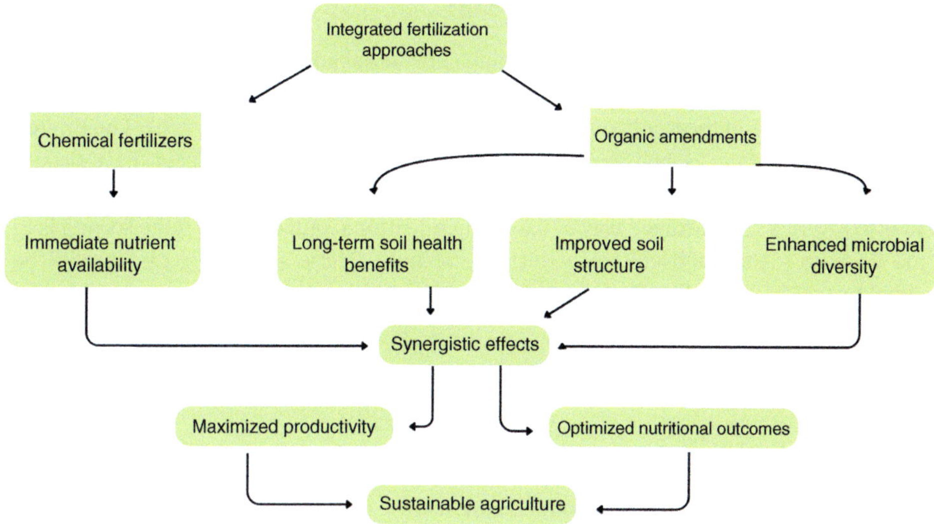

Figure 11.3 Synergistic benefits of integrating chemical fertilizers with organic amendments for sustainable agriculture.

11.1.7 Precision Agriculture and Nutrient Management

As the contemporary agenda on environmentally friendly farming develops, technologies supporting precision delivery of water and nutrients are gaining ground as the nonnegotiable technologies, fulfilling the dual objective of enhancing nutrient use efficiencies and maintaining the productivity of crop plants (Wang et al., 2024b).

In tandem with the above, the focus on microbial fertilizers has also gained the interest of considerable academic interest as a potential alternative to chemical fertilizers, which would offer not only greater agricultural yields but also fewer negative health and agricultural environmental consequences (Wang et al., 2024b). Biological formulations of fertilizers rely on the positive microorganisms to increase the availability of nutrients as well as improve the soil health, and increase the biosynthesis of active bio-compounds, increasing the nutritional value of crops.

11.1.8 Silicon-Enhanced Sustainable Crop Production

What about the possibility of establishing viable silicon cycles in natural ecosystems, a technique that not only shows greater sustainability, but also results in more resilient crops to drought and pest pressures? Silicon supplementation is a very young field of study in agriculture, but the evidence available highlights significant benefits in both plant health and nutritional composition. According to the results of the experiments carried out, silicon-supplemented fertilization procedures enhance the mechanical strength, increase disease resistance, and encourage the production of desirable components in edible plant parts.

11.1.9 Research Gaps and Future Directions

Within the last few years, we have been seeing substantial advances in the art of fertilizer, and in the understanding of nutrient-quality interrelationships. The latter nevertheless exist notwithstanding these advances; there remain several key research gaps. The limited quantities of field data, as well as the absence of standardized testing policies, have slowed the wider implementation of innovative fertilization methods. The first step to advance the field is to invent a standardized way of assessing nutritional outcomes, and the second is to come up with a thorough system of testing the field. These are efforts that cannot be avoided to achieve the full potential of precision agriculture.

11.2 The Science of Crop Nutritional Quality

11.2.1 Background

In the modern sphere of discussion of food systems, the nutritional quality of crops is at the interference of agricultural productivity and human health. Given the fact that growing global populations are forced to use a planetary surface that has a limited supply of arable land, the need to intensify the nutrient density in the crops has now become more feverish. Existing research makes clear that such an aim could only be achieved in a multidisciplinary setting that links plant physiology, soil science, genetics, and nutritional biochemistry, as one recent review refers to a supposedly more appropriate vision as a nutrition-first lens (Avnee et al., 2023).

What needs to be highlighted is that crop nutritional quality cannot be equated to the concentration of nutrients only; it includes bioavailability, storage and processing stability of foods, and synergies in the body. Recent studies have also shown that the historical approach to farming, whereby the main thrust was yield maximization, had in effect corroded the nutritional value of most staple foods to the point where what researchers now termed as hidden hunger, or, in the words of researchers, circumstances of being calorically satisfied but with micronutrient needs unmet. Estimates indicate that around 2 billion individuals are victims of chronic malnutrition or what researchers refer to as hidden hunger, which is related to a variety of diseases, such as mental deterioration, stunted development, and high death rates (Sheoran et al., 2022).

11.2.2 Fundamental Principles of Crop Nutritional Enhancement

11.2.2.1 Bioavailability and Nutrient Density Optimization

The precise understanding of underlying science that decides the nutritional value of crops will depend on demystifying how plants form, transport, and concentrate necessary nutrients in their edible portions. Biofortification, in its turn, denotes the controlled increase in the concentration and/or bioavailability of micronutrients in core crops, an event that can resolve the ailment of micronutrient shortage on a global scale (Kozłowska & Szostak-Węgierek, 2022). Where the strategy surely aims at increasing absolute quantities of nutrients, it also aims at optimizing the quantity of nutrients in which they are stored in terms of the chemical form better suited to place them into to achieve maximal absorption and use in the human digestive system (Table 11.1).

Table 11.1 Genetic and environmental determinants influencing crop nutritional quality.

Category	Determinant	Description	Impact on nutritional quality
Genetic determinants	Quantitative trait loci (QTL)	Complex genetic architecture controlling nutritional traits	Respond dynamically to environmental conditions
	Regulatory genes	Key genes controlling the synthesis and accumulation of essential nutrients	Provide targets for conventional breeding and biotechnological interventions.
	Genome-wide association studies (GWAS)	Breakthrough tool for identifying genetic variants	Enable identification of nutritional trait-associated genes
	Next-generation sequencing (NGS)	Advanced genomic technology	Accelerates crop improvement efforts
	Genome editing (GE)	CRISPR/Cas9 and other editing technologies	Reduces time and resources for crop improvement
Environmental determinants	Soil health parameters	Organic matter content, microbial diversity, mineral availability	Particularly sensitive influence on nutritional quality
	Rhizosphere microbiome	Beneficial microorganisms in the root zone	Known role in crop nutrient dynamics
	Climate conditions	Temperature, precipitation, and atmospheric CO_2	Affects nutrient synthesis and accumulation
	Soil pH and structure	Chemical and physical soil properties	Influences nutrient bioavailability
	Water management	Irrigation and drainage practices	Affects nutrient uptake and transport

Nutrient bioavailability is affected by complex interactions of many compounds with plant tissues, which is demonstrated by recent progress in nutritional enhancement. As an illustration, phytic acid is proven to inhibit the absorption of iron and zinc in the intestines, and vitamin C can increase iron bioavailability. This very understanding of the subtleties of relationships encourages the investigator to abandon the limited approaches to research of individual components and take into account the whole nutritional matrix (Table 11.1).

11.2.3 Genetic and Environmental Determinants

There is cumulative evidence that nutrition traits in crops are the result of a genomic structure full of quantitative trait loci whose expression is dependent on soil, climate, and growth factors (Table 11.1). Recent genomic research studies have given a separation of major regulatory genes that coordinate in the production and accumulation of necessary nutrients, hence providing accurate targets to both conventional breeding approaches and to modern biotechnological tinkering. In addition to this, the new advances of genome-wide association studies (GWAS), next generation-sequencing (NGS), and genome editing (GE) have transformed crop-development projects (Table 11.1).

Empirical evidence also indicates that the nutritional quality of crops is more sensitive to indicators of soil health, such as the presence of organic matter, soil microbial diversity, and availability of minerals (Table 11.1). Microbiome also provides a potential prospect of growing nutrient-dense crops due to the established effect of rhizosphere processes on nutrients. These understandings have generated fresh interest in regenerative agriculture, where the reinvigoration of soil ecological health is brought to the forefront as the primary method of nutritional improvement.

11.2.4 Contemporary Approaches to Nutritional Enhancement

11.2.4.1 Biofortification Strategies

In the modern crop-science vocabulary, biofortification has become the panacea of possible approaches to the systematic improvement of crop nutritional identity. More importantly, the discipline involves three major approaches, that is, conventional plant breeding, agronomic, and genetic engineering, which have their benefits and drawbacks. Any attempt to enhance the nutrition of the diet can certainly not be without a stringent comprehension of these variations, especially when the target population, the environmental circumstances, and the available resources are taken into consideration.

The traditional breeding schemes are based on the methodical discovery and mixing of genetic variations that inherently appear in the enhancement of desirable nutrients. Having said that, this method is inevitably slower than the alternatives possible with the application of biotechnology, but it tends to lead to varieties that will be more readily accepted by consumers and by the regulatory bodies that will be required to approve them. Agronomic biofortification, in its turn, implies using blends of fertilizers that are rich in micronutrients, also aimed at raising the content of available minerals throughout the cultivation process, which offers a more timely solution, but, perhaps, is less sustainable in the long term.

11.2.5 Advanced Genetic Technologies

Due to the collaboration of modern genetic engineering and the study of plants, the face of nutritional improvement has significantly transformed. In the specific case of the CRISPR/Cas9 platform, the timeframe in which new plant varieties have been generated relative to traditional breeding has been reduced, not only limiting the amount of time to which it is subject but also the economic limitations that have long been inherent in crop enhancement and health. Genome-editing methods allow a researcher to alter the encoding of specific genes without disrupting the overall genomic structural framework by allowing him or her to modify precisely the metabolic pathways controlling nutrient synthesis and storage.

These developments notwithstanding, attempts to achieve enhanced crop production, increased nutritional content, and increased bio-production are still upset by mounting environmental pressures and the slow rates of adopting new technologies. However, recent findings in molecular plant breeding have led to a series of foods with significantly higher nutritional values, without a concomitant loss in agronomic performance, and environmental performance has affirmed the world of plant phenotyping and genomic marker analysis.

11.2.6 Mechanisms of Nutrient Accumulation

11.2.6.1 Metabolic Pathway Engineering

In the past decades, there has been a growing insight into the biochemical pathways involved in the production of nutrients, thus providing an opportunity to carry out intervention at specific nutritional levels. By regulating essential biosynthetic pathways, researchers have already managed to increase vitamin levels, increase minerals, and enhance protein quality. The most common types of these enhancements include increasing the expression of key enzymes or regulatory proteins that control rate-limiting functions in the target biosynthetic pathways.

Since these metabolic circuits are so interconnected, it is most appropriate to evaluate the potential implications of genetic manipulation by use of advanced computational models. In this context, systems biology has emerged as an essential science that gives researchers the predictive capacity that can be used to design nutritional-enhancement interventions that would avoid adverse effects on the growth and development of the plants.

11.2.7 Nutrient Transport and Storage Optimization

We should not forget the idea that effective transport of nutrients produced at the sites of synthesis to storage organs is another pillar of nutrition improvement. Recent empirical studies have identified the individual transporter proteins that transport minerals, vitamins, and other nutrients in the plant bodies. Any manipulation of their expression and/or activity (their activity) has a direct and strong effect on the nutritional profile of harvested crops.

During the same time, storage systems are a very important determinant of the nutritional quality of postharvest as well as the succeeding processing. The understanding of cellular compartmentalization of nutrients has served to provide the framework of approaches that have been implemented to increase stability and even bioavailability in processed foods.

11.3 Next-Generation Fertilizers: Innovations for Enhancing Crop Nutritional Profiles

11.3.1 Enhanced-Efficiency Fertilizers: The Foundation of Modern Nutrient Management

The use of enhanced-efficiency fertilizers (EEFs) is a significant change to the modern agricultural nutrient management practice that combines physicochemical processes of fertilizer release with control of biological conversion of nitrogen in the soil (Lam et al., 2022). Through such innovations, the issue of nutrient loss is being confronted, and at the same time, nitrogen use efficiency is also improved, which is also currently explored to be insufficient at the global level. As empirical results show, the social returns, or rather social returns to the prevention of the costs related to nitrogen pollution, are significantly greater than the initial investment amount, hence promoting further adoption of the same practice in farming enterprises aimed at the pursuit of environmental protection and increased profitability (Ali et al., 2025; Beig et al., 2022).

11.3.2 Bionanofertilizers: Merging Biotechnology with Nanotechnology

Integrating biotechnology with nanotechnology holds great promise for revolutionizing agriculture and addressing both current and future challenges (Shanmugavel et al., 2023). Fellow scholars, it is time to take a break and examine an up-and-coming, yet auspicious transformation to modern farming, the rise of bionanofertilizers, the crossing of nanotechnology and biotechnology to produce environmentally friendly alternatives to traditional agrochemical fertilizers (Arora et al., 2024) (Figure 11.4). These systems combine nanoparticle components along with symbiotic microorganisms and therefore extend nutrient absorption and restore soil health. Empirical findings propose improvement of up to 30% in nutrient use efficiency, 20% in crop yields compared to monolithic systems of chemical agents, with the capability of substituting up to 50% of the conventional regimen (Fu et al., 2023) (Figure 11.4). These bionanoformulations will react to specific plant physiographical demands, including response to severe growth stages, due to precision-based delivery systems.

11.3.3 Controlled-Release Fertilizer Technologies

Controlled-release fertilizers (CRFs) have evolved far from the earlier generation of polymer-coated fertilizers of yesterday and now feature advanced stimuli-responsive strategies (Govil et al., 2024). Fertilizers based on the technology of nanotechnology may help to improve the nutrition of crops, but with a lesser impact on the environment, as the devices can regulate the release of nutrients. Such complex systems utilize a package of

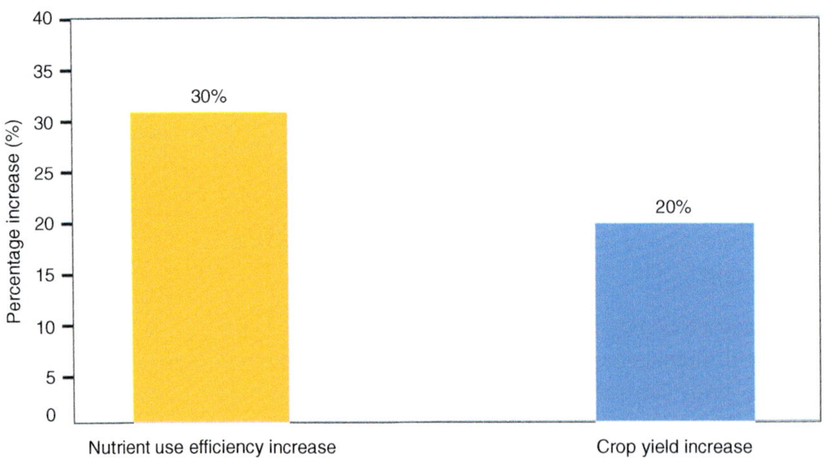

Figure 11.4 Potential increases in nutrient use efficiency and crop yield with bionanofertilizers, modified from Martinez et al. (2017).

release triggers such as soil temperature, moisture content, and pH levels (Shen et al., 2023). The consequent dynamic fertilization plans adjust according to the crop needs in real time. In modern CRFs, the prospects of loss of nutrients are closed, and release patterns are engineered at the scale of plant physiology.

11.3.4 Smart Fertilizer Systems and Precision Agriculture Integration

The so-called smart fertilizer system has been the general term applied in the current agricultural scenario, where the Internet of Things (IoT) and advanced fertilizer compositions are combined, thus leading to comprehensive nutrient management systems. They are seen to regulate soil properties actively (organic carbon content, nutrient availability, and water retention), controlling what goes on in the soil at the same time. Soil stress indicators are observed by the use of real-time sensor networks that measure the status of the soil nutrients continuously. These data are used to specifically increase desired nutrients, such as iron, zinc, and essential amino acids, throughout the growth of crops.

11.3.5 Agronomic Biofortification Through Advanced Fertilization

Biofortification offers a practical and affordable way to deliver essential micronutrients to populations with limited access to diverse diets or other nutritional interventions (Bouis & Saltzman, 2017). We must first explain the jargon that we are going to use: agronomic biofortification alludes to a strategic implementation of selective mineral fertilizers to the ground or directly to the crops. The aim is to increase both the concentration and bioaccessibility of certain nutrients in the product obtained by harvest. Taking data reported and synthesized by (Verma et al., 2023), at least 450 biofortified versions of 12 crops had been released in 41 countries by 2024. The rate of adoption of this innovation in the world is very impressive.

In history, agronomic biofortification arose as a simple fortification approach to micronutrients. However, recent studies have brought the field closer to high-tech multi-nutrient add-on strategies, considering the interaction of nutrients and the bioavailability of nutrients while keeping the acceptability of sensory properties unscathed.

11.3.6 Sustainable Nitrogen Management Technologies

At a more recent scale, strategic use of slow-release/controlled-release nitrogenous fertilizer has shown its ability to increase crop production and reduce environmental limitations (in agriculture) (Liu et al., 2024a). The most important part of this accomplishment is the record of nitrogen use efficiency regulation, a long-standing constraint in agricultural systems, affecting the whole world. In this context, nano-nitrogen-based fertilizers utilize the aid of nanocarrier systems to encapsulate nitrogen compounds, as a measure to induce reduced, demand-driven release patterns. The increased protein level and general nutritional value of crops yielded as a result are both visible and quantifiable.

11.4 Fertilizer-Induced Modulation of Crop Functionality

11.4.1 Physiological Mechanisms of Nutrient–Crop Interactions

Fertilizers are catalysts or agents that alter the functioning of crops, triggering complicated physiological cascades controlling growth, development, and metabolic activities. Effective and planned supply of key macro elements—nitrogen (N), phosphorus (P), and potassium (K)—initiates integrated molecular activities that eventually have a bearing on the productivity and potential of crops (Yousefi & Douna, 2023). Recent research shows the modulation caused by fertilizers occurs in a complex network with energy metabolism regulation, hormone signaling, and stress responses at the cellular level (Jiaying et al., 2022).

Nitrogen is the most influential nutrient in determining the physiology of crops. Adequate nitrogen determines the protein synthesis, chlorophyll biosynthesis, and photosynthetic ability, and thus morphology and physiological processes of a plant through the hormonal route and metabolic-mediated changes in the conversion of organic acids. The type of form of nitrogen applied, namely as nitrate (NO_3^-) or ammonium (NH_4^+), produces different physiological effects; that is, plants treated with nitrate usually have improved vegetative growth than the plants treated with ammonium.

Similarly, albeit non-inferior, phosphorus plays an equally significant role in determining the way plants respond to environmental stress. It plays a pivotal role in the development of roots, chloroplasts, and overall plant metabolism by mediating energy transfer and involvement in nucleic acid metabolism. Because of the low bioavailability of phosphorus to most soils, the trace nutrient has been a challenge to the productivity of crops, particularly in stressful environments where efficient use of energy is very critical.

11.4.2 Molecular Responses to Fertilizer Application

Our multi-omics approach, a combined work over the last 10 years, has revealed in hitherto unequaled depths the complexity of fertilizer-induced changes in crop functionality in terms of their molecular basis. Once plants experience shortages in the key macronutrients, primarily nitrogen, phosphorus, and potassium, a severe rise in the production of reactive oxygen species occurs, leading in turn to the activation of oxidative stress responses, as a result of which cellular metabolism is completely restructured. The manifestation of these perturbations is in the form of altered patterns of gene expression, readjusted speed of protein synthesis, as well as recalibrated metabolic pathways.

The results of nitrogen and potassium deficiency are more extreme, developmental stands, as compared to those of phosphorus levels; roots atrophy significantly, morphologically, as well as the chlorophyll synthesis process is hindered; elements that jointly affect the efficiency of photosynthesis and the capability of carbon sequestration. The key bottleneck occurs in the chloroplast: a disturbance in its structure and functioning has a direct effect on the photosynthetic process (Yahaya et al., 2023). New interrogations of genomes have also revealed particular transcriptional regulators that are sensitive to the fertilizer supply, thus explaining the time-course of how crops adjust to changes in nutrient availability. Equal interest are the details that we now know about the complex regulatory loops that control the interaction between nitrogen and phosphorus availability. These responsive signaling pathways compose environmental as well as developmental programs, and finally conclude the crop performance on various fertilizer regimes (Saurabh et al., 2024). Collectively, these facts highlight the demand to design accurate fertilization procedures that factor in genetic foundations and the ever-changing conditions of the environment, in case we are ever to salvage a situation of maximum crop functionality in the 21st century.

11.4.3 Fertilizer Effects on Crop Growth Coordination

Fertilizer use is also an effective manipulation tool in terms of crop growth and metabolism, hence providing a formal transformation in the relationship between sustainable agricultural production. Modern Green Revolution varieties of the major cereals demonstrate peculiar reactions to fertilizer treatments, an effect revealing the engineered features of these varieties meant to increase nutrient use efficiency (Nakachew et al., 2024). Nevertheless, in a bid to achieve increased productivity by intensifying the use of fertilizers, there is the unavoidable risk thereof developing environmental hazards, first among them all being depletion of the quality of soil and with that, the ensuing issue of nitrogen pollution.

Recent research thus focuses on the methods that are geared toward maximizing the efficiency of nutrient usage whilst maintaining high yields. The empirical evidence suggests that when a balanced fertilization procedure is followed, it can enhance the crop yield by up to 30.9% and the nutritional value of crops such as cereals, vegetables, and oil crops as well (Ishfaq et al., 2023). This twofold thrust in yield/quality is testimony to the complex mutualism between nutrient provision and the functional capacity of crops.

This is also subject to temporal considerations, where precision-timed applications always exhibit better performance as compared to conventional methods. Applications of advanced placement systems, especially banding and side-dressing, provide the best nutrient supply in the root zone and limit losses by leaching or volatilization (Fang et al., 2024).

11.4.4 Integrated Approaches for Enhanced Crop Functionality

In the modern-day science of sustainable agricultural production, fertilizer use as an agricultural device acts as a tool to adjust crop development and metabolism. Present-day Green Revolution varieties of cereal crops, for example, have specific reactions to added fertilizer, and these reactions indicate the intentional development of these cultivars to maximize nutrient use efficiency (Nakachew et al., 2024). However, the universal increase of fertilizer intensity required to achieve productivity is bound to jeopardize the long-term sustainability of the environment, especially in the two processes of soil quality deterioration and intensification of nitrogen pollution.

It follows that the current studies on the criterion focus on scientific research that aims at both optimizing the utilization of nutrients and sustaining crop performance. Under empirical research, it was found that sticking to a carefully balanced system of fertilization has an opportunity to increase the harvest by 30.9%, and, at the same time, to increase the level of nutritional value of cereals, leafy vegetables, and oil culture (Fageria et al. 2008). This twofold advance in both quality and quantity of the yield proves the complexity of the symbiosis between the nutrient provision and the functional properties of crops being planted. These are additionally moderated by time: accuracy-timed fertilizer enhanced routinely exercising greater affirmation than conventional procedures. The method of precision placements that emerge, especially banding and side-dressing, preferably supplying some nutrients to the root zone area with the least leakage and volatilization (Tyagi et al., 2022).

11.5 Nutrient Biofortification: Fertilizers for Addressing Micronutrient Deficiencies

11.5.1 The Hidden Hunger Crisis and Fertilizer Solutions

A new agricultural strategy that addresses the issue of global micronutrient deficiencies (MNDs) affecting billions of individuals is nutrient biofortification. We will use fertilizers specially formulated to plant staple crops and give them a higher supply of minerals and address what has been characterized by scholars as a form of hidden hunger: when people receive an adequate amount of calories, yet they still cannot get the essential amounts of vitamins and minerals (Guardiola-Márquez et al., 2022). The fundamental premise of biofortification with fertilizers is easy to understand, namely specific uses of micronutrient-fortified fertilizers in crop production. One method of enhancing the accessibility of nutrient-rich foods (micronutrient-rich foods) at the farm level would be through agronomic biofortification harnessed by mineral and organic nutrient supplementation, boosting the concentration of micronutrients in the staple crops (Manzeke-Kangara et al., 2023). Biofortification offers a sustainable and cost-effective approach to combating malnutrition (Ofori et al., 2022). Recent researches point out that the method is capable of sufficiently targeting multiple MNDs simultaneously, particularly essential micronutrients such as zinc, iron, and selenium.

11.5.2 Strategic Implementation and Measurable Outcomes

Biofortification measures implemented through fertilizers target MNDs, which more than two billion people across the world grapple with. It is estimated that nearly 2 billion individuals suffer from chronic malnutrition, commonly known as hidden hunger, with its outcomes such as cognitive deterioration, poor growth, and increased mortality (Koç & Karayiğit, 2022). The modern composition of fertilizers is developed to provide crops with necessary micronutrients, including iron, zinc, and selenium, as well as a range of different vitamins, currently in their development stages (Figure 11.5).

Biofortification with fertilizers has a more effective impact than the simple addition of nutrients in fertilizers (Figure 11.5). More than a quarter of the world population is afflicted by such MNDs, which is also known as the hidden hunger. Agronomic biofortification aims at enhancing the concentration of a desired mineral in food crops and supplementing human mineral dietary intake (Teklu et al., 2023). These require strict timing and nutrient ratio addition of fertilizers in the soil, and favorable soil conditions to enhance healthy intake and availability of the micronutrients in any harvested foods (Figure 11.5). Experimental research proves that various application modes are highly effective in biofortification. Studies conducted in Ethiopia have identified that a combination of soil and foliar

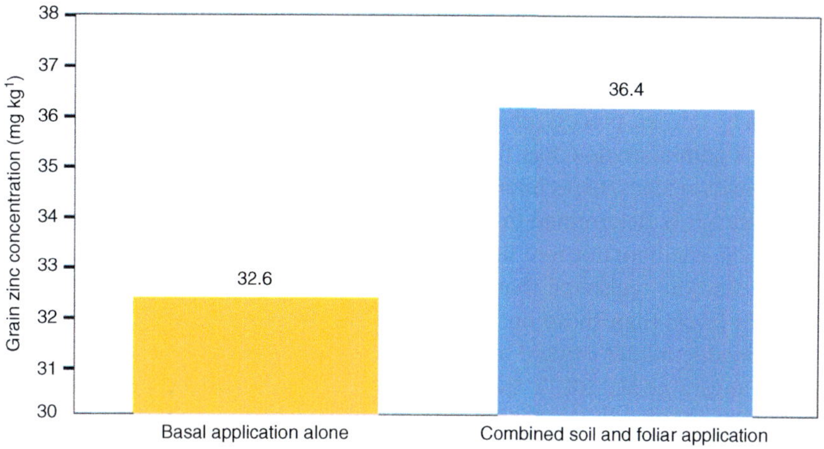

Figure 11.5 Effect of application method on grain zinc concentration, showing higher zinc levels with combined soil and foliar application compared to basal application alone.

zinc application provided better grain micronutrient levels compared to applying a single application (Manzeke-Kangara et al., 2023). Combining both basal and foliar zinc fertilizers resulted in a mean grain zinc concentration of 36.4 $mg\,kg^{-1}$, which is an improvement over treatment with sole basal zinc, where levels of 32.6 $mg\,kg^{-1}$ were achieved—a significant increase in the nutritional quality (Figure 11.5).

11.6 Nutrient Management Strategies for Improving Crop Quality

11.6.1 Precision Agriculture and Quality-Focused Nutrition

Present nutrient management processes aim at maximizing the use of fertilizers to increase crop yields and the quality of the produce. Albeit striving to remain highly effective in terms of increasing the yield, the different traditional methods of fertilization do not always prove sufficient to respond to the rising demand for nutrient-dense foods in the processes of global food systems. Precision fertilization techniques that consider crop requirements, soil variability, and environmental conditions are therefore the basis of modern nutrient management. The use of fertilizer has succeeded in 30.9% (CI: 28.2–33.7%) in increasing the yield of crops on average and enhancing the nutritional value of all nutritionally relevant elements (Ishfaq et al., 2023). These integrated strategies show how proper management of fertilizer could be used to achieve both productivity and nutritional aims.

11.6.2 Integrated Systems and Site-Specific Management

The practicable mixture of inorganic and organic fertilizer sources would be an especially worthwhile direction on the way to increasing the value of crops. The modern research proposes a combined approach that can bring chemical and organic fertilizers together with the possibility of supporting sustainable agricultural practices (Sande et al., 2024). This type of a combined system will take advantage of the immediate nutrient availability and convenience unique to synthetic fertilizers, combined with the soil-health provision and the long, steady release process characteristics of organic amendments. On-site-dependent fertilizer administration has further developed into an important aspect of quality-oriented nutrition plans. Experimental data are supportive when the role of decomposition in the landscape is determined to define the decisive impact on the effectiveness of fertilization and the resultant nutrient-uptake regimes. A study done in the highlands of Ethiopia, as an illustration, registered that plots with foot slope positions recorded more grain micronutrient levels than those on hillslope or in the mid-slope area, which was explained by soil organic matter content and preferred nutrient availability at lower altitudes (Manzeke-Kangara et al., 2023). Even the scheduling and technique of fertilizer application are decisive factors in the quality of the crops. Micronutrient foliar sprays are effective in increasing grain content of minerals; however, field limitations—such as crop height, cost, and the possibility of leaf burn—need to be carefully considered before adoption in a smallholder system of production is achieved. The controlled manipulation of macronutrient inputs can also be considered vital. It was demonstrated empirically that nitrogen application can regulate micronutrient concentrations in grain, with reduced rates of nitrogen often leading to increased grain concentrations of zinc and other micronutrients. The discovery highlights the intricate interaction between nutrients and the requirements of balanced fertilization regimens.

11.7 Impact of Fertilizer on Plant Secondary Metabolites

11.7.1 Bioactive Compounds and Nutritional Enhancement

Plant secondary metabolites represent a conglomeration of chemical components, which do not directly participate in the primary process of growth and development but are essential to defensive systems and have high nutritional and medicinal importance to be consumed by human beings. The cultivation processes, such as the crop-fertilization process, substantially affect the production and concentration of these bioactive compounds. The secondary metabolites form a diverse category, including the phenolic compounds, alkaloids, terpenoids, and flavonoids. These types of compounds promote antioxidant properties, give agricultural products very specific flavors, and provide health benefits to the products. Such comprehensive awareness about the impacts of fertilizer management on secondary metabolite formation is thus needed to optimize the productivity and nutritional profiles of the crops.

11.7.2 Macronutrient Effects on Metabolite Biosynthesis

The fertilization with nitrogen has complex effects on the secondary metabolites. Under most conditions, moderate levels of applied nitrogen trigger the appearance of biosynthesized secondary metabolites that accumulate in the amino acid pathway and may reduce concentrations of the carbon-based products like phenolics. This has led to the metabolic trade-offs, which the plants have to strike between the processes that promote their growth and the processes that enhance their defense. Availability of phosphorus also has a dramatic impact on secondary metabolite patterning linked with energy metabolism and stress response. Sufficient phosphorus nutrition is the basis of the formation of substances important in the transfer of cellular energy and is likely to increase the concentrations of some flavonoid substances and other phenolics that enhance plant stress resistance and quality.

11.7.3 Micronutrient Catalysis and Targeted Enhancement

Micronutrient fertilization, especially zinc, iron, and manganese fertilization, has been demonstrated to substantially increase secondary metabolite synthesis. These nutrients act as coenzymes to the wide and diverse group of enzymes that intercede in primary and secondary metabolic pathways. As an example, zinc deficiency significantly decreases the enzyme activity involved in the biosynthesis of phenolic compounds, and this leads to low antioxidant properties of the harvested products. The interaction between the fertilization strategy and the environmental stress has a determining effect on secondary metabolite generation. Under moderate environmental stress, plants often have increased levels of protective secondary metabolites when subjected to suitable nutrition. Such a relationship indicates that balanced fertilizer practices can have the potential to improve the production of health-promoting compounds, as well as sustain crop productivity. Organic fertilizers have also proven themselves to be especially effective in boosting the level of secondary metabolites in crops. This is because the slow-release nature of nutrients derived via organic sources and the various bioactive features of organic amendments have the potential to activate secondary metabolism pathways of biosynthesis. Moreover, the organic fertilizers provide favorable soil microorganisms that can affect plant metabolism and the synthesis of secondary compounds. The time of fertilizer application, depending on the plant development state, very significant impact on the accumulation pattern of secondary metabolites. ANPs are used in applications that match the times when secondary metabolites are being actively made and, therefore, are in higher concentration in the products harvested. These development relationships are vital in providing insights on what has to be done to optimize fertilization in such a way that crops' nutritional and functional characteristics are enhanced. Recent studies also show that some secondary metabolic profiles can be regulated through specific fertilization strategies. For example, selenium fertilization not only fixes the selenium deficiency of humans but also amplifies the generation of selenium-enclosed ameliorates, together with other organo-selenium compounds that bear the great antioxidant qualities. In the same way, silicon supplements can enhance the yield of phenolic compounds and enhance the plant tolerance to stresses, and also might have the nutraceutical potential in enhancing harvest products.

11.8 Precision Fertilization: Tailoring Nutrient Supply for Optimal Quality

11.8.1 Introduction to Precision Fertilization

To start with a critical one, precision fertilization constitutes a paradigm shift currently underway within the domain of contemporary agronomy and highlights the strategic use of nutrients based on site-specific crop needs and soil conditions (Xing et al., 2024). This novelty has already taken up a leading position in the debate about sustainable agriculture since it pursues the goals of both increasing the yield and quality of crops and the rational use of resources and reducing the environmental consequences of agricultural activities (Lu et al., 2022). The single principle that governs precision fertilization is the tailored application of nutrients to specific field zones, consequently increasing nutrient use efficiency, and, accordingly, augmenting crop quality in general.

11.8.2 The 4R Principle in Precision Fertilization

In modern agronomic literature, precision fertilization has its focal point revolving around the so-called 4R principle, namely, right source, right rate, right time, and right place (Figure 11.6). This expression describes different interactions of soil, plants, and other broader ecosystem components as established in various research articles, like those executed by Maestrini and Basso (2018). Ensuring that the 4R framework is at the center of managerial decisions, practitioners strive to achieve two goals at the same time: not only to increase nutrient uptake but also to reduce losses and pollution (Figure 11.6). That way, fertilizer applications become economically viable and ecologically safe, and ultimately increase the quality of crops by guaranteeing nutrient availability throughout the growing phase (Xing et al., 2024).

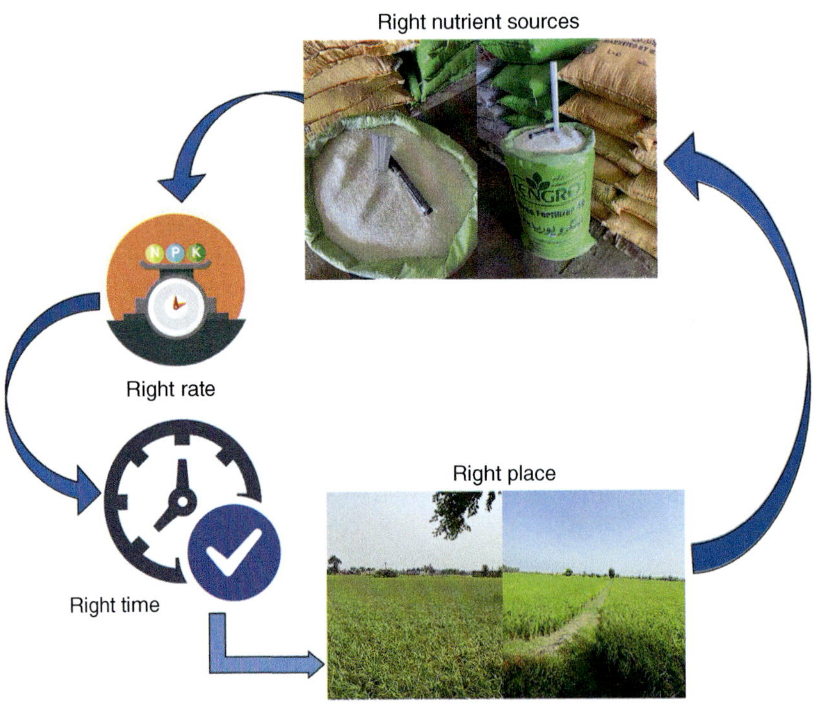

Figure 11.6 Principles ensuring fertilization effectiveness through proper time and applications.

11.8.3 Variable Rate Technology (VRT) Applications

VRT has radically transformed precision fertilizing in that agricultural manufacturers and manufacturers can now carefully apply various amounts of fertilizer to the diverse areas of a field as per the crop type and field requirements of the specific region (Pawase et al., 2024). It enables the optimization of fertilizer application whereby the correct quantity and form of nutrients are applied to every section of a field based on soil-test information, yield maps, and satellite- or drone-based sensors, thus identifying spatial variability in the nutrient levels in the soil and crop health (Tenreiro et al., 2023) (Figure 11.7). Modern-day all-embracing assessments show that VRT is capable of increasing the crop yield by 15–20% at the same time reducing the total capital investment by 25–30% and improving the level of operational efficiency by 20–25% by implementing machine-learning algorithms and AI-powered crop monitoring systems (Pawase et al., 2024) (Figure 11.7).

11.8.4 Data-Driven Nutrient Management

Recent precision-fertilization systems are based on a strong data-collection and analytical base to guide fertilizer application (Xing et al., 2024). New tools and techniques are used in precision agriculture to capture, analyze, and make sense of various information flows that improve crop-disease management practices and increase productivity and quality, as well as decrease input, particularly fertilizer, reliance (Shi et al., 2020) (Figure 11.8). These systems combine soil-testing, plant-tissue, yield mapping, and remote-sensing information to create detailed nutrient-management strategies that closely match emerging crop nutrient needs and physiologies (Figure 11.8).

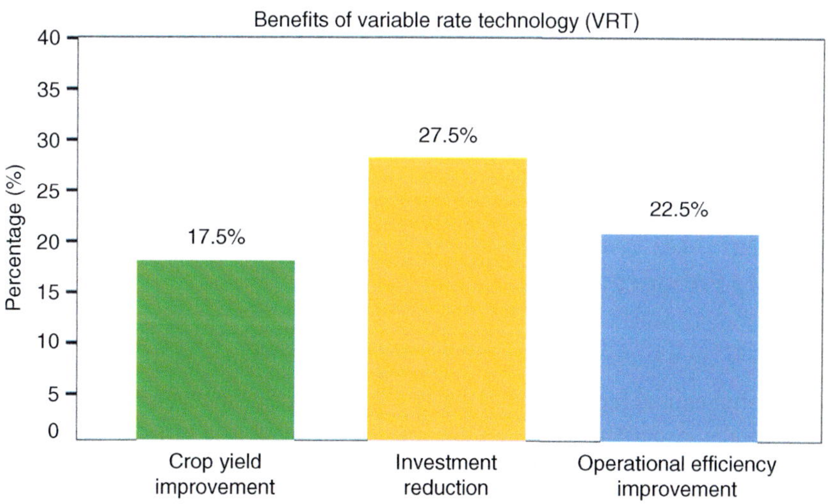

Figure 11.7 Benefits of variable rate technology in crop yield, investment, and operational efficiency.

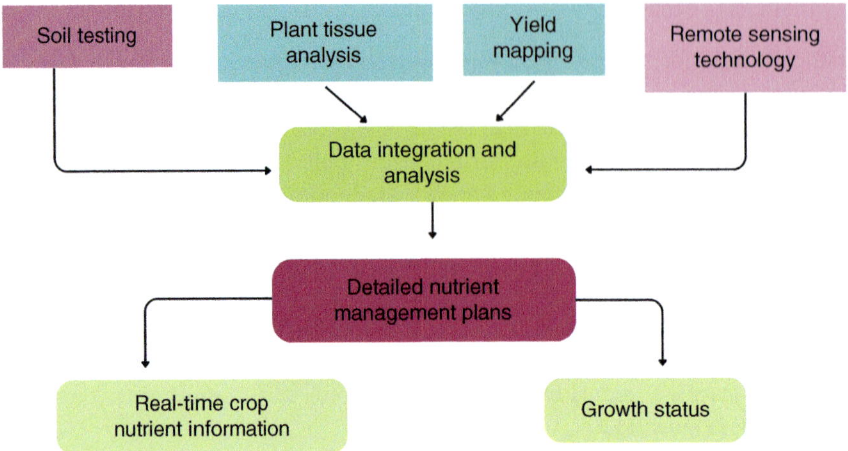

Figure 11.8 Framework for developing nutrient management plans using integrated data sources and analysis.

11.8.5 Environmental and Economic Benefits

The use of precision fertilization measures would be an effective tool for improving environmental stewardship and economic productivity (Figure 11.9). When used in conjunction with spatially variable nutrient management plans, these strategies produce several benefits. The most observed is the ability to regulate fertilizer application using a variable-rate application, hence, maximizing net returns, especially on farms with uneven topography and variable water regime, where evapotranspiration diversities lead to unique yield potential (Tenreiro et al., 2023). With nutrient supply precision to match the localized crop requirement needs, the farmers can reduce losses of fertilizer through the fertilizer wastes, reduce nutrient leaching through runoff, raise nutrient utilization, such as efficiency, crop quality, reduce the cost of production, and achieve optimal input–output productivity ratios (Shi et al., 2020) (Figure 11.9).

11.9 Environmental and Health Impacts of Fertilizer-Induced Quality Enhancements

11.9.1 Environmental Degradation and Ecosystem Disruption

A systematic use of chemical fertilizers in the quest to increase crop levels has brought about environmental consequences that are threatening to reach the brink of disaster. Fertilizers recurrent application of fertilizer changes the pH of the soil, increases the ability of a soil to resist pests, and promotes the acquisition of acid crusts, which in turn reduces organic matter and limits the growth of plants (Chittora et al., 2023). These transformations detrimentally affect the soil-structure integrity in the agricultural ecosystem, and this leads to the development of sustainability issues, which in the long

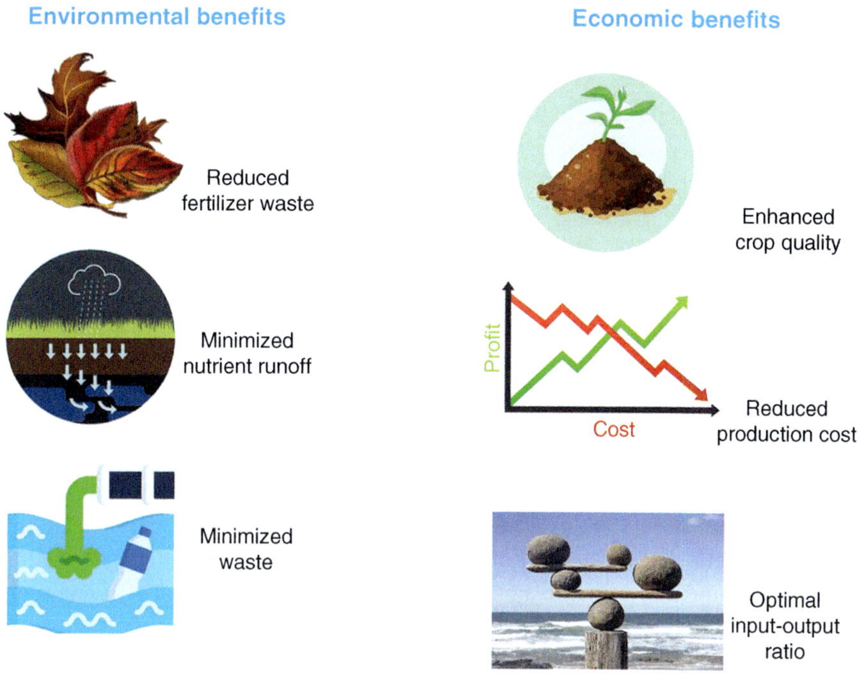

Figure 11.9 Dual benefits of precision fertilization and optimization of inputs and outputs. *Source:* Dimitri Otis / Pixels / Public domain.

term affect both the biodiversity and the soil vitality. Fertilizer drainage also imposes significant effects on water systems. The resulting eutrophication process takes the form of algae growth, oxygen depletion in surface water, transmission of pathogens and nitrates into drinking water, and other foul smells and gases into the air (Berg, Meehan, & Scherer, 2023). Such impacts disrupt aquatic food chains and create dead zones where marine life is absent, thus reorganizing not only ecosystem processes but also water biodiversity in watersheds.

11.9.2 Climate Change and Greenhouse Gas Emissions

Fellow employees, we are here in the first place to agree that the fertilizer use is another prominent source of worldwide greenhouse gas emissions through various metabolic pathways. Overdependence on the reactive nitrogen-rich compounds imposes a threefold burden on the soil, water, and air, where fertilizer-based nutrients enter surface water systems, are released to the troposphere, and get sorbed to pollutants in the particulate matter, leading to the lack of environmental sustainability (Tyagi et al., 2022). This dual-emission difference poses a great challenge to the climate-mitigation plans in the agronomic production system. The regional imbalances of fertilizer use inflate the environmental effects mentioned above. Even though extensive optimization of fertilizer application over the past decades significantly increased worldwide food security, the current literature records ineffective patterns of application and strong regional variations, which yield harmful environmental consequences and mineral imbalances in soils (Penuelas et al., 2023). Policy-wise, such results highlight the urgency to incorporate environmental externalities into fertilizer sustainability calculations in both a macro and micro levels of governance.

11.9.3 Water Quality Deterioration and Contamination

One of the most intractable environmental health issues that has been associated with high rates of fertilizer usage is contamination of groundwater. Specifically, the issue of nitrate contamination, which is of growing international importance, comes with grave responsibilities concerning both public and environmental health due to the ease with which soils can transport nitrate ions into the groundwater, with this being a source of long-term contamination (Islam et al., 2025). The substances introduced are slow-moving through soil layers, thus making the problem resistant to remediation and having lasting effects on whole communities, at least in agricultural regions. This dilemma is gradually increasing in terms of its geographical coverage, especially in developing countries, where fertilizer use has increased without proper monitoring systems. The persistent growth of nitrate contamination of ground and surface waters is why it is urgent to address both the environmental and health consequences of this process.

11.9.4 Human Health Implications

First-hand contact with the fertilizer-related pollutants represents a significant community-health risk, especially to susceptible groups. The empirical fact-based evidence proves that significant health-related consequences, such as methemoglobinemia (blue baby syndrome), associated gastrointestinal and neurological diseases, and the increased risks of cancer, are correlated with high concentrations of nitrate in drinking water (Chaudhary et al., 2025). The most vulnerable groups, including fetuses and pregnant women, as well as infants, are most prone to experiencing these impacts. Pathophysiologically, methemoglobinemia is the most severe expression of nitrate toxicity, and the effects are particularly severe in infants, at less than six months of age, whose gastrointestinal systems are too immature to adequately break down these compounds. On a global scale, the tendency toward a continuous rise of nitrate residues in foodstuffs and water, which has been observed, should be drastically curbed at the moment by authorities in the field of public health and policymakers in the agricultural industry.

11.10 Future Directions: Advancing Fertilizer Technologies for Nutritional Quality Enhancement

The current debates in the fertilizer design present day are toward the precision of the strategies to promote efficiencies in the use of nutrients, as well as optimization of micronutrients, and reduction of environmental losses. An emerging body of evidence defines some routes to research, with particular focus on the creation of new formulations designed to have controlled-release properties, the combination of nanotechnology with bio-based systems, and the timing of the technologies with precision-agriculture tools. Together, they want to increase agricultural yields and, at the same time, enhance nutritional density.

11.10.1 Enhances Controlled Release Formulations

CRFs have been engineered to align nutrient availability with the corresponding plant nutrient uptake cycle. As Mansouri et al. (2023) report, stimuli-responsive hydrogels and metal-organic frameworks now afford the capacity to detect environmental signals such as soil moisture and to release nutrients in tandem. Nutrient-supplying fertilizers that can support nutrient supply during the vegetative and grain-filling stages have consistently increased grain nutrition with minimal loss of nutrients by volatilization. Disintegrating hydrogel materials upon the initiation of root exudates or a change of pH offer a potentially attractive direction toward the precision of the site of application. Current evidence indicates that CRFs achieving an initial release rate of less than 15% and a final release rate that exceeds 75% within 40–360 days represent the most effective profile for practical field deployment (Mansouri et al., 2023).

11.10.2 Integration of Nanotechnology and Bio-based Inputs

Fellow researchers, I would like to frame our discussion by recent developments in the study of nanofertilizers. The literature (Nazir et al., 2024; Zou et al., 2021) indicates that these advanced nutrient carriers, particulate materials with mean diameters less than 100 nm, significantly elevate uptake and translocation efficiency in cereal crops. A peer-reviewed experiment in Scientific Reports (2024) reported that wheat seeds exposed to nano-zinc oxide achieved a 91% enhancement in zinc use efficiency relative to seedlings treated with bulk zinc sulfate. The excellent adhesive qualities, as well as the controlled diffusion rate of the nanogenic formulation, are also worthy of merit. More importantly, the narrowed down application rates make nanofertilizers substantially cost-effective and environmentally friendly in the context of their traditional analogs.

In parallel, there has been the initiation of an inquiry related to the synergistic mixture with biofertilizer, like beneficial rhizobacteria. Empirical evidence suggests that these mixtures not only intensify root colonization and nutrient mobilization but also preserve microbial diversity within the rhizosphere (Zou et al., 2021). Altogether, the findings reiterate that nanofertilizer technology is a revolutionary process that is bound to balance the nutrient demands and sustainability principles.

11.10.3 Synergy Between Nanotechnology and Controlled-Release System

The present research in soil science also attaches more and more importance to nanostructured CRFs, that is, the fertilizer granules with the nanoparticles being encapsulated. These nanotechnology hybrids combine controlled-release formulations with pinpoint accuracy of the nanomaterials and the longevity of the nutrient delivery provided by traditional CRFs. Mansouri et al. (2023) report that nanocomposite CRFs prepared from silica and biochar matrices afford prolonged nutrient availability while simultaneously sustaining soil structure. The long period of the nutrition release (10–20 days) is particularly useful in crops with a lengthy growing period. The authors also indicate that the process of chemically coating nanoparticles with hydrophobic substances like trimethoxysilane would eliminate leaching and enhance the uptake of micronutrients in grain tissues in the stage of reproduction.

11.10.4 Coupling Fertilizer Innovation with Precision Agriculture

An effective direction toward the development of smart fertilizers is their combination with modern real-time monitoring tools. The precision agriculture platform supported by IoT can rely on soil-based sensor networks and aerial photography to combine into spatial differences in fertilizer application. Nguyen et al. (2024) report that zinc foliar treatments targeted via remote sensing in wheat fields enhanced grain zinc content by 10% relative to uniform application. Provided that the precision delivery creates cuts in excess inputs, it enhances the nutrient consistency of grains. This kind of integration can also be used in predictive modelling of optimum nutrient mixture and application periods relative to particular crop developmental stage, subject to local environmental context. Therefore, formulations in the future will be more calibrated to coincide with points in crop development. For example, selenium biofortification in rice is optimized during the grain-filling stage, whereas wheat demonstrates a superior response to early root-zone zinc application followed by foliar top-ups (Nazir et al., 2024). It has been proposed that precision is achieved through smart carriers, through which nutrient release occurs when and where it is biologically required, for example, enzyme-responsive nanoparticles that degrade in the presence of specific root exudates. This intervention minimizes the risk of crop contamination, besides reducing the risks of environmental pollution through reduced residual nutrients.

11.10.5 Soil-Health-Compatible Nanocarriers

One of the most important questions that can be pursued nowadays in agricultural science relates to the construction of nano-delivery systems that would be synchronized with the activity of soil microbial communities. Recent investigations involving mesoporous silica nanoparticles loaded with iron and zinc demonstrate negligible perturbation of beneficial rhizobacterial populations within the rhizospheres of maize and barley (Chaudhary et al., 2025). Chitosan and cellulose are biodegradable coverings that enhance the degradation of the nanoparticles so that they can be absorbed into the soil matrices gradually. Such an arrangement conserves nutrient supply, as well as the integrity of microbes, and contributes to long-term soil fertility. The overall impact of all these interventions has been the manufacture of micronutrient fortified foodstuffs and making them safe to consume by people. Studies by Nature Scientific Reports (2024) further demonstrate that smart fertilizer systems have yielded marked enhancements in protein concentration, iron bioavailability, and phytochemical content of grains and vegetables. These progresses hold great promise in reducing the effects of hidden hunger, where there exists an imbalance of micronutrients that plague populations that rely on staple cereals.

11.11 Conclusion: Fertilizers as Key Drivers of Agricultural and Nutritional Transformation

In recent research works, the value of fertilizers has moved beyond their traditional attribution of a productivity driver; fertilizers have been considered a mandatory tool to achieve victory over food input as well as food content. New fertilizer developments are rejuvenating the agricultural performance by catalyzing the precise, energy-efficient, and physiological nutrient deposition. This chapter looks into several scientific solutions that are indicative of how the next generation of fertilizers can lead to the reduction of the problems associated with sustainable farming, without at the same time compromising the nutritional value of crops. CRFs present a huge step ahead in nutrient management, allowing a controlled release of nutrients to be applied at a level to match the dynamics of crop uptake. The formulations reduce the number of applications, minimize losses of nutrients, and increase the efficiency of nutrient uptake. The advantages are demonstrated by new-generation controlled-release systems, hydrogels that are eco-friendly and capable of prolonging nutrient availability in concert with simultaneous moisture retention, thereby facilitating superior nutrient partitioning in grain tissues during critical growth phases (Song et al., 2024). Parallel advances center on stimuli-responsive coatings that react to soil moisture cues or root exudates to release nutrients precisely when crop demand is highest (Mansouri et al., 2023). A new aspect of innovation is nanotechnology. The nano-formulated fertilizers, mainly zinc oxide nanoparticles, enhance nutrient uptake by expanding the surface area and intracellular displacement. A recent study revealed that wheat crop conditioned with nano-ZnO showed markedly enhanced zinc content and superior morphological traits relative to plants treated with conventional zinc sulfate (Nazir et al., 2024). The increased absorptive power, as explained by the study, has a massive potential to improve the MND issue worldwide.

The importance of beneficial soil microbes in the productivity of plants cannot be underestimated. Arbuscular mycorrhizae and rhizobacteria observe nutrient status and, in case of nutrient-limited conditions, they become solubilizers and mobilizers of nutrients. This association, when used together with advanced fertilizers, gives a synergetic effect that increases nutrient availability and the ability to withstand stress in growth environments. This plan is especially beneficial in low or eroded soils where the traditional fertilizers tend to be less productive. In recent years, it has managed to gain attention due to the variety of roles that these hydrogel-based carriers play (they moderate the rate at which nutrients are discharged and increase water use efficiency, a factor that is imperative in drought-prone locations). Such hydrogels are generally biologically degradable with little ecotoxicological impact, thus being in line with the current sustainable use of inputs. Song et al. (2024) reported that starch-based hydrogels loaded with nutrients catalyzed improved plant growth and structural enhancements in the rhizosphere. Their higher capacity to retain water stabilizes nutrient transport through delicate phases of plant development, thus enhancing the concentration of nutrients in the tissues, which can be eaten. The case with nano-biofertilizers, in which nanomaterial platforms are combined with bioinoculants, is a similar development. Such a hybrid formulation promotes a strong colonization of the microbes on the rhizosphere and also enhances the bioavailability of crucial nutrients, including phosphorus and iron. Mirbakhsh (2023) underscored that nano-formulation strategies can be engineered to liberate nutrients in response to plant physiological cues, thereby elevating nutrient use efficiency while curbing environmental contamination.

Advances in fertilizer technology form part of the essential elements in climate-resistant agriculture. Synergistan implementation of nano-urea, for example, in combination with conservation couples like zero tillage, significantly increases the efficiency with which the nitrogen is used and at the same time, results in the reduction of greenhouse-gas emissions, which is the simultaneous achievement of both productivity and sustainability. In addition, this is a more effective way of reducing the unnecessary use of nutrients and runoff of fertile nutrients that tend to be used in the ordinary system of farming. These breakthroughs have a direct positive impact on human nutrition, increasing the micronutrient content of staple crops: rice, wheat, and maize, thus reversing the problem of so-called hidden hunger, in which over two billion people suffer globally. Recent research has found that application of improved fertilizers on these crops has seen a measurable increase in zinc and iron absorption, thus opening an opportunity for an extensive intervention in nutrition-sensitive agriculture. To conclude, the fertilizer innovations cannot be ignored with regard to the twofold goals of improving agricultural yields and strengthening human health. Advancements, including but not limited to controlled release materials, nano-enabled carriers, bio-based compositions, and microbial incorporations, are developed to maximize the efficiency of nutrient use, mitigate environmental distress, and improve the micronutrient amount of food crops. These shifts have foreseen a positive course to a world where farming serves not only the needs of food security but also the wider themes of health and sustainability for the whole population.

11.11.1 Conclusion

The fertilizers are an irreplaceable force of agricultural and nutritional revolution, changing in a new form and becoming a highly advanced instrument of food improvement and safety. Higher rates of fertilizer technology, which include controlled-release ones, integration of nanotechnology, and bio-based ones, are transforming nutrient delivery to maximize its effectiveness and reduce environmental impact. Such innovations can play a fundamental role in meeting worldwide food needs and in countering the problem of hidden hunger by drastically enhancing the nutritional quality of staple crops. Sustainable farming of the future depends on the exploration and upcoming usage of these intelligent, precision-targeted fertilizer plans, which would make the earth healthier and make the world better-nourished.

References

Abdoli, M. (2020). Effects of micronutrient fertilization on the overall quality of crops. In: *Plant micronutrients: Deficiency and toxicity management* (ed. T. Aftab and K.R. Hakeem), 31–71. Cham: Springer International Publishing.

Ali, A., Jabeen, N., Farruhbek, R. et al. (2025). Enhancing nitrogen use efficiency in agriculture by integrating agronomic practices and genetic advances. *Frontiers in Plant Science* 16: 1543714. https://doi.org/10.3389/fpls.2025.1543714.

Arora, P.K., Tripathi, S., Omar, R.A. et al. (2024). Next-generation fertilizers: The impact of bionanofertilizers on sustainable agriculture. *Microbial Cell Factories* 23 (1): 254. https://doi.org/10.1186/s12934-024-02485-1.

Avnee, S., Sood, S., Chaudhary, D.R. et al. (2023). Biofortification: An approach to eradicate micronutrient deficiency. *Frontiers in Nutrition* 10: 1233070. https://doi.org/10.3389/fnut.2023.1233070.

Barłóg, P., Grzebisz, W., and Łukowiak, R. (2022). Fertilizers and fertilization strategies mitigating soil factors constraining efficiency of nitrogen in plant production. *Plants* 11 (14): 1855. https://doi.org/10.3390/plants11141855.

Beig, B., Niazi, M.B.K., Sher, F. et al. (2022). Nanotechnology-based controlled release of sustainable fertilizers: A review. *Environmental Chemistry Letters* 20 (4): 2709–2726. https://doi.org/10.1007/s10.11-022-01409-w.

Berg, M., Meehan, M., and Scherer, T. (2023). *Environmental implications of excess fertilizer and manure on water quality*. NDSU Agriculture Extension Publications.

Bouis, H.E. and Saltzman, A. (2017). Improving nutrition through biofortification: A review of evidence from HarvestPlus, 2003 through 2016. *Global Food Security* 12: 49–58. https://doi.org/10.1016/j.gfs.2017.01.009.

Bouis, H., Foley, J., Lividini, K. et al. (2024). Biofortification: Future challenges for a newly emerging technology to improve nutrition security sustainably. *Current Developments in Nutrition* 8 (12): 104478. https://doi.org/10.1016/j.gfs.2017.01.009.

Chaudhary, I.J., Chauhan, R., Kale, S.S. et al. (2025). Groundwater nitrate contamination and its effect on human health: A review. *Water Conservation Science and Engineering* 10 (1): 33. https://doi.org/10.1007/s41101-025-00359-y.

Chen, Y., Zhang, B., and Annis, G.B. (2024). Plant biotechnology and genetics for sustainable agriculture and global food security. *Frontiers in Plant Science* 15: 1479632. https://doi.org/10.3389/fpls.2024.1479632.

Chittora, D., Parveen, T., Yadav, J. et al. (2023). Harmful impact of synthetic fertilizers on growing agriculture and environment. *Global Journal of Pharmaceutical Sciences* 11 (1): 555804.

Fageria, N.K., Baligar, V.C., and Li, Y.C. (2008). The role of nutrient efficient plants in improving crop yields in the twenty first century. *Journal of Plant Nutrition* 31 (6): 1121–1157. https://doi.org/10.1080/01904160802116068.

Fang, J., Chen, B., Wang, F. et al. (2024). Nitrogen, phosphorus, and potassium co-limitation in terrestrial ecosystems: A global meta-analysis. *Plants, People, Planet* 6 (6): 1329–1340. https://doi.org/10.1002/ppp3.10563.

Fu, Y., Mason, A.S., Song, M. et al. (2023). Multi-omics strategies uncover the molecular mechanisms of nitrogen, phosphorus and potassium deficiency responses in *Brassica napus*. *Cellular & Molecular Biology Letters* 28 (1): 63. https://doi.org/10.1186/s11658-023-00492-4.

Govil, S., Long, N.V.D., Escriba-Gelonch, M., and Hessel, V. (2024). Controlled-release fertiliser: Recent developments and perspectives. *Industrial Crops and Products* 219: 119160. https://doi.org/10.1016/j.indcrop.2024.119160.

Guardiola-Márquez, C.E., Santos-Ramírez, M.T., Segura-Jiménez, M.E. et al. (2022). Fighting obesity-related micronutrient deficiencies through biofortification of agri-food crops with sustainable fertilization practices. *Plants* 11 (24): 3477. https://doi.org/10.3390/plants11243477.

Ishfaq, M., Wang, Y., Xu, J. et al. (2023). Improvement of nutritional quality of food crops with fertilizer: A global meta-analysis. *Agronomy for Sustainable Development* 43 (6): 74. https://doi.org/10.1007/s13593-023-00895-x.

Islam, A., Sarkar, B., Shit, P.K., and Gazi, H.A.R. (2025). Groundwater nitrate contamination in agroecosystems: A comprehensive review of detection methods, pollution mechanisms, and policy implications. *Progress in Physical Geography: Earth and Environment* 49 (3): 281–317. https://doi.org/10.1177/03091333241233955.

Kaushal, S. (2024). Harmonizing agriculture: Nurturing soil through micronutrients management and texture dynamics. *International Journal of Plant & Soil Science* 36 (6): 188–199.

Koç, E. and Karayiğit, B. (2022). Assessment of biofortification approaches used to improve micronutrient-dense plants that are a sustainable solution to combat hidden hunger. *Journal of Soil Science and Plant Nutrition* 22 (1): 475–500.

Kozłowska, A. and Szostak-Węgierek, D. (2022). Targeting cardiovascular diseases by flavonols: An update. *Nutrients* 14 (7): 1439.

Lam, S.K., Wille, U., Hu, H.W. et al. (2022). Next-generation enhanced-efficiency fertilizers for sustained food security. *Nature Food* 3 (8): 575–580. https://doi.org/10.1038/s43016-022-00542-7.

Liu, J., Wang, D., Yan, X. et al. (2024a). Effect of nitrogen, phosphorus and potassium fertilization management on soil properties and leaf traits and yield of *Sapindus mukorossi*. *Frontiers in Plant Science* 15: 1300683.

Liu, Y., Lan, X., Hou, H. et al. (2024b). Multifaceted ability of organic fertilizers to improve crop productivity and abiotic stress tolerance: Review and perspectives. *Agronomy* 14 (6): 1141. https://doi.org/10.3390/agronomy14061141.

Lu, Y., Liu, M., Li, C. et al. (2022). Precision fertilization and irrigation: Progress and applications. *AgriEngineering* 4 (3): 626–655. https://doi.org/10.3390/agriengineering4030041.

Jiaying, M., Tingting, C., Jie, L., Weimeng, F.U., Baohua, F., Guangyan, L., Hubo, L., Juncai, L., Zhihai, W., Longxing, T. and Guanfu, F. (2022) 'Functions of nitrogen, phosphorus and potassium in energy status and their influences on rice growth and development', *Rice Science*, 29(2),166–178. https://doi.org/10.1016/j.rsci.2022.01.005

Maaz, T.M., Dobermann, A., Lyons, S.E., and Thomson, A.M. (2025). Review of research and innovation on novel fertilizers for crop nutrition. *npj Sustainable Agriculture* 3 (1): 25. https://doi.org/10.1038/s44264-025-00066-0.

Madhupriyaa, D., Baskar, M., Sherene Jenita Rajammal, T. et al. (2024). Efficacy of chelated micronutrients in plant nutrition. *Communications in Soil Science and Plant Analysis* 55 (22): 3609–3637.

Maestrini, B. and Basso, B. (2018). Drivers of within-field spatial and temporal variability of crop yield across the US Midwest. *Scientific Reports* 8 (1): 14833. https://doi.org/10.1038/s41598-018-32779-3.

Mansouri, H., Ait Said, H., Noukrati, H. et al. (2023). Advances in controlled release fertilizers: Cost-effective coating techniques and smart stimuli-responsive hydrogels. *Advanced Sustainable Systems* 7 (9): 2300149.

Manzeke-Kangara, M.G., Amede, T., Bailey, E.H. et al. (2023). Landscape and micronutrient fertilizer effect on agro-fortified wheat and teff grain nutrient concentration in western Amhara. *Agronomy* 13 (10): 2598.

Martínez-Moreno, F., Igartua, E., and Solís, I. (2017). Barley Types and Varieties in Spain: An Historical Overview. *International Journal of Agriculture and Natural Resources* 44: 12–23. https://doi.org/10.7764/rcia.v44i1.1638.

Mirbakhsh, M. (2023). Role of nano-fertilizer in plants nutrient use efficiency (NUE)—a mini-review. *arXiv preprint* arXiv:2305.14357.

Nakachew, K., Yigermal, H., Assefa, F. et al. (2024). Review on enhancing the efficiency of fertilizer utilization: Strategies for optimal nutrient management. *Open Agriculture* 9 (1): 356.

Nazir, M.A., Hasan, M., Mustafa, G. et al. (2024). Zinc oxide nano-fertilizer differentially affects morphological and biochemical traits and enhances zinc biofortification in wheat (Triticum aestivum L.). *Scientific Reports* 14: 63987.

Nguyen, T.T., Buresh, R.J., and Pham, V.H. (2024). Meta-analysis on smart fertilizer efficiency and soil carbon retention in South and Southeast Asia. *Soil & Tillage Research* 229: 105093. https://doi.org/10.1016/j.still.2023.105093.

Ofori, K.F., Antoniello, S., English, M.M., and Aryee, A.N. (2022). Improving nutrition through biofortification—a systematic review. *Frontiers in Nutrition* 9: 1043655.

Pawase, P.P., Nalawade, S.M., Walunj, A.A. et al. (2024). Comprehensive study of on-the-go sensing and variable rate application of liquid nitrogenous fertilizer. *Computers and Electronics in Agriculture* 216: 108482. https://doi.org/10.1016/j.compag.2023.108482.

Peñuelas, J., Coello, F., and Sardans, J. (2023). A better use of fertilizers is needed for global food security and environmental sustainability. *Agriculture & Food Security* 12 (1): 1–9.

Sande, T.J., Tindwa, H.J., Alovisi, A.M.T. et al. (2024). Enhancing sustainable crop production through integrated nutrient management: A focus on vermicompost, bio-enriched rock phosphate, and inorganic fertilisers—a systematic review. *Frontiers in Agronomy* 6: 1422876. https://doi.org/10.3389/fagro.2024.1422876.

Saurabh, K., Prakash, V., Dubey, A.K. et al. (2024). Enhancing sustainability in agriculture with nanofertilizers. *Discover Applied Sciences* 6 (11): 559. https://doi.org/10.1007/s42452-024-06267-5.

Schaller, J., Webber, H., Ewert, F. et al. (2024). The transformation of agriculture towards a silicon improved sustainable and resilient crop production. *npj Sustainable Agriculture* 2 (1): 27. https://doi.org/10.1038/s44264-024-00035-z.

Shanmugavel, D., Rusyn, I., Solorza-Feria, O., and Kamaraj, S.K. (2023). Sustainable SMART fertilizers in agriculture systems: A review on fundamentals to in-field applications. *Science of the Total Environment* 899: 165542. https://doi.org/10.1016/j.scitotenv.2023.165542.

Shen, M., Liu, S., Jiang, C. et al. (2023). Recent advances in stimuli-response mechanisms of nano-enabled controlled-release fertilizers and pesticides. *Eco-Environment & Health* 2 (3): 161–175. https://pmc.ncbi.nlm.nih.gov/articles/PMC10702921/.

Sheoran, S., Kumar, S., Ramtekey, V. et al. (2022). Current status and potential of biofortification to enhance crop nutritional quality: An overview. *Sustainability* 14 (6): 3301. https://doi.org/10.3390/su14063301.

Shi, Y., Zhu, Y., Wang, X. et al. (2020). Progress and development on biological information of crop phenotype research applied to real-time variable-rate fertilization. *Plant Methods* 16 (1): 11. https://doi.org/10.1186/s13007-020-0559-9.

Song, Y., Ma, L., Duan, Q. et al. (2024). Development of slow-release fertilizers with water-retention function using eco-friendly starch hydrogels. *Molecules* 29 (20): 4835. https://doi.org/10.3390/molecules29204835.

Taklu, D., Gashu, D., Joy, E.J.M. et al. (2023). Effectiveness of agronomic biofortification strategy in fighting against hidden hunger. *Agronomy* 13 (8): 2173. https://doi.org/10.3390/agronomy13082173.

Tenreiro, T.R., Avilés, F., Gómez, J.A. et al. (2023). Opportunities for variable rate application of nitrogen under spatial water variations in rainfed wheat systems—an economic analysis. *Precision Agriculture* 24 (2): 853–878. https://doi.org/10.1007/s11119-022-09977-1.

Tyagi, J., Ahmad, S., and Malik, M. (2022). Nitrogenous fertilizers: Impact on environment sustainability, mitigation strategies, and challenges. *International Journal of Environmental Science and Technology* 19 (11): 11649–11672. https://doi.org/10.1007/s13762-022-04027-9.

Verma, K.K., Song, X.P., Degu, H.D. et al. (2023). Recent advances in nitrogen and nano-nitrogen fertilizers for sustainable crop production: A mini-review. *Chemical and Biological Technologies in Agriculture* 10 (1): 111. https://doi.org/10.1186/s40538-023-00488-3.

Wang, T., Xu, J., Chen, J. et al. (2024b). Progress in microbial fertilizer regulation of crop growth and soil remediation research. *Plants* 13 (3): 346. https://doi.org/10.3390/plants13030346.

Xing, Y. and Wang, X. (2024). Precise application of water and fertilizer to crops: Challenges and opportunities. *Frontiers in Plant Science* 15: 1444560. https://doi.org/10.3389/fpls.2024.1444560.

Xing, Y., Zhang, X., and Wang, X. (2024). Enhancing soil health and crop yields through water-fertilizer coupling technology. *Frontiers in Sustainable Food Systems* 8: 1494819. https://doi.org/10.3389/fsufs.2024.1494819.

Yahaya, S.M., Mahmud, A.A., Abdullahi, M., and Haruna, A. (2023). Recent advances in the chemistry of nitrogen, phosphorus and potassium as fertilizers in soil: A review. *Pedosphere* 33 (3): 385–406. https://doi.org/10.1016/j.pedsph.2022.12.004.

Yousefi, H. and Karimi Douna, B. (2023). Risk of nitrate residues in food products and drinking water. *Asian Pacific Journal of Environment and Cancer* 6 (1): 69–79. http://waocp.com/journal/index.php/apjec/article/view/1036.

Zou, Y.N., Zhang, F., Srivastava, A.K. et al. (2021). Arbuscular mycorrhizal fungi regulate polyamine homeostasis in roots of trifoliate orange for improved adaptation to soil moisture deficit stress. *Frontiers in Plant Science* 11: 600792. https://doi.org/10.3389/fpls.2020.600792.

12

Regulatory Frameworks and Policy Implications for Next-Generation Fertilizers

Muhammad Anas[1,2], *Aliza Falak*[2], *Waseem Ahmed Khattak*[2], *Sudenaz Soylu*[3], *Javed Iqbal*[4], *and Banzeer Ahsan Abbasi*[5]

[1] *Department of Soil Science & Plant Nutrition, Selçuk University, Konya, Türkiye*
[2] *Department of Plant Science, Faculty of Biological Sciences, Quaid-i-Azam University, Islamabad, Pakistan*
[3] *Department of Field Crops, Faculty of Agriculture, Selcuk University, Konya, Türkiye*
[4] *Department of Botany, Bacha Khan University, Charsadda, Pakistan*
[5] *Department of Botany, Rawalpindi Women University, Rawalpindi, Pakistan*

12.1 Introduction

12.1.1 Overview of Fertilizer Regulation

Fertilizer regulation describes the series of policies, codes of conduct, and institutional frameworks aimed at monitoring production, distribution, quality control, and environmental safety of fertilizers (Simtowe & Expert, 2015). The underlying aim of these regulations is to keep the quality of nutrients in fertilizers up to the set standards, ensure the absence of dangerous elements in them, and application methods that do not pose a threat to the well-being of human beings and the environment. Regulatory systems are different in different countries and regions, yet they share some basic features: registration of products, convincing documents, and testing performance. An example is that in India, the Essential Commodities Act prescribes the stimulation of all fertilizer interventions through the Fertilizer Control Order (FCO) (Yadav, 2023). It also fixes the nutrient content limits per brand of fertilizer and requires the samples to be tested as assurance of quality. Environmental agencies like the Environmental Protection Agency (EPA) and the United States Department of Agriculture (USDA) control the use of fertilizers, emphasizing the safety of the environment and agricultural effectiveness in the United States. In the European Union, a unified regulation system that covers conventional and organic fertilizers, biostimulants, and microbial inputs has been introduced by Fertilizing Products Regulation (FPR). Despite these systems, the majority of regulations have been formulated bearing in mind conventional chemical fertilizers, mainly nitrogen (N), phosphorus (P), and potassium (K) compounds (Ashitha & Rakhimol, 2021). Consequently, they tend to be inadequate to meet the special attributes of the new generation of fertilizers. As an example, the nanofertilizers (fertilizers with nanoscale-engineered nutrients) involve altered processes of risk assessment because of unique physical, chemical, and biological relationships with the soil and plants. The same can be said about polymer-coated fertilizers and biofertilizers, which are difficult to classify, test in performance, and continue monitoring after a prolonged period of use in an environment. The manufacturers are confronted with uncertainties regarding approval procedures and the fear of adoption of new products by the farmers without an updated regulatory framework that could adapt itself to these changes (Eastwood & Renwick, 2020).

12.1.2 Importance of Policy in Sustainable Fertilizer Use

The policy is a very essential tool in directing fertilizer application toward sustainability (Ejedegba, 2024). It influences the effort that is done in producing, distributing, and utilizing fertilizers through subsidies, taxes, licenses, educational campaigns, and many others. A well-designed fertilizer policy may reconcile agricultural productivity and environmental protection, resource productivity, and socioeconomic fairness. Among various important contributions of fertilizer policy, an important one can be considered to be handling environmental externalities. Misuse and lack of control on the use of fertilizers add to the concerns about the environment like leaching of nitrates, eutrophication of rivers and lakes, and the release of nitrous oxides, one of the severe greenhouse gases. States such as Denmark, Germany, and Netherland have led in nutrient management practices that include limiting the level of fertilizer application, nutrient budgets of farms, and promotion of environmentally friendly inputs with the help of tax preferences or eco labels. These policy tools minimize the risk of environmental degradation by ensuring better and accurate management of nutrients (Oenema et al., 2003). Besides the stewardship of the environment, policy is significant in promoting innovation. Just to introduce a successful new fertilizer into the market involves a lot of research work, testing, and approval of the product, which may be costly and time-consuming. Policy frameworks with clear and consistent, science-based information will eliminate ambiguity in the minds of the innovators as well as those who invest. Fast-tracked approval mechanisms, pilot exemptions, and government research and trial subsidies will be effective in arguably speeding up the creation and transfer of the next-generation fertilizers (NGFs). Another way in which governments can enhance innovation is by using public–private partnerships, research grants, and agricultural innovation platforms (Nikolaevich et al., 2019).

In addition, policy has an impact on accessibility and fairness in the market. Smallholder farmers are the main producers of agricultural output in most developing countries, and they do not always have access to improved inputs, technical knowledge, and capital. Policies on the use of fertilizers should be inclusive so that technological advancement does not lead to benefiting some parties, namely large-scale agribusinesses. Such tools as fertilizer vouchers, targeted subsidies, microcredit, and farmer field schools have the potential of filling that gap, resulting in the adoption of efficient fertilizers by many farmers (Alawode, 2025). It is also important to have been the case good agricultural extension services to train the farmers about proper and secure use of new fertilizer products. Another imperative consideration is policy coherence. The functioning of fertilizer policy must not be in isolation, but it must be subordinated to large-scale strategies connected with climatic changes, the health of the soil, food security, and local development in the countryside

(Hassan, 2024). As an example, a policy that encourages farmers to utilize more fertilizer to achieve higher yields should also consider the effect this has on soil biodiversity, the quality of water, and land productivity in the long run. Incorporating the fertilizer governance into a comprehensive policy integrated in a systems perspective policy guarantees that the economic, environmental, and social objectives are also simultaneously pursued (Sukprasert & Phadungkit, 2024).

12.1.3 Emerging Needs in the Context of Next-Generation Fertilizers

Emergence of the next generation of fertilizers poses both challenges and benefits that the existing methods of control and policies are not that time to tackle most of the time. The nanofertilizers, microbial consortia, polymer-coated granules, and biodegradable nutrient carriers are generally structured to augment nutrient use efficiency (NUE), decrease the rate of application, and reduce environmental harm (Singh et al., 2025). They are quite new and complicated and should involve entirely different methods of risk evaluation, product typology, and performance testing. The most urgent requirement is the new definitions of regulations. According to the traditional systems of regulations, fertilizers are labeled solely based on their nutrient and chemical composition. This is not a sufficient classification when there is a product that has several functions, as is the case with nanofertilizers or biofertilizers, which can contain chemical, physical, and biological ingredients (El-Ghamry et al., 2018). The regulatory agencies should thus embrace a new classification that incorporates new functions, delivery mechanisms, and environmental interactions. One of the other immediate needs is the formulation of specialized testing and evaluation guidelines. Unlike traditional fertilizers, where nutrient release rates and effects on crops are defined as the primary criteria of their choice, NGFs might demand a different type of testing, which means nanotoxicology, tests of biodegradability, and the impact of fertilizing on the soil microbiome (Arora et al., 2024). The procedures of standardized tests will not only streamline the approval by the regulators but also promote comparability of the product and consumer confidence.

The standardization that builds harmonization across international borders is also important. The absence of standards to form NGFs on the global level acts as a trade barrier, raises the demands of the multinational manufacturers, and hinders the spread of sustainable technologies worldwide. An effort to come up with definitions, testing procedures, and labeling standards accepted globally should be led by institutions such as the FAO, ISO, and Codex Alimentarius (Fortin, 2023). Also, regulatory approval of products that are proven to be well tested in other countries can be enhanced through mutual recognition agreements (MRAs) between countries. The adoption of digital technologies, namely blockchain, the Internet of Things (IoT), and artificial intelligence (AI), in fertilizer regulation can revolutionize traceability, surveillance, and policy enforcement.

In that case, it is possible to revolutionize traceability, surveillance, and enforcement of policies used with fertilizers. As an example, farmers and regulators can be provided with immediate access to a product, composition, origin, and use instructions by implementing smart labels attached to real-time databases. Monitoring tools can also facilitate an overview of nutrient application patterns nationwide or even regionally, so that data may be used to make policy and compliance decisions. Finally, there is a need to have institutional capacity-building. Many regulatory agencies, particularly in developing nations, do not have the infrastructure, expertise, and resources to assess sophisticated fertilizer preparations (Benson & Mogues, 2018). International training programs, research networks, and technical support through international organizations such as global capacity building could augment the capability of national agencies to react to the emerging likelihood of the fertilizer environment. It might also be advisable to develop regional centers of excellence, which would pool their resources and harmonize practices in the countries with similar agroecological conditions (Mishenin et al., 2021).

12.2 Global Regulatory Landscape

The shift toward the globalization of agriculture and the rapidly changing technology in fertilizers requires that the regulatory regime follow suit to guarantee the safety of products on the market, as well as the environment, and equitable access to the marketplace (Pinstrup-Andersen, 2002). The question of fertilizer regulation is no longer arising as a European or a national concern but has been made a global concern, especially about NGFs. These are nanofertilizer, slow- and controlled-release products, and biofertilizer, and all of them have their own special needs that few regulatory regimes are designed to address (Sales et al., 2024). Regulatory approaches and frameworks in the different countries of the world are gradually becoming more diverse—the different countries have a well-developed regulatory framework (which is more of a strict regulation), and others still have fragile and outdated regulations. In this part, study the variations between the developed and the developing countries, how the demands of unified norms are pressing up against each other, and how the international trade regulation and external trade/import policy are influencing the flow of fertilizer products on the global platform.

12.2.1 Developed vs. Developing Country Frameworks

Fertilizers and their regulation have a comparatively different mark in terms of dominance, extent, and level of enforcement among developed and developing nations, as well as the level of incorporating the theme of sustainability and innovation into the regulation process (Ashford & Hall, 2011). Within the developed world, including North America, Europe, and certain regions in East Asia, the regulation of fertilizers is usually extensive and well organized, and their implementation is likely to be taken seriously (Novotny et al., 2010). These systems are sometimes science-founded, having elaborate procedures of product registration, quality assurance, risk analysis, and post-market surveillance mechanisms. The Fertilizing Products Regulation (EU FPR) of the European Union can be described as a progressive regulatory framework covering not only mineral fertilizers but also organic fertilizers and bio-based fertilizers, plant biostimulants, and soil improvers (Athanasoulia et al., 2021). It imposes rigorous testing, impartial conformity assessment, and a harmonized CE clearly marked as being safe and performing in all the member countries. Likewise, fertilizable drugs in the United States are carried out on the federal and state levels. Regulation of fertilizers with pesticidal claims or otherwise potentially hazardous to the environment is regulated by the EPA, and the Association of American Plant Food Control Official (AAPFCO) assists with the harmonization of regulatory state requirements mainly by providing model legislative texts and best practice recommendations (Zhang et al., 2021).

Most developing nations, on the other hand, do not enjoy the kind of institutional and regulatory sophistication. Regulation of fertilizer can mainly be restricted to examination of the basic nutrient content and simple quality assurance (Wang et al., 2008). As such, for instance, in many African and South Asian countries, the enforcement is poor with low laboratory infrastructure, downsizing of staff, and tight budgets. Such lapses lead to the growth of fake or poor-quality fertilizers, hence causing a lack of confidence and productivity among farmers (Michelson et al., 2022). Also, the legislations of most developing nations have yet to be amended that reflect the products of the next generation and resulting in ambiguity of regulations and non-incorporation in the market by innovative fertilizers. Moreover, fertilizer subsidy and its distribution in most developing countries may be regulated by the state, leading to bureaucratic bottlenecks and political disturbances. Such systems tend to increase or even abuse the use of some types of fertilizer, mainly nitrogen, and restrict the use of balanced and environmentally correct inputs. In this regard, regulatory reform is not limited to product safety but also strives to establish an enabling environment to integrate soil fertility management (ISFM) and climate-smart agriculture. To overcome this gap, there is a rising demand for international collaboration, capacity building, and harmonization initiatives by donors (Michelson et al., 2022). International bodies and countries with developed economies can greatly contribute to aiding the developing nations through technical training, laboratory equipment, policy making, and online regulation applications. Finally, it is necessary to have a fair and equitable development of regulatory potential so that farmers worldwide can have access to fertilizers that can be characterized as safe, effective, and sustainable (Pretty, 2008).

12.2.2 Harmonization of Standards

In a globalized economy, the problem of harmonization of fertilizer laws among nations and regions is becoming of great necessity. Since fertilizer products, particularly new generation products, have a transborder level, the lack of uniformity in the national regulation translates into significant barriers to trade, high costs of compliance, and suppression of innovation. Harmonization has been coined to mean bringing preferences such as definition, classification rules, testing techniques as well as labeling norms to a mutually agreed common ground to enable mutual recognition and eliminate fragmentation of regulations (Nakagawa, 2011). Regional alliances and international organizations provided the initiative in recent years in the direction of harmonization. The East African Community (EAC) has come up with a harmonized fertilizer regulatory regime in a bid to guarantee the quality of the product as well as promote regional trade (Turley et al., 2022). In the same manner, Economic Community of West African States (ECOWAS) has come up with regional fertilizer regulations through which they harmonize the specifications of the major fertilizer types and encourage cross-recognition of fertilizer authorizations. Such frameworks are of special significance to the problem of fake fertilizers, most of which find a place in

African markets because of poor national control and divided regulation. On international level, there are organizations like Food and Agriculture Organization (FAO), International Fertilizer Association (IFA), and Codex Alimentarius Commission that are engaged in the development of voluntary international codes and technical standards (van der Meulen & Wernaart, 2019). These are to streamline governments of countries to follow harmonized and scientifically sound regulations. Another critical platform is the ISO Technical Committee 134 on fertilizers and soil conditioners, which offers a standardized approach to sampling and testing as well as labelling of products that are the same and reliable in different jurisdictions. Nonetheless, harmonization is going through a number of obstacles. First, they are not institutionally aligned (the countries are not compatible in terms of their administrative organization, regulatory ideology, and political priorities). Second, these regulatory sovereignties are not readily willing to surrender independence in regulation by some nations, also it could also be protectionist agendas among some countries that do not want to accommodate foreign regulations, just to name the obvious. Third, harmonization entails not only high levels of investment but also capacity-building, particularly in countries with low incomes, where they do not have basic regulatory infrastructure. To address the challenges, MRAs among nations and regional centers of excellence may become useful. MRAs enable countries to recognize the approved regulation by other countries if they all have similar safety and efficacy standards. Regional centers can be used to offer training, technical assistance, and common testing. Harmonization is not only beneficial to the manufacturers since they will reduce duplication of efforts but also beneficial to the governments in enhancing efficient control as well as building confidence in farmers of similar quality of the product.

12.2.3 Trade Regulations and Export/Import Policies

The trade in fertilizers happens internationally and is controlled by a set of national policies, bilateral economic treaties, and multilateral systems like those of the World Trade Organization (WTO) (Baldwin, 2016). Such trade rules determine the kind of fertilizers that can be imported, exported, and the terms and prices. In case of regular fertilizers such as urea, DAP (di-ammonium phosphate), and potash, the policies of the trade usually center on tariffs, quotas, and strategic reserve (Subash et al., 2020). But in the case of NGFs, the trade rules are rather underdeveloped, which causes market uncertainty in official global supply chains. There is a big problem of having inconsistent documentation and classification of advanced fertilizers (Martínez-Blanco et al., 2014). Most of its customs agencies are using old codes of Harmonized System (HS), which fail to align with adequate distinction between traditional and innovative products. Consequently, nanofertilizers, biofertilizers, and controlled-release formulations can be delayed, misclassified, or even refused entry at the borders since such products are not clearly defined in terms of regulation. It is particularly flawed in the case of start-ups and small enterprises that attempt to enter the foreign market with new, sustainable technology. The other area of concern is the application of nontariff barriers (NTBs) to curtail imports; that is, the application of licensing schemes, quality certifications, and phytosanitary approvals. These measures are generally given good reasons based on safety or environmental reasons, but this can be differentiated or can be a discriminatory practice, thereby going against the

WTO principles. A country might not grant a trader a license to use a nanofertilizer that has already been licensed in other countries simply because they must carry out multiyear field tests just to protect themselves against nanofertilizer.

There is an urgent need to revise and harmonize the trade policies to enhance efficiency in the trade process that would take into consideration the changing characteristics of the products in the fertilizer industry (Pekdemir, 2018). These are the development of new HS codes, alignment of labeling requirements, and digitized document systems to simplify the process of customs clearance. The modernization of trade infrastructure can be achieved with the help of organizations like the WTO, UNCTAD, and World Customs Organization (WCO) that can assist through technical help devices that facilitate the lack of practice in trade. In addition, nations are expected to strive to have bilateral and regional trade agreements by way of agreeing on bilateral and regional mutual recognition of fertilizer standards and their certification (https://www.wto.org/english/res_e/booksp_e/world_trade_report15_e.pdf). Such agreements can lower the costs of transactions, increase access to the market, and make investment in green technology-based fertilizer more friendly. It may give as an example the African Continental Free Trade Area (AfCFTA), which has the potential to harmonize the rules of fertilizer trade on the continent, increasing the self-sufficiency and innovation of their production across the region (Mballa, 2022). At last, the export and import policies should be able to strike a balance between economic goals and environmental and population health. The export of advanced fertilizers is phenomenal as it will help national industries, but it is also necessary that the products should be safe at their place of origin and the international market. High-quality and sustainable imports must be promoted through importations and not as protectionist measures. In a globalized world, there is a need to have a transparent, evidence-based, and internationally harmonized trade policy toward responsible upsizing of NGFs (Atkins & Macpherson, 2022).

12.3 National Fertilizer Acts and Standards

12.3.1 India: Fertilizer Control Order (FCO)

The fertilizer industry of India is controlled through the FCO that was adopted in 1985, under the Essential Commodities Act of 1955. Fertilizers in the country are controlled in terms of their manufacture, sale, and distribution, including their quality and price, by the FCO (Ray, 2016). It lays down requirements regarding different fertilizer products, makes it compulsory to have specific packaging and labeling systems, along with defining the procedures for approving the product. All fertilizers, including imported and locally made ones, face a requirement that they must be marketed upon meeting required nutrient levels and physical characteristics as prescribed in terms of the FCO. Pricing is also determined by the FCO, which offers registration of new fertilizers where multi-season agronomic tests are mandatory, and the data on bio-efficacy. The FCO has recently extended its scope to include biofertilizers and organic fertilizers, but guidelines on the nanofertilizers and other new generation products are still under development (Subramanian et al., 2025). It is the mandate of the laboratories attached to the State Agriculture Departments to carry out quality tests; however, it is not equally enforced because of the limited capacity.

The fertilizer policy of India has a severe impact on the dynamics of regulations due to subsidies (Birner et al., 2011). In the Nutrient-Based Subsidy (NBS) scheme, the government offers direct subsidy on urea and fixed subsidy on P&K fertilizers that result in excess use of nitrogen-rich inputs and unbalanced fertilization. Although FCO is a very important regulatory body, reform is expected to transform it into an environmental stability and fertilizer reform technology. Some of the suggestions to make the FCO more amenable to contemporary agricultural realities include digitization of the mechanisms that ensure compliance, integrating them with an already existing program called soil health card, and creating expedited lanes of approval for the next generation of fertilizer.

12.3.2 USA: EPA and USDA Guidelines

In America, the regulation of fertilizers is decentralized, with some efforts being implemented by both the state and federal governments. The EPA may have a part in regulating fertilizers, which may contain dangerous chemicals or act as pesticides, namely under the Federal Insecticide, Fungicide, and Rodenticide Act (FIFRA) (Omoyajowo, 2024). U.S. Department of Agriculture (USDA) impacts the use of fertilizer indirectly by having programs that provide conservation, research, and sustainable practices assistance and support. Product registration, labelling, and inspection are matters of agencies at the state level, frequently supported by the AAPFCO (Keener, 2019). AAPFCO has come up with model laws definitions, which most of the states incorporate to regulate the composition of fertilizers, claims, and their safety uniformly. Being at the same time decentralized and coordinated, such a framework enables flexibility and regional adjustment; however, it can also lead to discrepancies in the assessment of the new fertilizer technologies (Wang, 2014). The U.S. regulatory system helps to encourage innovation with its comparatively unambiguous registration of conventional fertilizers, but the status of so-called second-generation fertilizer types (nanofertilizers, biofertilizers, and biostimulants) is unclear. Assessment of nanomaterials has been issued by the EPA, and still, no standardized assessment of the nanofertilizer has been established. Biostimulants could be classified into various categories depending on the claims, which implies ambiguity among stakeholders in the industry (Du Jardin, 2015). To counter this, the USDA and EPA have been collaborating with stakeholders in bringing in greater clarity, particularly where the current farm bill of 2018 inspired the establishment of uniform definitions relating to plant biostimulants. In general, the U.S. regulatory environment is strong, but higher levels of harmonization between the states and new guidelines on novel inputs are two important areas that need improvement (Pekdemir, 2018).

12.3.3 EU: Fertilizing Products Regulation (FPR)

The Fertilizing Products Regulation (EU FPR), Regulation (EU) 2019/1009, is one of the most far-reaching and innovative regulatory schemes in the world of fertilizers (Stubenrauch, 2022). The FPR came into force on July 16, 2022, and replaces the outdated Fertilizer Regulation (EC) No. 2003/2003 and substantially extends the regulatory scope to cover organic fertilizers, soil improvers, growing media, plant biostimulants, and NGFs. The CE marking scheme of fertilizer products that pass EU-wide hurdles on safety, efficacy, and environmental issues is one of the huge innovations of the FPR (Athanasoulia et al., 2021). The product should be assessed by a notified body as being in conformity, and evidence produced that it meets harmonized EU standards of contaminant limits and agronomic performance to achieve CE marking. It is also important that the FPR will take into consideration various criteria of the emerging technologies, such as controlled-release fertilizers, biostimulants, and microbial additives. It requires the full disclosure of ingredients and mode of action and promotes life-cycle environmental analyses. The FPR also creates limitations on pollutants like cadmium and heavy metals, thereby establishing the view of the EU toward environmental sustainability. But the high regulatory standards are associated with problems. The prices and procedures associated with the CE certification can put off small and medium enterprises (SMEs) from venturing into the business. The FPR is strongly considered as a reference model in uniting the principles of safety, innovation, and environmental ambitions in the management of fertilizer. It establishes a precedent for other regions seeking to control next-generation fertilizers in a harmonized and scientifically based way (Metzger & Kusch, 2024).

12.3.4 China: Ministry of Agriculture and Rural Affairs (MARA)

In China, the Ministry of Agriculture and Rural Affairs (MARA) oversees fertilizer control with support of other agencies, including the State Administration for Market Regulation (SAMR) (Liang et al., 2022). The fertilizer market of the world is largely dominated by China, as it is the largest consumer and producer of fertilizer, and thus its regulatory system plays a very critical role in the global fertilizer market. Conventional products like urea and superphosphate are strictly controlled under national standards (GB standards), which entail the requirement of nutrient content, labeling, and storage, among others. In the past years, China has intervened in focusing on green agriculture and sustainable intensification, and this has led to a revision of fertilizer regulations (Xie et al., 2019). MARA came up with the policy, Zero Growth of Fertilizer Use by 2020; the objective of the policy is to stabilize fertilizer application but increase efficiency of fertilizer use with the help of soil testing and formula-based fertilizer. Such campaigns have been quantitatively effective in decreasing the total fertilizer input without a kill in productivity. In the case of NGFs, the Chinese regulatory environment is still in the making (Lam et al., 2022). There is little guidance released by the government against slow-release and controlled-release fertilizers, and biofertilizers are covered under microbial product standards. However, a single approval process does not exist regarding nanofertilizers, and a regulatory situation relating to biostimulants and multifunctional inputs remains to be vague. China is investing and a great promotion of research and innovations, and there are several state initiatives boosting fertilizer research. There must be legal clarification to promote commercialization faster. MARA has indicated an interest in the establishment of risk-based tiers approval of innovative fertilizers. Now that China is consolidating its leadership in sustainable agriculture, the next important step to pursue would be to modernize its regulatory framework to understand emerging technologies (Qin et al., 2022).

12.3.5 Africa: Regional and National Fertilizer Acts

The setting of the African regulatory environment is fragmented and variable because it accommodates different levels of national capability, market maturity/commitment, and institutions to regulate fertilizer (Saad, 2014). Though there are several relatively well-developed fertilizer registration and quality control systems, as in South Africa and Morocco, there are significant numbers of other countries with weak enforcement policies, poor regulation, and poor testing facilities. To counter such differences and enable trades to be done within these regions, unsuccessful organizations, such as the African Union, ECOWAS, COMESA, and EAC, have come up with harmonized standards of fertilizer regulations. The fertilizer regulation (C/REG.13/12/12) of ECOWAS harmonizes nutrient details, labeling, and packaging of fertilizers in an area comprising 15 states of West Africa (Keyser et al., 2015). On the same note, EAC has introduced standard regulations of fertilizer product approval, which are facilitated by community labs and collective advancements of certification. There is also prevalent counterfeiting of fertilizers in most African markets because of laxity in their inspection and enforcement. There is little access for farmers to information concerning product quality and little knowledge on how to apply products as well. Besides, underfunding of national regulatory agencies, insufficient numbers of personnel, and poor coordination are not uncommon. NGFs are gradually gaining entry through African markets, with the level of regulation being insufficient (Hoeschle-Zeledon, 2019). They have little or no national policies that talk of biofertilizers, nanofertilizers, or controlled-release formulations. Small-scale demonstrations like pilot projects and donor-sponsored programs have implemented advanced fertilizers in some areas, but the expanded use has been impeded by the absence of legal structures and market incentives. The enhancement of regulatory institutions, investment in laboratory power, and the utilization of regional cooperation are the most important things that could enhance the management of fertilizers in Africa (Sommer et al., 2013). The harmonized regulations will not only lead to better product quality and confidence among the consumers of a given product but also facilitate smallholder farmers to get access to sustainable fertilizers on the continent.

12.4 Certification and Labeling of Next-Generation Fertilizers

12.4.1 Requirements for Slow/Controlled-Release Products

The breakthrough that came with fertilizer-release technology is slow- and controlled-release fertilizers (SRFs and CRFs), whose nutrients are not instantly released, but in a regulated, timed, and crop-specific way that matches nutrient requirements as the plant consumes them. This minimizes the loss of nutrients in terms of leaching, volatilization, and denitrification, hence increasing NUE and environmental influence (Govindasamy et al., 2023). Regulatory frameworks require those tests to be conducted in the laboratory to mimic field conditions and measure what percentage of nutrients is released in what period (e.g., 30, 60, or 90 days) (Dimkpa et al., 2017). CRFs in the EU have certain standards that they should satisfy before being released into the market. When compared to regular fertilizers and biostimulants, they are regulated under the requirements of the FPR. The second significant requirement is the coating material, as well as its environmental profile. Most CRFs are covered with synthetic polymers; this triggers the issue of pollution with microplastics (Bhattacharjee et al., 2025). This has made some jurisdictions require that the materials used to coat the products be biodegradable and nontoxic, particularly for the products marketed as friendly to the environment. Before approvals, some standardized risk assessments and biodegradation tests are becoming a mandatory requirement. In addition, labeling should contain conditions of release and the coating formula, storage conditions, and preferred instructions for using it. Such contradictory or deceptive statements of controlled-release performance may lead to recalls of products or denial of registration. Considering its growing use, most nations are revising their country rules to add codes of classification, performance standards, and compatibility examination processes to SRFs and CRFs. Such regulatory transparency will be crucial in the growth of market confidence and sustainable use in the various agroecological zones (Kumar & Khan, 2024).

12.4.2 Guidelines for Nanofertilizers and Biofertilizers

The next field of agricultural inputs is nanofertilizers and biofertilizers delivering specific nutrients and maximal contact between plants and microbes, respectively (El-Ghamry et al., 2018). Their new ways of action and diverse compositions require separate control regulations to guarantee quality, safety, and efficacy. There is a substantial dearth of international uniformity between the regulations that have been put out by various countries, even though some countries have started the process of drafting rules. Nanofertilizers are nutrients that have been engineered to operate at the nanoscale (generally 1100 nm) and, as a result, have special characteristics related to their large surface area, slow-release, and bioavailability (Yadav et al., 2023). The possibility of toxicity, bioaccumulation, and fate in the environment must be covered by regulatory requirements. The European Union, as an example, has a mandatory identification of nanomaterials that should be indicated specifically in the labeling and be analyzed in special risk assessments (Rauscher et al., 2017). In India, FCO has already listed some preliminary requirements with regard to zinc-based nanofertilizers, where it must be subjected to multi-locational field trials, shelf-life studies, and assessment of environmental risks before its registration. Microbial inoculants such as Rhizobium, Azospirillum, and phosphate-solubilizing bacteria that are described as biofertilizers are controlled with respect to strain authenticity, number of viable cells, contamination, and shelf life (Mishra & Barolia, 2020). Most countries have registered biofertilizers as a class of their own in the regulations of agricultural products or microbial products. The criterion of quality is usually entailed in the confirmed number of colony-forming units (CFUs), composition mix of the carrier material, and lowest levels of efficacy under the local soil and crop conditions. There are more problems in the classification of emerging hybrid products, for example, nano-enabled biofertilizers. Regulators should determine the legal status of such products are to fall within the microbial, chemical, and nanomaterial regulations. Agencies are being advised to implement tier risk-based systems, enhance laboratory testing facilities, and work internationally on the standardization of standards (Ali et al., 2024).

12.4.3 Environmental Safety Labeling

Environmental safety labeling is one of the most essential tools of agricultural sustainability, transparency, adequate fertilizer handling, and ecosystem protection as the status of environmental safety improves (Sinha et al., 2009). The labels are the link between the manufacturers, regulators, and end users to provide them with important information about the composition of the product, dangers in its use, environmental safety, and handling instructions. NGFs are especially important to be labelled with environmental protection indications; their composition and action can be more complicated than that of traditional NPK-fertilizers (Anas et al., 2025). Fertilizers could either be polymer-coated fertilizers or nano-formulated, such that they may degrade in soil and may have to provide detailed information about biodegradability, ecotoxicity, or residual behavior in the soil and water systems (Jayasoorya & Kumar, 2024). To this effect, such countries as EU countries need to be compliant with the chemical labeling needs (e.g., REACH and CLP), and they need a combination of pictograms, hazard statement, and precautionary measures to be included in products. A number of jurisdictions are turning toward eco-labeling programs of low-impact fertilizers. These voluntary certification schemes evaluate lifecycle effects and could comprise standards like low cadmium level, carbon footprint, power-efficient manufacturing, and sustainable packaging (Plassmann, 2018). A case in point is the EU Ecolabel, which can be applied to fertilizer products with high standards of sustainability. Labeling also must take into consideration that international transport regulations should be followed, particularly by products where the substance it carries is subject to hazardous material codes. Also, clear usage instructions are very important, that is, how and when, and the conditions under which the product should be stored, and which products it should be used with. If not provided, there is a possibility that it will be misused. Blockchain and QR, SDS, and product performance analytics are opening these advantages in a transparent way, with dynamic access to certificates and safety data sheets (SDS) and analytics. Through the changing regulations, safety labeling will be at the core of instilling accountability as well as educated utilization of fertilizers (Ochieng, 2015).

12.5 Risk Assessment and Regulatory Testing Protocols

12.5.1 Ecotoxicology Standards

The regulatory framework of fertilizers is necessarily important because the development of fertilizers in the new generation, as nanofertilizers, biofertilizers, and controlled-release products, requires ecotoxicology standards (Ochieng, 2015). These norms are expected to analyze the risks of fertilizers on both the flora and fauna by considering the harmful impacts of fertilizers on nontarget species in the terrestrial and the water environment, including soil microorganisms, earthworms, insects, aquatic invertebrates, and vertebrates like fish and amphibians. Acute and chronic toxicity tests are covered under the scope of ecotoxicology testing. Organization for Economic Cooperation and Development (OECD) has put in place guidelines, which put a specific path to achieve the impact of some chemical substances other than key indicator species such as *Daphnia magna and Eisenia fetida*. Ecotoxicological risk assessments are also compulsory in the EU for all CE-marked fertilizers and biostimulants for which possible releases to the environment are of a major volume and are controlled under the FPR (Athanasoulia et al., 2021). New issues regarding nanomaterials have increased the significance of the generation of special nano-ecotoxicological procedures, which presently means that nanoparticles behave differently from their macro-form. To give an example, they could bioaccumulate more easily or produce reactive oxygen species that are destructive to aquatic organisms. To provide consistency and comparability, the majority of the regulatory agencies demand standardized methodologies of test, such as LC50 (lethal concentration), NOEC (No Observed Effect Concentration), and EC50 (effective concentration) values, and propose models of environmental fate modeling (Wolf & Segner, 2023). These standards of ecotoxicology are frequently applied in gauging whether a product requires extra environmental labels or an environmental limitation on the use of the product. Standardization of these protocols globally is of priority as variations in such requirements made by various jurisdictions may block international product approvals. Ecotoxicology will remain part of the foundation of safe and sustainable agricultural advancement as fertilizer technologies develop (Nadarajan & Sukumaran, 2021).

12.5.2 Soil and Water Impact Testing

The major environmental compartments that are altered due to the use of fertilizers include soil and water, and thus testing of impacts on such systems is a very important regulatory requirement. In contrast with traditional fertilizers, the products of the next generation potentially have a new mode of interaction with soil and water-modifying its cycling nutrients, the structure of microorganisms, and even physicochemical characteristics (McCartney & King, 2011). Thus, the regulators are increasingly demanding more thorough impact analysis of a product prior to its approval. Soil impact testing aims at identifying the effect of a fertilizer on the pH, cation exchange capacity, organic matter percentages, microbial biomass, and enzyme activity in soil. The beneficial microbes might be increased with the use of biofertilizers, and slow-release formulations should not inhibit the local population of microbes and leave toxic leftovers (Chaudhary et al., 2020). Tests are usually carried out simulating field conditions to measure changes in soil structure, compaction, and moisture retention with time. Water impact testing will analyze the possibilities of leaching, runoffs, and pollution of surface and ground (Dandautiya et al., 2018). Significant indicators are the nitrate mobility, the phosphate mobility, and the heavy metal content, as well as the presence of synthetic compounds or the metabolites of degradation. Such tests are especially significant for fertilizers with a polymeric coating or nanomaterials, which can remain in the waterways. Prediction of fertilizer transport and transformation has become standardized in Europe and North America in the form of Pesticide Root Zone Model (PRZM) and Groundwater Loading Effects of Agricultural Management Systems (GLEAMS). Cumulative effects may also be required in long-term field trials or mesocosm applications, as these will be necessitated by its regulators. Finally, testing of soil and water makes sure that fertilizers contribute not only to the productivity of the crops but also aid in applying healthy elements in the environment as well. These tests form part of integrated product assessments and guide the input on usage instructions, buffer zones, and conditional approvals of a product according to regional risk profiles (Galert & Hassold, 2021).

12.5.3 Regulatory Data Requirements for Approval

To qualify and be registered to be legally marketed any fertilizer product, it should meet the specific regulatory data requirements, which indicate quality, efficacy, safety, and environmental acceptability of the given product. They vary with jurisdiction but normally contain fundamental categories: composition, agronomic performance, toxicological safety, environmental impact, and label accuracy (Winder et al., 2005). The initial element is elaborate chemical and physical characterization that encompasses nutrients, solubility, size of granules, pH, moisture content, and containment of either heavy metals or pollutant materials. Within the EU, products that require CE certification under FPR are required to provide a technical dossier, including conformity assessment results made by a notified body. Then, field trial efficacy information is necessary. These experiments ought to show that the fertilizer enhances the growth of plants, uptake of nutrients, or the health of soil, in particular circumstances (Singh & Ryan, 2015). They normally must replicate the trials, have them peer-reviewed, and carry them out in various agroecological zones to guarantee generalization. Agronomic multi-locational trials and seasonal field trials are a precondition to FCO approval in India. The third pillar is made up of toxicological and ecotoxicological studies that need to certify that the product will not present any major risk to man, animals, or the ecosystem. In case of nanofertilizers, regulators usually demand more information on particle size distribution, reactivity on the surface, and bioaccumulation ability (Fatima et al., 2021). Lastly, the labeling and use directions need to be verified, including any statements on the nutrient level, rate of application, or environmental advantages. Profiteering and exaggerated products may be sanctioned by the regulating authorities and even suspended or withdrawn. Due to duplication and cost-cutting motives, most countries are examining the possibility of mutually recognized data and risk-tiered pathways (Elston et al., 2018). These would mean that well-travelled products would be subjected to more squelching, whereas low-risk or organic inputs would have a faster process of approval.

12.6 Future Directions and Policy Recommendations

The development of technologies in fertilization and the growth in the need for sustainable agriculture would require the fundamental restructuring of regulating and policy regimes. Most of the modern regimes are device-specific to the traditional types of fertilizers and may not be adequately geared to evaluate or regulate NGFs, like nanofertilizers, biofertilizers, and the matter of controlled-release formulations. To make sure that these innovations are safely assimilated in agricultural systems, government should consider some sort of forward-looking or adaptive policy. This part provides the most important policy recommendations that focus on friendly regulation to innovation and integrating into the models of the circular economy, as well as cooperation on capacity building at the regional levels.

12.6.1 Adaptive Regulatory Models for Innovation

The processes of innovation in fertilizer technologies are much faster than the traditional cycles of regulation. To keep up with it, states should embrace flexible regulatory systems, the ones that adjust to scientific awareness and market forces. These models must consist of a tiered system of risk-based assessment in which low-risk inputs, such as some biofertilizers, can experience an accelerated approval, whereas elevated-risk products, such as nanoparticle-based fertilizers, may be put under a more extensive examination. This stratification decreases bottlenecks to achieve safe and low-risk innovations at the same time, guaranteeing the safety of the people and the environment. Regulatory sandboxes constitute another such important aspect because, through them, the new fertilizer products can be tested in specifically provided limited real-life environments and provisionally approved. Such a strategy enables regulators to gather evidence and develop guidance that can be useful for early-stage innovation. The statistics should also develop the internal systems of horizon scanning to ensure that these bodies are able to monitor the emerging technologies with a view to preparing appropriate policy response mechanisms. Academia and industry cooperation are important. The public-private collaboration should facilitate a co-development of the standards, exchange of safety data, and decrease workload on the individual innovators. In addition, the computerization of the process of authorization, which means an online submission, dossiers that can be traced, and automated review systems, can simplify compliance and enhance transparency. Adaptive regulatory governance is not the ideal form of policy; it is a requirement of inclusive and science-based agricultural change.

12.6.2 Integrating Circular Economy Principles

The ultimate solution is a sustainable fertilizer policy that would be built on the principles of a circular economy and maximize the utilization of resources, minimize waste, and recycle or generate materials. In the area of fertilizers, this can be in terms of nutrient recycling, to recapture phosphorus, nitrogen, and potassium as a part of organic wastes supplied as livestock manure, food waste, crop wastes, and wastewater treatment by-products. The promotion of measures that can help, such as the use of organic waste-derived fertilizers, composts, as well as digestates, can be used to reduce the need for energy-intensive synthetic fertilizers as well as greenhouse gas emissions to a large extent. EU, as an example, under its Circular Economy Action Plan, has discouraged end-of-waste criteria on organic fertilizers, and according to the plan, certain recovered materials are getting the chance to reenter the market in a legal and safe way. Recent governments need to install an incentive system like taxes or subsidies for example, with circular fertilizer is purchased. These policies have the potential to promote innovations in nutrient recovery technology and promote decentralized biogas systems and composting. Additionally, using the closest existing concept, fertilizer standards, in recognizing the circular economy goals, will be effective in redefining what can be considered as fertilizer waste, as well as biologically enriched products. The other opportunity is lifecycle assessment (LCA) integration of the fertilizer policy. Regulatory bodies may require or encourage LCAs to assess the environmental implications of products on their life cycle, starting with manufacture until it has reached its end.

Integrating LCA measures in the process of certification or eco-labeling would help steer manufacturers and farmers toward a less impactful option. However, in the end, circularity must not be an extra principle of future fertilizer policy, but a fundamental one: the unity of agronomic efficiency and ecological integrity.

12.6.3 Regional Collaboration and Capacity Building

Regulation typically brings about limitations in technical infrastructure, human capital, and institutional coordination, particularly in the developing world, in the case of fertilizers. Regional cooperation and strengthening are one of the best approaches in conquering these deficits. International alliances may be used to bring about standardization, minimize duplication of rules, and ease accessibility of sustainable fertilizer products in the market. Important steps regional economic communities such as ECOWAS, EAC, COMESA, and ASEAN have taken include achieving harmonization of fertilizer policies, through standardization of fertilizer specifications, labelling, and quality control procedures. These frameworks allow MRAs so that, besides producers of fertilizers in one member country being able to sell them in the entire region, trade is promoted, and regulatory expenses are decreased. Capacity building still lies at the core. Governments and international agencies ought to make investments in the investigation of regional testing laboratories, training centers, and knowledge hubs of the fertilizer enrolments, inspection, and supervision. These investments not only assist in adopting a harmonized standard but also enhance access to high-quality input by farmers. Multilateral actions like the Africa fertilizer and soil health summit have been a central component in such support to the local agencies and compatibility of data platforms across borders globally, funded by donors. Regional cooperation can also be boosted by the use of digital tools. Redundancy and counterfeiting will be avoided by a common, real-time updated digital registry of registered fertilizers maintained on a country-to-country basis. There can be regional networks of scientists, policymakers, and industry experts that can transfer data, assess risks, and help come up with regulatory frameworks for next-generation products. Export. Regional cooperation is not just a political dream, but a practical concern to handle the cross-border agriculture issues and gain an equal right to access to sustainable fertilizers by different kinds of economies.

12.7 Conclusion

The NGFs have given hope of fixing the environmental issues, as the use of the NGFs in the agricultural process has been developing rapidly. Their general application is, however, at present not made possible by the regulatory regimes, which were mainly adopted to suit the conventional fertilizers. It is also in this chapter that issues related to the utmost necessity of regulatory modernization to be able to fit the distinctive features of NGFs, including nanofertilizers and biofertilizers. Current systems will have to undergo changes in the form of the latest definitions; common procedures associated with testing and thorough evaluation of risk. A good example is the European Union FPR which not only has captured the LCAs and eco-labeling, but it has done so to achieve effectiveness and environmental safety. The models of this kind emphasize the necessity to harmonize the policies to be used with the innovative character of these products, encouraging their safe usage. There is a major issue that is presented by the worldwide difference in regulatory standards. Whereas the developed states such as those in the EU and the United States have drawn up stringent structures, most of the developing nations experience old-fashioned policies and poor enforcement. This divides farms and forms a barrier to trade, thus restricting access to sustainable technologies by smallholder farmers. To cope with this, harmonization across countries is important, to achieve which special bodies like the FAO and ISO. These gaps can be bridged by MRAs, and regional cooperation so that the norms of safety and performance of NGFs are the same wherever they are. Through such working relationships, these initiatives can promote the free movement of trade and promote the use of NGFs in various agricultural regimes. The other key players in terms of NGFs promotion are the economic and policy tools. Market change can be achieved by restructuring the fertilizer subsidies to an equilibrium that favors sustainable fertilizers, introducing carbon credits as a factor of reduced withdrawal, and the creating incentives that will encourage market innovation in the industry. Also, digital, such as blockchain, can provide a greater level of traceability and compliance, which gives farmers and regulators real-time results on how the product performs and how it affects the environment. The steps not only facilitate the change toward sustainable practices, but they also guarantee a supply-chain accountability. Finally, NGFs can be made successful due to farmer-centric strategies placing emphasis on access and education. All this information being propagated by these technologies needs to be readily available to the smallholder farmers who are the backbone of agriculture in many parts of the world. There was a need to have policies that incorporate specific subsidies, training programs, and extension services that close the gap between innovations and practical applicability. With the combination of these measures and other sustainability objectives, including UN SDGs, NGFs will be able to transform agriculture, increase production, and reduce its environmental impacts. The way toward such a more sustainable future will need adaptive regulations, international cooperation, and integrative policies to fulfill the full potential of NGF.

References

Alawode, A. (2025). Evaluating agricultural subsidy reforms and their effects on smallholder farmer income and efficiency. *International Journal of Advance Research Publication and Reviews* 2 (5): 180–201.

Ali, A., Hamza, A., and Saleh, H. (2024). Risk management strategies in medical laboratory practice. *International Journal of Medical Laboratory Research* 34: 274.

Anas, M., Falak, A., Hassan, S. et al. (2025). Microbial interactions and bacterial responses to metal stress in plants: Mechanisms, adaptations, and applications for sustainable agriculture. *Journal of Crop Health* 77 (1): 36.

Arora, P.K., Tripathi, S., Omar, R.A. et al. (2024). Next-generation fertilizers: The impact of bionanofertilizers on sustainable agriculture. *Microbial Cell Factories* 23 (1): 254.

Ashford, N.A. and Hall, R.P. (2011). The importance of regulation-induced innovation for sustainable development. *Sustainability* 3 (1): 270–292.

Ashitha, A. and Rakhimol, K. (2021). Fate of the conventional fertilizers in environment. In: *Controlled release fertilizers for sustainable agriculture* (ed. F.B. Lewu, T. Volova, S. Thomas, and R. K.R), 25–39. Elsevier.

Athanasoulia, I.-G., Novakovic, J., & Kyriazi, M. (2021). D 6.1-Inventory and integrated assessment of the EU regulations related to the production and application of bio-based fertilisers.

Atkins, J. and Macpherson, M. (2022). *Extinction governance, finance and accounting: Implementing a species protection action plan for the financial markets*. Taylor & Francis.

Baldwin, R. (2016). The world trade organization and the future of multilateralism. *Journal of Economic Perspectives* 30 (1): 95–116.

Benson, T. and Mogues, T. (2018). Constraints in the fertilizer supply chain: Evidence for fertilizer policy development from three African countries. *Food Security* 10 (6): 1479–1500.

Bhattacharjee, L., Gopakumar, A.N., Beheshtimaal, A. et al. (2025). Mechanisms of microplastic generation from polymer-coated controlled-release fertilizers (PC-CRFs). *Journal of Hazardous Materials* 486: 137082.

Birner, R., Gupta, S., and Sharma, N. (2011). *The political economy of agricultural policy reform in India: Fertilizers and electricity for irrigation*, vol. 174. Intl Food Policy Res Inst.

Chaudhary, I.J., Neeraj, A., Siddiqui, M.A., and Singh, V. (2020). Nutrient management technologies and the role of organic matrix-based slow-release biofertilizers for agricultural sustainability: A review. *Agricultural Reviews* 41 (1): 1–13.

Dandautiya, R., Singh, A.P., and Kundu, S. (2018). Impact assessment of fly ash on ground water quality: An experimental study using batch leaching tests. *Waste Management & Research* 36 (7): 624–634.

Dimkpa, C., Bindraban, P., McLean, J.E. et al. (2017). Methods for rapid testing of plant and soil nutrients. In: *Sustainable agriculture reviews* (ed. E. Lichtfouse), 1–43. Springer.

Du Jardin, P. (2015). Plant biostimulants: Definition, concept, main categories and regulation. *Scientia Horticulturae* 196: 3–14.

Eastwood, C.R. and Renwick, A. (2020). Innovation uncertainty impacts the adoption of smarter farming approaches. *Frontiers in Sustainable Food Systems* 4: 24.

Ejedegba, E.O. (2024). Advancing green energy transitions with ecofriendly fertilizer solutions supporting agricultural sustainability. *International Research Journal of Modernization in Engineering Technology and Science* 6: 2582.

El-Ghamry, A., Mosa, A.A., Alshaal, T., and El-Ramady, H. (2018). Nanofertilizers vs. biofertilizers: New insights. *Environment, Biodiversity and Soil Security* 2 (2018): 51–72.

Elston, T., MacCarthaigh, M., and Verhoest, K. (2018). Collaborative cost-cutting: Productive efficiency as an interdependency between public organizations. *Public Management Review* 20 (12): 1815–1835.

Fatima, F., Hashim, A., and Anees, S. (2021). Efficacy of nanoparticles as nanofertilizer production: A review. *Environmental Science and Pollution Research* 28 (2): 1292–1303.

Fortin, N.D. (2023). Global governance of food safety: The role of the FAO, WHO, and Codex Alimentarius in regulatory harmonization. In: *Research handbook on international food law* (ed. M.T. Roberts), 227–242. Edward Elgar Publishing.

Galert, W. and Hassold, E. (2021). Environmental risk assessment of technical mixtures under the European registration, evaluation, authorisation and restriction of chemicals—a regulatory perspective. *Integrated Environmental Assessment and Management* 17 (3): 498–506.

Govindasamy, P., Muthusamy, S.K., Bagavathiannan, M. et al. (2023). Nitrogen use efficiency—a key to enhance crop productivity under a changing climate. *Frontiers in Plant Science* 14: 1121073.

Hassan, B. (2024). The impact of agricultural policy on global food security. *Frontiers in Agriculture* 1 (2): 325–355.

Hoeschle-Zeledon, I. (2019). Africa research in sustainable intensification for the next generation: Sustainable intensification of key farming systems in east and southern Africa.

Jayasoorya, R. and Kumar, P. (2024). Utilization of biodegradable carrier-based nano herbicide formulations for sustainable weed management in agriculture. *Frontiers in Agronomy* 6: 1497041.

Keener, K.M. (2019). Food regulations. In: *Handbook of farm, dairy and food machinery engineering* (ed. M. Kutz), 15–44. Elsevier.

Keyser, J.C., Eilittä, M., Dimithe, G. et al. (2015). *Towards an integrated market for seeds and fertilizers in West Africa*. Washington: World Bank Group.

Kumar, R. and Khan, A.K. (2024). Supply chain sustainability and SDG compliance: Challenges and opportunities for agro-based industries. *Sustainable Development Goals & Business Sustainability* 442: 290.

Lam, S.K., Wille, U., Hu, H.-W. et al. (2022). Next-generation enhanced-efficiency fertilizers for sustained food security. *Nature Food* 3 (8): 575–580.

Liang, J., Yang, X., Jiao, Y. et al. (2022). The evolution of China's regulation of agricultural biotechnology. *aBIOTECH* 3 (4): 237–249.

Martínez-Blanco, J., Lehmann, A., Muñoz, P. et al. (2014). Application challenges for the social life cycle assessment of fertilizers within life cycle sustainability assessment. *Journal of Cleaner Production* 69: 34–48.

Mballa, M. O. (2022). Achieving integration through the african continental free trade area (AfCFTA): Between promises and challenges.

McCartney, M. P., & King, J. (2011). Use of decision support systems to improve dam planning and dam operation in Africa.

Metzger, A. and Kusch, C. (2024). Innovation policy beyond patents: A case study on the development of climate-friendly fertilizers. *GRUR International* 73 (8): 742–750.

Michelson, H., Gourlay, S., & Wollburg, P. (2022). Non-labor input quality and small farms in sub-Saharan Africa.

Mishenin, Y., Yarova, I., and Koblianska, I. (2021). Ecologically harmonized agricultural management for global food security. In: *Ecological intensification of natural resources for sustainable agriculture* (ed. M.K. Jhariya, R.S. Meena, and A. Banerjee), 29–76. Springer.

Mishra, B.K. and Barolia, S.K. (2020). Quality assessment of microbial inoculants as biofertilizer. *International Journal of Current Microbiology and Applied Sciences* 9 (10): 3715–3729.

Nadarajan, S. and Sukumaran, S. (2021). Chemistry and toxicology behind chemical fertilizers. In: *Controlled release fertilizers for sustainable agriculture* (ed. F.B. Lewu, T. Volova, S. Thomas, and R. K.R), 195–229. Elsevier.

Nakagawa, J. (2011). *International harmonization of economic regulation*. Oxford University Press.

Nikolaevich, M., Fedorovich, O., Vladimirovna, N., and Igorevna, D. (2019). Public and private partnership: Innovation-driven growth of agriculture at the regional level. *Journal of Environmental Management & Tourism* 10 (39): 1435–1444.

Novotny, V., Wang, X., Englande, A.J. Jr. et al. (2010). Comparative assessment of pollution by the use of industrial agricultural fertilizers in four rapidly developing Asian countries. *Environment, Development and Sustainability* 12 (4): 491–509.

Ochieng, R. (2015). Towards a regulatory framework for increased and sustainable use of bio-fertilizers in Kenya.

Oenema, O., Kros, H., and de Vries, W. (2003). Approaches and uncertainties in nutrient budgets: Implications for nutrient management and environmental policies. *European Journal of Agronomy* 20 (1–2): 3–16.

Omoyajowo, K. (2024). Sustainable environmental policies: An impact analysis of US regulations on pesticides and chemical discharges. *Available at SSRN 4943299.*

Pekdemir, C. (2018). On the regulatory potential of regional organic standards: Towards harmonization, equivalence, and trade? *Global Environmental Change* 50: 289–302.

Pinstrup-Andersen, P. (2002). Food and agricultural policy for a globalizing world: Preparing for the future. *American Journal of Agricultural Economics* 84 (5): 1201–1214.

Plassmann, K. (2018). Comparing voluntary sustainability initiatives and product carbon footprinting in the food sector, with a particular focus on environmental impacts and developing countries. *Development and Policy Review* 36 (4): 503–523.

Pretty, J. (2008). Agricultural sustainability: Concepts, principles and evidence. *Philosophical Transactions of the Royal Society, B: Biological Sciences* 363 (1491): 447–465.

Qin, T., Wang, L., Zhou, Y. et al. (2022). Digital technology-and-services-driven sustainable transformation of agriculture: Cases of China and the EU. *Agriculture* 12 (2): 297.

Rauscher, H., Rasmussen, K., and Sokull-Klüttgen, B. (2017). Regulatory aspects of nanomaterials in the EU. *Chemie Ingenieur Technik* 89 (3): 224–231.

Ray, A. (2016). *Distribution Channel & Customer Assessment On Iffco Fertilizers In Puri District Of Odisha* (Doctoral dissertation, Orissa University of Agriculture and Technology.

Saad, M. S. (2014). The impact of economic reforms on human security: The case of post-conflict Rwanda, 1996–2013.

Sales, H.B., Carolino, d.S.A., Nunes, d.A.R.Z. et al. (2024). Advances in agricultural technology: A review of slow-release nanofertilizers and innovative carriers. *Communications in Soil Science and Plant Analysis* 55 (12): 1849–1882.

Simtowe, F., & Expert, F. P. (2015). Support for the establishment of a regional fertilizer policy and regulatory framework in east and southern Africa. *An Assessment of National Fertilizer Policies, Regulations and Standards for Ethiopia. African Fertilizer and Agribusiness Partnership, Productivity Prosperity Partnership*, 1–34.

Singh, B. and Ryan, J. (2015). Managing fertilizers to enhance soil health. *International Fertilizer Industry Association, Paris, France* 1: 1–24.

Singh, K., Singh, R., Nwosu, N.J. et al. (2025). Optimizing nutrient delivery in agronomic crops: A review of enhanced efficiency fertilizers. *Journal of Plant Nutrition and Soil Science* 11: 15680.

Sinha, R., Herat, S., Valani, D., and Chauhan, K. (2009). The concept of sustainable agriculture: An issue of food safety and security for people, economic prosperity for the farmers and ecological security for the nations. *American-Eurasian Journal of Agricultural & Environmental Sciences* 5 (Suppl): 1–4.

Sommer, R., Bossio, D., Desta, L., Dimes, J., Kihara, J., Koala, S., Mango, N., Rodriguez, D., Thierfelder, C., & Winowiecki, L. (2013). Profitable and sustainable nutrient management systems for East and Southern African smallholder farming systems: Challenges and opportunities: A synthesis of the Eastern and Southern Africa situation in terms of past experiences, present and future opportunities in promoting nutrients use in Africa.

Stubenrauch, J. (2022). Innovative phosphorus governance: How to address recurring regulatory shortfalls—the example of Germany, costa Rica and Nicaragua. In: *International yearbook of soil law and policy 2020/2021* (ed. H. Ginzky, E. Dooley, I.L. Heuser, et al.), 435–462. Springer.

Subash, S., Jhajhria, A., & Pal, S. (2020). Trade and investment policy for overseas acquisition of fertilizers and raw materials: Role of the government.

Subramanian, K., Rajkishore, S., Praghadeesh, M. et al. (2025). Regulatory frameworks, guidelines, and policies to promote nanofertilizers: Current status in India and abroad. In: *Nanofertilizers in agriculture* (ed. A.K. Patra), 113–154. Springer.

Sukprasert, K. and Phadungkit, W. (2024). Integrating sustainable development frameworks into agricultural policies: A policy analysis perspective. *Contemporary Issues in Behavioral and Social Sciences* 11 (2): 1–12.

Turley, L., Bermúdez, S., Voora, V. et al. (2022). *Voluntary sustainability standards in east Africa*. International Institute for Sustainable Development (IISD).

van der Meulen, B. and Wernaart, B. (2019). Food and agriculture organization (FAO) and Codex Alimentarius commission. In: *Research Handbook on the European Union and International Organizations*, 82–100. Edward Elgar Publishing.

Wang, J. (2014). Decentralized biogas technology of anaerobic digestion and farm ecosystem: Opportunities and challenges. *Frontiers in Energy Research* 2: 10.

Wang, Z., Li, S., and Malhi, S. (2008). Effects of fertilization and other agronomic measures on nutritional quality of crops. *Journal of the Science of Food and Agriculture* 88 (1): 7–23.

Winder, C., Azzi, R., and Wagner, D. (2005). The development of the globally harmonized system (GHS) of classification and labelling of hazardous chemicals. *Journal of Hazardous Materials* 125 (1–3): 29–44.

Wolf, J.C. and Segner, H.E. (2023). Hazards of current concentration-setting practices in environmental toxicology studies. *Critical Reviews in Toxicology* 53 (5): 297–310.

Xie, H., Huang, Y., Chen, Q. et al. (2019). Prospects for agricultural sustainable intensification: A review of research. *Land* 8 (11): 157.

Yadav, P. (2023). A study of nutrient based subsidy policy of fertilizers in india and its impact on drivers of buying behaviour of farmers.

Yadav, A., Yadav, K., and Abd-Elsalam, K.A. (2023). Nanofertilizers: Types, delivery and advantages in agricultural sustainability. *Agrochemicals* 2 (2): 296–336.

Zhang, M., Wang, C., Zhang, R. et al. (2021). Comparison of the guidelines on good agricultural and collection practices in herbal medicine of the European Union, China, the WHO, and the United States of America. *Pharmacological Research* 167: 105533.

13

Optimizing Next-Generation Fertilizer Use Through Integrated Green Manure and Reduction Strategies for Sustainable Agriculture

Tauqeer Ahmed Qadri[1], *Muhammad Rizwan Khan*[2], *Muhammad Anas*[2,3], *Amjid Khan*[2], *Waseem Ahmed Khattak*[2], *Shabbar Abbas*[1], *Javed Iqbal*[4], *and Banzeer Ahsan Abbasi*[5]

[1] *Department of Biosciences, COMSATS University Islamabad, Islamabad, Pakistan*
[2] *UNESCO-UNISA Africa Chair in Nanoscience and Nanotechnology (U2ACN2), College of Graduate Studies, University of South Africa, Muckleneuk Ridge, Pretoria, South Africa*
[3] *Department of Soil Science & Plant Nutrition, Selcuk University, Konya, Türkiye*
[4] *Department of Botany, Bacha Khan University, Charsadda, Pakistan*
[5] *Department of Botany, Rawalpindi Women University, Rawalpindi, Pakistan*

13.1 Introduction

13.1.1 Background and Significance

The extensive adoption of synthetic fertilizers has transformed current agriculture. From the creation of the Haber-Bosch process at the beginning of the 20th century, which made it possible to fix atmospheric nitrogen industrially, agricultural output experienced an unprecedented growth (Anas & Quraishi, 2025; Prasad & Shivay, 2021). The Green Revolution of the mid-20th century continued to deepen fertilizer dependency by encouraging high-yielding crop varieties that rely heavily on chemical fertilizers and pesticides to produce the food to feed the world's growing population (Chawla et al., 2023). However, this dependence on conventional fertilizers has been quite costly to the environment and the economy. An excessive use of nitrogenous and phosphatic fertilizers has resulted in nutrient runoff and leaching, thus contributing to the eutrophication of freshwater and marine ecosystems. The usage of nitrogen fertilizers is also directly related to the emission of nitrous oxide (N_2O), a powerful greenhouse gas (GHG) that is close to 300 times more harmful than carbon dioxide (Filonchyk et al., 2024). Besides environmental degradation, these practices have degraded soil quality, disrupted microbial communities, and caused reduced nutrient use efficiency, which has reduced long-term agricultural sustainability and increased input costs to farmers. These are very critical issues in times of climate change, limited resources, and the need to provide for an increasing population with decreasing arable land. In this regard, input-intensive to knowledge-intensive, sustainable nutrient management systems have become not only desirable but necessary (Maja & Ayano, 2021).

13.1.2 Emergence of Sustainable Agriculture

Sustainable agriculture has turned out to be the answer to the ecological and agronomic issues that bedevil conventional farming practices. This system-based approach combines ecological concepts with contemporary agronomy to guarantee continued productivity and conservation of resources, as well as sustainability of the environment. It focuses on practices used to rebuild and sustain soil health, which include organic amendments, crop rotation, conservation tillage, and cover cropping (Anas et al., 2025; Cárceles Rodríguez et al., 2022). Agricultural conservation that requires minimal soil disturbance, permanent soil cover, and crop diversification has become popular because it can restore soil organic carbon, moderate water flows, and stimulate microbial activity. These enhancements help in enhancing resilience against drought, pests, and climate variability. Further, such practices lessen the reliance on synthetic inputs by utilizing natural processes like nitrogen fixation and nutrient cycling (Francaviglia et al., 2023). Sustainable agriculture is consistent with the global priorities such as the United Nations Sustainable Development Goals (SDGs), namely, the zero hunger (SDG 2), sustainable land management (SDG 15), and climate action (SDG 13). As more people become aware of agriculture's environmental footprint, governments and institutions globally are now pushing for integrated soil fertility management (ISFM) frameworks that combine organic and inorganic inputs for improved efficiency and less emissions (Atapattu et al., 2024).

13.1.3 Importance of Next-Generation Fertilizer Strategies

Being able to keep both farming and the environment safe now depends on the use of next-generation fertilizers. Such fertilizers are called controlled-release fertilizers (CRFs), nanofertilizers, and biofertilizers, and they work as creative methods for giving nutrients to plants with less waste (Arora et al., 2024). Using polymer coatings or chemical inhibitors, CRFs help the nutrients be supplied to the crops as they are needed and lessen the chances of losing them as gases or chemicals. Due to the nanotechnology used in producing them, nanofertilizers have a larger surface, dissolve more easily, and target plants better, making plants take up more nutrients and reducing contamination of the environment (Babu et al., 2022).

Using *Rhizobium*, *Azospirillum*, PSB, and mycorrhizal fungi as biofertilizers enhances both the soil's condition and the plant's nutrition by improving nitrogen fixation and phosphorus release in the soil (Nosheen et al., 2021). Some issues with these advanced fertilizers are that they are costly to produce, many farmers are unaware of them, and they work differently in different locations. Following this, blending the new fertilizers with the tried-and-true soil practices of using green manure (GM) improves the use of nutrients and strengthens the environment (Arora et al., 2024).

When you use GM, fast-growing or legume plants are added to the soil, which helps increase organic matter, provides more nitrogen, and supports a greater variety of microbes. As a result, the soil is made more suitable for roots, holds more water, keeps weeds under control, and provides a good environment for microbes. If GM is used together with reduced next-generation or synthetic fertilizers, it leads to higher crop yields, less GHG emissions, and keeps the soil healthy (Kocira et al., 2020). In addition, an integrated approach fits with agroecology by copying the way nutrients are cycled and by protecting and supporting beneficial bacteria in the soil around the roots. So, it saves farmers money and helps reduce the risks to nature, which are very significant for people with limited resources (Rehman et al., 2022).

The main objective of this chapter is to discuss mixing GM with new fertilizers to enhance how efficiently nutrients are used in farming. Its objectives include reviewing developments, uncovering strategies that can be used together, and providing helpful guidelines for protecting the environment while still yielding crops.

13.1.4 Green Manure: Concept, Types, and Agroecological Functions

GM is a class of crops, usually fast-growing herbaceous plants, grown to add to the soil at a particular stage of their growth cycle, which is usually before flowering. The main objective of green manuring is to increase the soil fertility, the content of organic matter, and the overall health of the soil. When these crops are plowed under when they are green, they decompose and release nutrients, improve soil structure, and promote favorable microbial activities (Prajapati et al., 2023). Historically, the utilization of GMs has been in use for thousands of years and has been a part and parcel of sustainable land management practices in civilizations. In the old agrarian communities of China and India, the farmers had been traditionally using species such as *Sesbania aculeata*, *Crotalaria juncea* (sunhemp), and *Tephrosia purpurea* for restoring the fertility of the soil after several cropping cycles. Such practices were so ingrained in the indigenous knowledge systems and were in harmony with the principle of returning nutrients to the soil to maintain productivity (Rani et al., 2024). With the advent of the 20th century, when agriculture was industrialized and synthetic fertilizers became widely used, the green manuring practices decreased. However, the renewed interest in regenerative and climate-resilient agriculture has reignited interest in GM as a nature-based solution to restore health in soil (Iderawumi & Kamal, 2022).

13.2 Classification of Green Manures

The GM species can be divided into two broad categories depending on their botanical features and the origin of development. These classifications have an impact on their functional features and applicability in various agroecosystems (Scavo et al., 2022).

13.2.1 Leguminous vs. Nonleguminous Species

- **Leguminous Green Manures**: These species are of the Fabaceae (Leguminosae) family and are particularly prized for their nitrogen-fixing properties. In symbiotic associations with Rhizobium bacteria in root nodules, they fix atmospheric nitrogen (N_2) into ammonium forms that can be used by plants. Some of these are *Trifolium* spp. (clovers), *Vicia* spp. (vetch), *Medicago sativa* (alfalfa), and *Cajanus cajan* (pigeon pea). Apart from nitrogen fixation, leguminous GMs enhance soil structure and microbial diversity (Raza et al., 2020).

- **Nonleguminous Green Manures**: These include cereal and cover crops, which grow fast, such as *Secale cereale* (rye), *Fagopyrum esculentum* (buckwheat), *Brassica juncea* (mustard), and *Avena sativa* (oats). Even though they do not have the ability to fix atmospheric nitrogen, they are effective at biomass accumulation, addition of organic matter, weed suppression, and phytoremediation. Some species also demonstrate the allelopathic effects, which also help to control weeds and suppress pests (Quintarelli et al., 2022).

13.2.2 Indigenous vs. Improved Cultivars

- **Indigenous Cultivars**: Locally adapted species and landraces have low inputs and are thus very suitable for low-resource and traditional farming systems. Such species are usually resilient to climatic stresses and pests in the local area. The use of their helps to sustain agro-biodiversity and to conserve traditional knowledge systems (Ulian et al., 2020).
- **Improved Cultivars**: Improved GM cultivars that are developed through selective breeding or biotechnological interventions are designed for such characteristics as high biomass yield, rapid decomposition, resistance to disease, and increased nutrient content. Although they provide additional agronomic benefits, their adoption might depend on the cost and availability of seeds and their agronomic compatibility with local practices (Singh et al., 2020).

13.2.3 Mechanisms of Soil Enrichment

GMs enhance soil enrichment by direct and indirect means that enhance nutrient availability, soil structure, and biological activity (Khan et al., 2020).

13.2.4 Nitrogen Fixation

Leguminous GMs increase soil nitrogen content by a large margin through biological nitrogen fixation. Rhizobium-based symbiosis enables these crops to take inert nitrogen from the atmosphere and make it bioavailable. This decreases the reliance on synthetic nitrogen fertilizers and also helps to meet the nutrient requirements of subsequent crops, especially in rotational systems (Muindi, 2021).

13.2.5 Organic Matter Incorporation

When incorporated into the soil, GM biomass breaks down and provides significant organic matter. This organic input helps to improve the texture of the soil, increases the amount of water held, increases CEC, and stabilizes aggregates. These enhancements are very instrumental in nutrient retention, soil aeration, and long-term soil fertility (Kamran et al., 2021).

13.2.6 Weed Suppression

GMs also effectively control weeds because they are able to form dense canopies that prevent sunlight penetration and, as such, limit germination of weed seeds. Additionally, other species (e.g., rye and mustard) exude allelopathic compounds that suppress the growth of other competing plant species. This natural weed control cuts the use of chemical herbicides and promotes cleaner production systems (Sultana et al., 2024).

13.3 Impacts on Soil Physical, Chemical, and Biological Properties

Integration of GMs brings about multidimensional enhancements in soil quality.

13.3.1 Physical Properties

GMs increase porosity, bulk density, and aggregate stability of soil. Improved soil structure allows for easy root penetration, more water infiltration, and less erosion risk. These impacts are especially important in degraded or compacted soils, where GMs can increase recovery and resilience rates (Prajapati et al., 2023).

13.3.2 Chemical Properties

The decomposition of GM pumps out critical macronutrients (N, P, K) and micronutrients (e.g., Mg, Zn, and Fe) into the soil. Organic acids formed during decomposition can make otherwise unavailable nutrients available, therefore increasing plant nutrient uptake. In addition, higher organic matter enhances the pH buffering and reduces nutrient leaching (Michael, 2021).

13.3.3 Biological Properties

GM inputs increase microbial diversity, activity, and biomass due to the availability of carbon-rich substrates for microbial metabolism. This microbial population increases enzymatic functions that are used in nutrient cycling, including nitrogen mineralization, phosphorus solubilization, and organic matter decomposition. Such dynamics of microbial life enhance the soil's ecological function and add to disease suppression and the system resilience (Tian et al., 2021).

13.4 Next-Generation Fertilizers: Innovations and Efficacy

13.4.1 Types and Technologies

The development of fertilizer technologies has resulted in the emergence of the next-generation fertilizers that are aimed at improving nutrient use efficiency (NUE) and reducing environmental impacts. These innovations include controlled-/slow-release fertilizers, nanofertilizers, and polymer-coated fertilizers (Lam et al., 2022).

13.4.2 Controlled-/Slow-Release Fertilizers (CRFs/SRFs)

CRFs and SRFs are meant for a slow release of nutrients to match the rate of plant uptake of nutrients. This is a controlled release that reduces the losses of nutrients by leaching and volatilization. CRFs are normally produced by encapsulating or coating fertilizer granules with materials that regulate the diffusion of nutrients like sulfur, polyolefins, or polyurethane. Recent developments are related to biodegradable and eco-friendly coatings, for example, the ones based on vegetable oils, starch, or chitosan to overcome environmental issues with synthetic polymers (Ganetri et al., 2021; Jariwala et al., 2022).

13.4.3 Nanofertilizers

Nanofertilizers are used in the process where nanoparticles are used to transfer nutrients to the plants more efficiently. Their high surface area to volume ratio provides for the increased absorption of nutrients and the controlled release (Anas & Quraishi, 2025). For example, chitosan-coated zinc nanoparticles have been found to delay the release of zinc, thus decreasing soil toxicity and restricting plants from accumulating too much. In addition, silica nanoparticles can adhere to soil, whereby they can retain important nutrients such as nitrogen, phosphorus, and potassium, thus enhancing the availability of nutrients for a long period of time (Al-Juthery et al., 2021; Daqa et al., 2023).

13.4.4 Polymer-Coated Fertilizers

In polymer-coated fertilizers, the granules of nutrients are protected by polymers that manage the speed of nutrient delivery. The coatings can be developed to respond to changes in temperature, pH, or humidity to ensure that nutrients are given to plants as required (Kassem et al., 2024). In this field, scientists have created types of smart coatings such as pH-sensitive, thermosensitive, enzymatic-sensitive, or self-healing to help improve and sustain fertilizer use (Mendonca Cidreira et al., 2025).

13.5 Role in Enhancing Nutrient Use Efficiency (NUE)

Using next-generation fertilizers is important for making better use of nutrients and for reducing their losses. For example, applying polymers to CRFs has helped nutrients reach plants at the right time, boosting yields and reducing nitrogen losses into the environment (Moradi et al., 2024). Similarly, because nanofertilizers are so small, they are taken up by plant cells and help the plants to use and absorb more nutrients. Technology allows farmers to use less fertilizer, less often, and it also helps the environment (Moradi et al., 2024). Comparative summary of next-generation fertilizers and green manure effects on soil fertility, microbial activity, and environmental outcomes are given in Table 13.1.

13.5.1 Environmental Footprint and Lifecycle Assessment

Concerns about fertilizer impact the environment primarily because of nutrient leakage into water, their role in global warming, and the use of valuable resources. New types of fertilizers address these problems by releasing their nutrients slowly and in low doses (Zheng et al., 2025). For example, research shows that CRFs decrease the amount of nitrogen that either leaches away or turns into a gas, thereby decreasing the release of GHGs and nutrient runoff. Besides, making biodegradable coatings for CRFs reduces the pollution from plastic and soil contamination (Jariwala et al., 2022). While producing these

Table 13.1 Comparative summary of next-generation fertilizers and green manure effects on soil fertility, microbial activity, and environmental outcomes.

Parameter	Controlled-release fertilizers (CRFs)	Nanofertilizers	Biofertilizers	Green manure	Reference
Nutrient release pattern	Slow, synchronized with crop demand	Controlled, efficient uptake	Gradual through microbial processes	Depends on the decomposition rate	Priya et al. (2024)
Nutrient use efficiency (NUE)	High	Very high	Moderate to high	Moderate, improves over time	Salim and Raza (2020)
Soil organic matter	No direct contribution	No direct contribution	Indirect via microbial stimulation	Significant increase	Bhattacharyya et al. (2022)
Microbial activity	Neutral to slightly positive	Neutral to variable	Strongly positive	Strongly positive	Ning et al. (2021)
Environmental impact	Low nutrient loss, biodegradable forms	Low emissions, minimal leaching	Eco-friendly	Sequesters carbon, reduces N_2O/CH_4	Sharma et al. (2022)
Cost	High	High	Low to moderate	Low	Nawadkar et al. (2024)
Adoption barriers	Cost, awareness	Cost, availability	Storage, shelf-life	Labor, training, policy support	Almutari (2023)
Best used in	High-input intensive systems	Precision and high-value crops	Organic, sustainable systems	All cropping systems with support	Rehman et al. (2022)

fertilizers is costly because of the advanced ways they are made, the positive effect on the environment can ultimately make up for the costs. Sustainable materials used for coating CRFs also play a role in a circular economy by reducing the use of resources that cannot be replaced and waste from production (Wu et al., 2021).

The synergistic application of next-generation fertilizers and GM represents a holistic method for nutrient management, aligning biological and technological approaches to optimize sustainability. This synergy is conceptually represented in Figure 13.1, which outlines the integrated pathways through which this input enhances nutrient cycling, soil health, and crop productivity while mitigating environmental degradation.

Figure 13.1 shows that the use of GM along with advanced fertilizers helps make nutrients more effective, increases the health of soil, supports a variety of microorganisms, and lessens the effect on the environment. Among other things, it points out nitrogen fixation, timed nutrient availability, and assistance to microbes, which lead to higher crop yields and greater resistance.

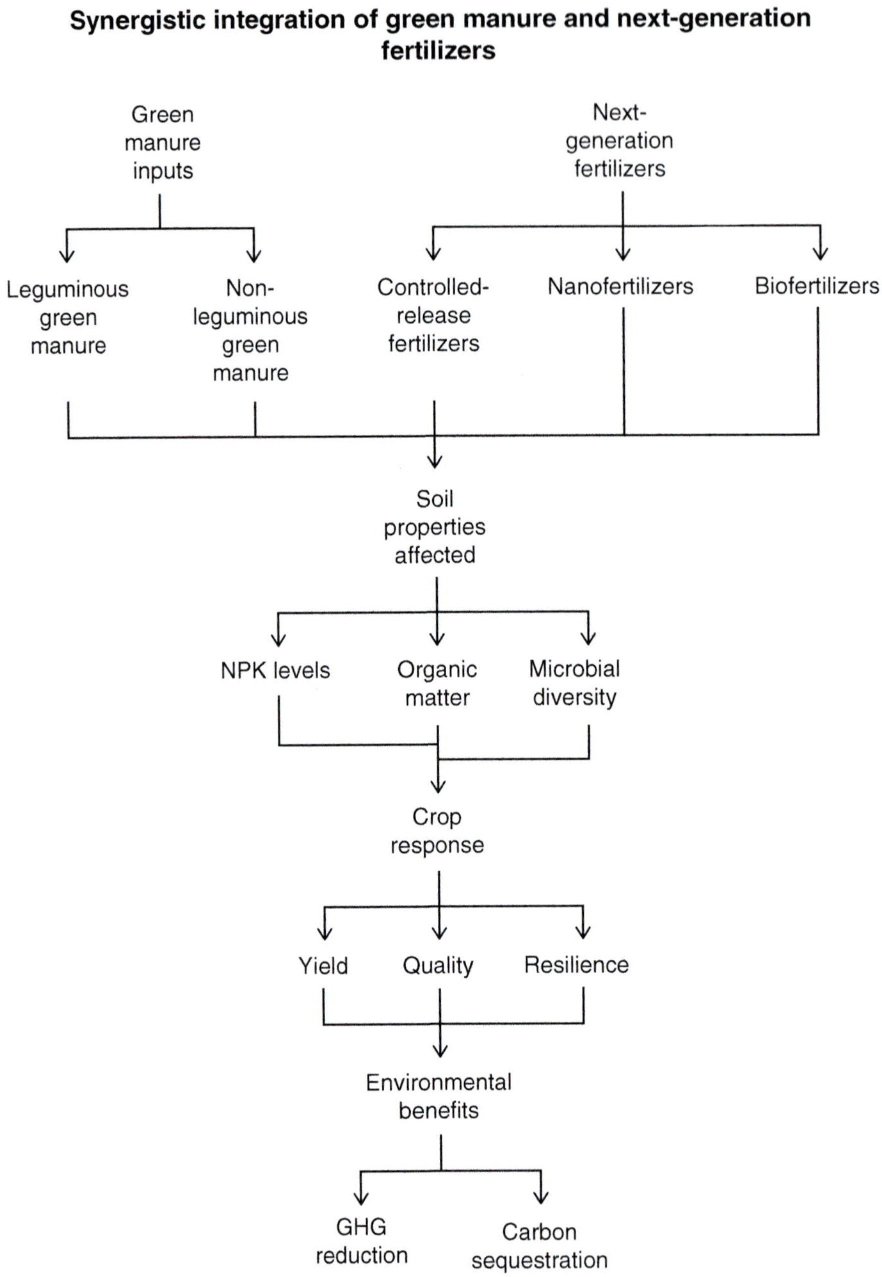

Figure 13.1 Conceptual framework illustrating the integrated use of green manure and next-generation fertilizers (CRFs, nanofertilizers, biofertilizers) for sustainable agriculture.

13.6 Challenges in Adoption and Cost–Benefit Analysis

In spite of the promising benefits of the next-generation fertilizers, there are a number of challenges that prevent them from being adopted on a large scale.

13.6.1 Production Costs

The manufacturing processes for CRFs and nanofertilizers are usually not different from the use of advanced technologies and materials; hence, the cost of production is higher than that of conventional fertilizers. For instance, the use of biodegradable polymers and nanomaterials may increase the cost of production, and this could hinder small-scale farmers from affording the production (Singh et al., 2024).

13.6.2 Raw Material Availability

Accessibility and cost of raw materials like polymers and urea may affect the production levels and prices of the next-generation fertilizers. Availability of raw materials fluctuates, which is a challenge to producers, and this may also alter the affordability and accessibility of these fertilizers (Prajapati et al., 2022).

13.6.3 Market Competition

Heavy competition from other slow-release fertilizers and conventional fertilizers can affect the market dynamics. Quality, efficiency, and sustainability differentiation are critical in the retention of a competitive edge in the market (Li & Li, 2024).

13.6.4 Adoption Rates

The uptake of polymer-coated fertilizers may be sluggish in areas with a low level of awareness or a lack of access to advanced agricultural inputs. Educating farmers and the availability of sustainable fertilizers are critical in market penetration (Kalita et al., 2024).

13.6.5 Cost–Benefit Analysis

Although the next generation of fertilizers has environmental advantages, the upfront costs could discourage their use. However, research has proven that the efficiency gains and the potential labor and fertilizer savings can compensate for the increased costs in the long run (Lam et al., 2022). For example, a study on polymer-coated urea showed that after a single application, 75% of the labor costs and 24–62% of the costs of agricultural inputs were saved, though the blended treatment was 13–14% more expensive than traditional fertilizers (Durham & Mizik, 2021).

13.7 Integrated Strategies: Green Manure and Fertilizer Reduction Synergies

13.7.1 Scientific Rationale and Agronomic Basis

The combination of GM with less chemical fertilizer application is an essential change to sustainable intensification of agriculture. The scientific basis of such an approach is the functional complementarity of organic and inorganic nutrient sources. GM, particularly leguminous species, increases soil nitrogen by biological nitrogen fixation, increases organic matter, and activates favorable microbial populations (Shrestha et al., 2021). These effects offset decreased external fertilizer inputs through maintaining soil fertility and enhancing the nutrient cycling efficiency. GM incorporation intensifies microbial interactions in the rhizosphere, thus increasing nutrient mineralization, enzymatic activities, and more synchronized nutrient availability, which matches the crop demand (Lyu et al., 2024). Furthermore, the mixture of organic and inorganic sources lessens the risk of nutrient loss, increases the cation exchange capacity, and enhances stable aggregate formation, thereby providing the optimal rhizosphere environment. This integrative approach fits into the larger objectives of ecological intensification and regenerative agriculture, which allow for productivity while maintaining the health of the soil and decreasing the environmental burden of agrochemical use (Liu et al., 2022).

13.7.2 Yield and Quality Enhancement in Major Crops

Many studies have shown that the practice of GM combined with a moderate amount of fertilizer can sustain or even increase crop yields and improve the quality of produce. In rice systems, Sesbania and Azolla as GM sources in combination with 50–75% of the recommended nitrogen rates have been demonstrated to give yields that are comparable to 100% chemical fertilizer applications (Kimani et al., 2021). Similarly, field studies in maize have also confirmed that the use of sunhemp or cowpea biomass and reduced nitrogen fertilizer is not only able to sustain grain yield but also improve kernel protein and micronutrient content (Dzvene et al., 2024).

In wheat production, the integration of leguminous GMs with 70% of conventional nitrogen input enhanced biomass production and grain yield, and enhanced better root development and increased chlorophyll content in plants (Bhardwaj et al., 2021). Research in pulse crops also suggests enhanced nodulation and biological nitrogen fixation under low nitrogen levels combined with GM residues. Such improvement is not only due to the contribution of GMs to nutrient supply but also because of their ability to alleviate soil pH, increase the microbial biomass nitrogen, and increase the availability of micronutrients such as zinc and iron that are essential for crop development and nutritional value (Barbieri et al., 2023). The GM application initiates a beneficial effect on soil properties and biological processes, which together enhance the nutrient dynamics and plant growth as in Figure 13.2.

The flowchart reveals that GM is involved in fixing nitrogen, making the soil richer, boosting microbes, suppressing weeds, and enhancing soil patterns, chemical and biological properties. They enhance the amount of nutrients in the soil, improve its condition, and make agriculture more sustainable.

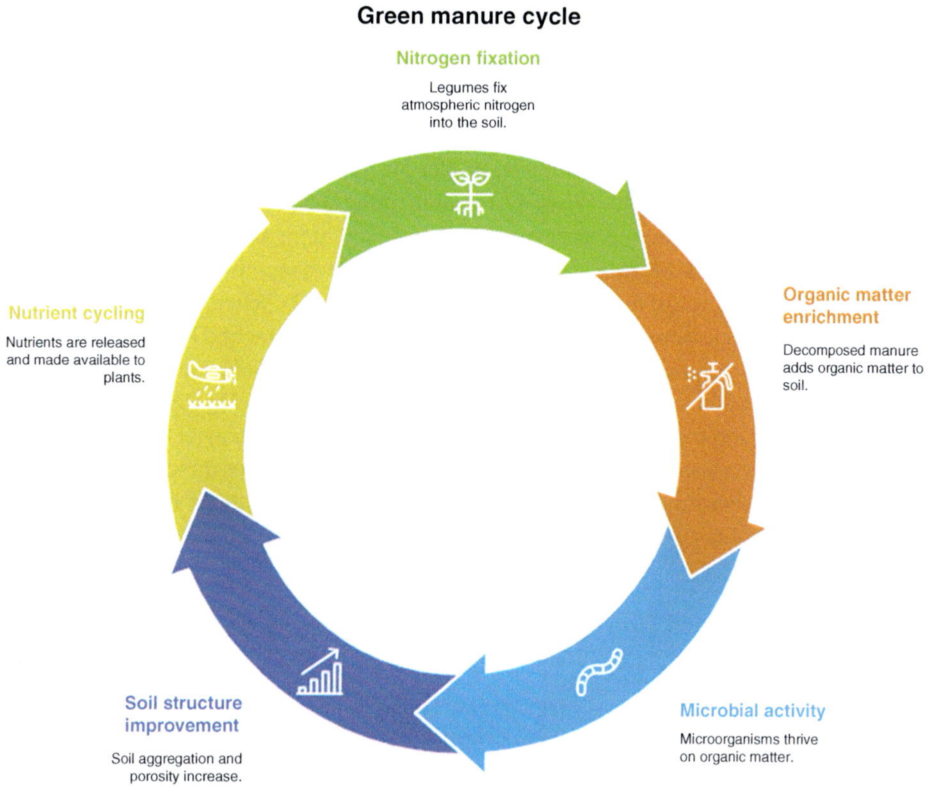

Figure 13.2 Mechanisms by which green manure enhances soil health and fertility.

13.7.3 Fertilizer Input Reduction Without Yield Penalty

Probably the strongest argument for the use of integrated approaches is the possibility of decreasing the application of synthetic fertilizers without compromising yields. Multi-season field experiments in different agroecological zones have generated strong empirical evidence (Selim, 2020). For example, incorporating GM in rice–wheat rotations enabled a 30–40% reduction in the amount of synthetic nitrogen applied without a significant reduction in yield. These findings are supported in tropical systems where the legumes such as Mucuna, crotalaria, and pigeon pea have been successfully incorporated in the cereal rotation to decrease the reliance on the urea-based fertilizers (Kumar et al., 2020b).

The secret of such success is the synchronization of GMs' nutrient release with the patterns of crops' nutrient uptake. Timely incorporation, for example, two to three weeks before sowing, ensures the mineralization of organic residues. In addition, GM inhibits weeds and enhances soil moisture retention, which are indirect ways of improving crop performance under low-fertilizer regimes (Ghaffar et al., 2022). Further economic analysis shows that farmers will pay off initial labor and biomass management costs and reap the benefits of lower fertilizer costs and improved resilience of soil in the long run, thus making their farming systems more cost-efficient in the long run (Vanlauwe et al., 2023).

13.7.4 Soil Health and Carbon Sequestration Benefits

When GM is used with lower fertilizer approaches, it helps to improve the condition of the soil and to store more carbon. GM delivers much labile carbon to the soil, boosting the activity of microorganisms and raising their overall carbon and nitrogen levels. As a result, the microbial loop becomes stronger, making it simpler for nutrients to be changed into forms useful for plants (Li et al., 2020b). When GM is used along with less fertilizer, the levels of dehydrogenase, phosphatase, and urease are higher in soil, demonstrating that the soil is working better biologically. More types of microbes contribute to recycling nutrients and help control the growth of harmful microbes (Asghar & Kataoka, 2022). When it comes to carbon, using GM for a long time improves the maintenance of organic matter and helps build higher levels of soil organic carbon, as well as decreasing climate change. Yadav et al.'s study found that bio-integrated systems in paddy and upland soils can store 0.3–0.6 Mg C per hectare each year, depending on the biomass added and tillage practices used (Yadav et al., 2024). As a result, soil can better resist storms and heavy rainfall because it becomes more stable, erosion is less likely, and it retains more water. All these changes make it possible for integrated nutrient management systems to maintain high productivity (Abdallah et al., 2021).

13.8 Microbial and Biogeochemical Interactions in Green Manure Systems

13.8.1 Green Manure Influence on Soil Microbial Diversity

The use of GM can significantly influence the microorganisms living in the soil. Studies have shown that GM crops such as ryegrass affect the types and activity of microbes that cycle carbon, nitrogen, and sulfur (Tao et al., 2022). Specifically, studies have found that the phylum Actinobacteria, which breaks down complex organic materials, is increasing in the soil. Some studies also suggest that GM crops can affect fungal diversity, leading to a decrease in the number of species present (Liao et al., 2021).

13.8.2 Nutrient Cycling and Rhizosphere Dynamics

When GM is added to the soil, it increases the movement of nutrients, mainly in the area around the roots and related to the activities of microbes. The presence of GM at a site increases the activity of microbes, which leads to the release of nitrogen and phosphorus (Paillat et al., 2020). It improves the amount of nutrients accessible to plants and encourages the growth of helpful microbes. GM may also change the composition of rhizosphere microbes, which helps nutrient cycling and supports growth-promoting populations (Kong & Liu, 2022).

13.8.3 Role of Mycorrhizae and PGPR in Enhancing Fertilizer Efficiency

Arbuscular mycorrhizal fungi (AMF) and plant growth-promoting rhizobacteria (PGPR) have important effects in improving nutrient absorption and fertilizer utilization. AMF enter into symbiotic relationships with the plant roots and increase the scope of their abilities to absorb water and nutrients, especially phosphorus, thereby increasing them

(Abkenar et al., 2024). PGPR such as Pseudomonas and Bacillus species can fix atmospheric nitrogen, solubilize phosphorus, and produce phytohormones, which lead to stimulation of growth in plants. Combined use of GM, AMF, and PGPR has been demonstrated to increase NUE so that at lower levels of chemical fertilizers, equivalent crop yields can be achieved (Etesami & Adl, 2020).

13.8.4 Green Manure and Microbial Carbon/Nitrogen Pathways

GM application affects soil microbial-mediated carbon and nitrogen routes. The incorporation of organic matter from GM increases microbial biomass carbon and nitrogen and consequently enhances enzyme activities that are responsible for nutrient cycling, like dehydrogenase and phosphatase. These changes help enhance the soil fertility and structure, as well as the carbon sequestration. Long-term application of GM has been linked to high values of soil organic carbon and nitrogen, thus suggesting its importance in sustainable soil management (Yadav et al., 2021).

13.9 Climate Resilience and Environmental Sustainability Through Green Manure Integration

13.9.1 Mitigation of Greenhouse Gas Emissions

The GM practices have come forth as a sustainable alternative to conventional fertilization practices, with potential advantages in the reduction of GHG emissions, mainly, nitrous oxide (N_2O) and methane (CH_4) (Daba et al., 2025). Farmers using milk-vetch rather than chemical fertilizer find that they emit less N_2O and do not see a drop in their yields. The drop in nitrogen is due to better nitrogen uptake and different changes in the soil's microorganisms (Yang et al., 2022). Yet, it is not as straightforward when it comes to GM and CH_4 emissions. The use of organic materials from GM food by methanogens may lead to more methane production in the anaerobic environment found in flooded rice paddies. Suggestions have been made to manage rice fields by draining excess water during the growing season, so CH_4 emissions are reduced and GM's benefits can be appreciated (Toma et al., 2021). In addition, integrating GM with less tillage techniques seems to reduce N_2O emissions in some soils (such as sandy loam) without affecting the harvested yields. Consequently, it is necessary to use management strategies tailored to each situation to get the best results from using GM plants (Shakoor et al., 2021).

13.9.2 Resilience to Abiotic Stresses

The adoption of GM in cropping systems increases resistance to abiotic stresses, such as drought, salinity, and high/low temperature. GM helps in soil structure improvement and improvement in organic matter content, which helps in increasing the water retention capacity. For example, GM-enriched soils have shown that they can hold up to 20 times their weight in water, acting as a buffer against drought (Villalobos-López et al., 2022). In saline environments, the associations between plants and mycorrhizal fungi, stimulated by

GM practices, are capable of increasing the plant's tolerance to salinity stress. These symbioses enhance nutrient uptake and aid in ion balance within the tissues of a plant and, therefore, avert salinity impacts (Khoza et al., 2025). In addition, GM practices can alter soil temperature and moisture regimes, which helps plants to be resilient to temperature extremes. The additional organic matter from GM provides insulation for soil temperatures and minimizes thermal stress on the roots of plants (KhokharVoytas et al., 2023).

13.9.3 Contribution to Sustainable Development Goals (SDGs)

The adoption of GM practices supports several United Nations SDGs, which support environmental sustainability and socioeconomic development.

SDG 2: Zero Hunger—GM helps to improve the soil fertility and crop yields, thus contributing to food security and sustainable agriculture (Malik, 2024).
SDG 13: Climate Action—GM practices contribute to the mitigation of climate change by the reduction of GHG emissions and improvement of carbon sequestration in soils (Arora & Mishra, 2023).
SDG 15: Life on Land—Better soil health and biodiversity because of GM integration enhances the conservation and restoration of terrestrial ecosystems (Arora & Mishra, 2024).

Moreover, GM practices support SDG 6: Clean Water and Sanitation through decreasing nutrient runoff, and improving water quality (van Koppen, 2024), and SDG 12: Responsible Consumption and Production through promoting sustainable land management practices (Hales & Birdthistle, 2023).

13.10 Economic and Policy Considerations in Green Manure Integration

13.10.1 Cost–Benefit Analysis of Integrated Practices

The introduction of GM into agricultural systems offers economic opportunities and challenges. Although initial adoption can be accompanied by the cost of labor, seed, and possible adjustments in short-term yields, long-term benefits tend to overshadow the expenses (Kawa, 2021).

13.10.2 Input Savings and Yield Gains

The inclusion of GM can bring dramatic decreases in the use of synthetic fertilizers, which means that farmers can save money. For example, studies have revealed that GM incorporation enhanced soil total nitrogen, soil available phosphorus, soil available potassium, and soil organic carbon sequestration rate by 13%, 8%, 8%, and 64%, respectively (Zhang et al., 2025). Consequently, GM inclusion enhanced the mean production of the subsequent season by 12%, whereby maize yields were significantly greater than those of rice and tobacco (Saquee et al., 2023).

13.10.3 Long-Term Soil Productivity

Besides the short-term benefits from yield, GM practices increase soil health through improvement in organic matter content, soil structure, and microbial activity. Such changes lead to soil fertility and productivity sustained in time without the need for external inputs and increased resilience of farming systems (Shahane & Shivay, 2021).

13.10.4 Incentives and Subsidy Models

Policies of the government and subsidy programs are very important in ensuring that GM practices are embraced.

13.10.5 India

India's spending on input subsidies has shot up in the recent past. As a percentage of agricultural output, the cost of India's agricultural input subsidies almost doubled to 11.6% in 2009–2010 from 6.0% in 2003–2004, with the increase largely due to large increases in the subsidies to fertilizer and electricity. However, there is an increasing understanding of the necessity to shift subsidies to sustainable practices such as the GM integration (Shah, 2021).

13.10.6 China

In China, the policy incentives for GM cultivation are not adequate; for example, there are no seed subsidies, technical training, and extension services. This dependence on spontaneous and experimental efforts of individual farmers makes it difficult for the technology of GM to be embraced on a large scale (Li et al., 2020a).

13.10.7 Sub-Saharan Africa

Agricultural input subsidies have been an important policy instrument in sub-Saharan Africa since the 1960s and 1970s. These subsidies have grown from demonstrating packs to big subsidy programs, which are mostly aimed at the small-scale farmers. Although these programs have been successful to some extent, the incorporation of GM in subsidy schemes might make them more effective as it will ensure sustainable farming practices (Yusuf, 2021).

13.11 Role of Extension Services and Farmer Training

Prominent extension services, farmers' training are vital in the successful adoption of GM practices.

13.11.1 Farmer-to-Farmer Extension

Traditional training models do not have a wide enough reach in such a way that many farmers are left out to lack information on sustainable practices such as GM. Farmer-to-farmer extension (F2F) has developed as an effective alternative. From studies, it has been found that

F2F more than doubled the probability of farmers' adoption of rice-GM rotation systems by 22.5% which confirmed its effectiveness (Sah et al., 2021).

13.11.2 Institutional Support

Interaction between the governments, research bodies, extension services, and farmer organizations is essential in developing an enabling environment for the GM practices implementation. This kind of cooperation can offer farmers enough knowledge, resources, and care that they need to implement GM effectively (Prajapati et al., 2025).

13.11.3 Policy Recommendations for Global Adoption

To ensure the acceptance of many of the GM practices, the following policy recommendations are suggested:

Reorient Subsidies: Move agricultural subsidies from traditional inputs toward sustainable practices, such as GM integration (Quiroga et al., 2024).
Enhance Extension Services: Invest in extension services and training programs that will equip the farmers with the knowledge and skills to implement GM (Osumba et al., 2021).
Support Research and Development: Fund research to establish region-specific GM practices and the long-term effects on soil health and productivity (Fan et al., 2025).
Promote Collaborative Approaches: Facilitate the collaborations amongst governments, research institutions, and farmer organizations to develop supportive networks for GM adoption (Pandey et al., 2022).
Monitor and Evaluate Programs: Develop a monitoring and evaluation system to gauge GM programs and guide policy changes (Pearce-Higgins et al., 2022).

13.12 Knowledge Gaps and Research Frontiers in Green Manure Integration

13.12.1 Long-Term Field Studies and Meta-analysis Needs

Although many research projects have revealed how GM can help the soil and crops, long-term field studies investigating the continued effects of GM are still lacking. To understand how GM affects these things over time, we need such research. Besides, current meta-analyses mainly study short-term results, making it difficult to use them when formulating long-term policies for agriculture. More research should be done that looks at GM crops in many different geographical, soil, and farming settings (Lei et al., 2022; Liang et al., 2022).

13.12.2 Standardizing Metrics for Soil and Crop Response

Different standards are used to measure soil and crop responses; it is hard to compare results from various studies. Because the methods to measure soil organic carbon, nutrients, microbes, and crop yield are not all the same, it is hard to compare results. Having agreed protocols for collecting and analyzing data is necessary to allow data from different studies to be compared. Using standard metrics would allow us to develop models that predict the advantages of GM integration for various situations (Siedt et al., 2021).

13.12.3 Integration with Precision Agriculture and Remote Sensing

Improvements in GM techniques now depend on precision agriculture and remote sensing. Crop health, water in the soil, and nutrients are monitored with satellite and drone images, which helps make GM applications more effective (Inoue, 2020). GM in agriculture can be used to make sure resources are used well and to reduce harm to the environment. However, very little is being done to combine these technologies practically. More research is needed to link GM with modern farming technologies to achieve better results and help farmers be more sustainable (Brookes & Barfoot, 2020).

13.12.4 Potential of Genetically Enhanced Green Manure Species

Genetically modified GM plants could be an important development for sustainable farming. Various changes in genes can increase growth, nitrogen retention, resistance to pests, and the ability to withstand different weather conditions (Thakur et al., 2023). Even today, many are concerned about the safety of genetically modified species for humans, nature, and the public. Studying GM crops is important to determine any long-term consequences they might have. Besides, it is important to develop rules and standards to ensure that these technologies are handled safely and ethically (Kumar et al., 2020a).

13.13 Conclusion

With climate change getting worse, the quality of soil declines, and food shortages increase, which is why sustainable farming is essential everywhere. Here, we considered how using both next-generation fertilizers and GM can increase the sustainability and output of farming. For some time now, overuse of conventional fertilizer has caused environmental harm by letting nutrients escape, polluting waters, and releasing GHGs. Slow-release, nanofertilizers, and biofertilizers could help handle the problem by reducing nutrient-related damage and making better use of fertilizers. Moreover, these products are most effective when used together with methods to improve the quality of soil and its microbes. Improving and enriching farm soil is best done with the help of leguminous and native GM crops. They improve the soil's health, control weeds, and help fix nitrogen. Because of next-generation fertilizers, the soil improves, and farmers can achieve stable and better harvests by using fewer artificial fertilizers. Multiple techniques applied to rice, maize, wheat, and legume crops are effective in reducing the amount of fertilizer and do not affect how much is harvested. Much like that, these systems support the environment by helping to decrease carbon emissions, N_2O, and CH_4, which support the targets of the United Nations and SDGs. The usefulness of these systems is increased thanks to helpful interactions among microbes in the rhizosphere. When GM is used, the number of microbes increases, and PGPR plus mycorrhizae help the crops access nutrients and make the soil healthier for farming. Although there are obvious advantages, difficulties in the economy and organizations prevent everyone from using them. Before farmers are able to use this technology, they might have to pay a lot, use few tools, and receive no training. These issues can be addressed by introducing policies, regional subsidies, and better extension services, mainly in developing areas. Other issues to consider in the study are testing for a long time, measuring soil and crop changes using standard methods, and how effective precision farming devices are. Modifying the genes in GM crops could result in better advantages depending on the environment. Basically, blending GMs with current fertilizers is a sustainable way for farmers to operate. Doing this helps us unite in science, policy, and practice to support food supply, the environment, and climate readiness in the future.

References

Abdallah, A.M., Jat, H.S., Choudhary, M. et al. (2021). Conservation agriculture effects on soil water holding capacity and water-saving varied with management practices and agroecological conditions: A review. *Agronomy* 11 (9): 1681.

Abkenar, M.B., Mozafari, H., Karimzadeh, K. et al. (2024). Arbuscular mycorrhizal fungi (AMF) and plant growth-promoting rhizobacteria (PGPR) as an alternative to mineral fertilizers to improve the growth, essential oil profile, and phenolic content of satureja macrantha L. *Journal of Crop Health* 76 (1): 347–356.

Al-Juthery, H. W., Lahmod, N. R., & Al-Taee, R. A. (2021). Intelligent, nano-fertilizers: A new technology for improvement nutrient use efficiency (article review). IOP conference series: Earth and environmental science. Iraq (Virtual Conference, 21–22 January 2021

Almutari, M.M. (2023). Synthesis and modification of slow-release fertilizers for sustainable agriculture and environment: a review. *Arabian Journal of Geosciences* 16 (9): 518.

Anas, M. and Quraishi, U.M. (2025). Green-fabricated MnO_2 nanoparticles function as dual nanofertilizers and chromium remediators, enhancing antioxidant pathways, ionomic networks, and physiological resilience in wheat. *Journal of Trace Elements in Medicine and Biology* 127661. https://doi.org/10.1016/j.jtemb.2025.127661.

Anas, M., Khattak, W.A., Fahad, S. et al. (2025). Mitigating heavy metal pollution in agriculture: A multi-omics and nanotechnology approach to safeguard global wheat production. *Journal of Hazardous Materials Advances* 17: 100584. https://doi.org/10.1016/j.hazadv.2024.100584.

Arora, N.K. and Mishra, I. (2023). Sustainable development goal 13: recent progress and challenges to climate action. *Environmental Sustainability* 6 (3): 297–301.

Arora, N.K. and Mishra, I. (2024). Life on land: progress, outcomes and future directions to achieve the targets of SDG 15. In: *Emerging technologies and marketing strategies for sustainable agriculture*, 1–7. Springer.

Arora, P.K., Tripathi, S., Omar, R.A. et al. (2024). Next-generation fertilizers: the impact of bionanofertilizers on sustainable agriculture. *Microbial Cell Factories* 23 (1): 254.

Asghar, W. and Kataoka, R. (2022). Green manure incorporation accelerates enzyme activity, plant growth, and changes in the fungal community of soil. *Archives of Microbiology* 204 (1): 7.

Atapattu, A.J., Ranasinghe, C., Nuwarapaksha, T.D. et al. (2024). Sustainable agriculture and sustainable development goals (SDGs). In: *Emerging technologies and marketing strategies for sustainable agriculture* (ed. J. Garwi, R. Masengu, and O.T. Chiwaridzo), 1–27. IGI Global.

Babu, S., Singh, R., Yadav, D. et al. (2022). Nanofertilizers for agricultural and environmental sustainability. *Chemosphere* 292: 133451.

Barbieri, P., Starck, T., Voisin, A.-S., and Nesme, T. (2023). Biological nitrogen fixation of legumes crops under organic farming as driven by cropping management: A review. *Agricultural Systems* 205: 103579.

Bhardwaj, A.K., Rajwar, D., Yadav, R.K. et al. (2021). Nitrogen availability and use efficiency in wheat crop as influenced by the organic-input quality under major integrated nutrient management systems. *Frontiers in Plant Science* 12: 634448.

Bhattacharyya, S.S., Ros, G.H., Furtak, K. et al. (2022). Soil carbon sequestration—an interplay between soil microbial community and soil organic matter dynamics. *Science of the Total Environment* 815: 152928.

Brookes, G. and Barfoot, P. (2020). Environmental impacts of genetically modified (GM) crop use 1996–2018: impacts on pesticide use and carbon emissions. *GM Crops & Food* 11 (4): 215–241.

Cárceles Rodríguez, B., Durán-Zuazo, V.H., Soriano Rodríguez, M. et al. (2022). Conservation agriculture as a sustainable system for soil health: A review. *Soil Systems* 6 (4): 87.

Chawla, R., Poonia, A., Samantara, K. et al. (2023). Green revolution to genome revolution: driving better resilient crops against environmental instability. *Frontiers in Genetics* 14: 1204585.

Daba, N.A., Huang, J., Shen, Z. et al. (2025). Green manure substitution for chemical nitrogen reduces greenhouse gas emissions and enhances yield and nitrogen uptake in ricerice cropping systems. *Field Crops Research* 322: 109715.

Daqa, W.M., Alshoaibi, A., Ahmed, F., and Rao, T.N. (2023). Potential applications of chitosan-coated zinc oxide nanoparticles for degrading pesticide residues in environmental soils. *Crystals* 13 (3): 391.

Durham, T.C. and Mizik, T. (2021). Comparative economics of conventional, organic, and alternative agricultural production systems. *Economies* 9 (2): 64.

Dzvene, A.R., Gura, I., Tesfuhuney, W. et al. (2024). Effect of intercropping maize and sunn hemp at different times and stand densities on soil properties and crop yield under in-field rainwater harvesting (IRWH) tillage in semi-arid South Africa. *Plant and Soil* 505 (1): 363–379.

Etesami, H. and Adl, S.M. (2020). Plant growth-promoting rhizobacteria (PGPR) and their action mechanisms in availability of nutrients to plants. In: *Phyto-Microbiome in stress regulation*, 147–203.

Fan, H., Miao, R., Guo, C. et al. (2025). Research on the effect of diversified cropping on crop quality: A review. *Agriculture* 15 (5): 456.

Filonchyk, M., Peterson, M.P., Zhang, L. et al. (2024). Greenhouse gases emissions and global climate change: Examining the influence of CO_2, CH_4, and N_2O. *Science of the Total Environment* 935: 173359.

Francaviglia, R., Almagro, M., and Vicente-Vicente, J.L. (2023). Conservation agriculture and soil organic carbon: Principles, processes, practices and policy options. *Soil Systems* 7 (1): 17.

Ganetri, I., Essamlali, Y., Amadine, O., Danoun, K., Aboulhrouz, S., & Zahouily, M. (2021). Controlling factors of slow or controlled-release fertilizers. In F. B. Lewu, T. Volova, S. Thomas, & R. K.R *Controlled release fertilizers for sustainable agriculture* (pp. 111–129). Elsevier.

Ghaffar, A., Rahman, M.H.U., Ahmed, S. et al. (2022). Adaptations in cropping system and pattern for sustainable crops production under climate change scenarios. In: *Improvement of plant production in the era of climate change* (ed. S. Fahad, M. Adnan, and S. Saud), 1–34. CRC Press.

Hales, R. and Birdthistle, N. (2023). The sustainable development goals–SDG# 12 responsible consumption and production. In: *Attaining the 2030 sustainable development goal of responsible consumption and production*, 1–10. Emerald Publishing Limited.

Iderawumi, A.M. and Kamal, T.O. (2022). Green manure for agricultural sustainability and improvement of soil fertility. *Farming and Management* 7 (1): 1–8.

Inoue, Y. (2020). Satellite-and drone-based remote sensing of crops and soils for smart farming—a review. *Soil Science and Plant Nutrition* 66 (6): 798–810.

Jariwala, H., Santos, R.M., Lauzon, J.D. et al. (2022). Controlled release fertilizers (CRFs) for climate-smart agriculture practices: a comprehensive review on release mechanism, materials, methods of preparation, and effect on environmental parameters. *Environmental Science and Pollution Research* 29 (36): 53967–53995.

Kalita, A., Elayarajan, M., Janaki, P. et al. (2024). Organo-monomers coated slow-release fertilizers: Current understanding and future prospects. *International Journal of Biological Macromolecules* 274: 133320.

Kamran, M., Huang, L., Nie, J. et al. (2021). Effect of reduced mineral fertilization (NPK) combined with green manure on aggregate stability and soil organic carbon fractions in a fluvo-aquic paddy soil. *Soil and Tillage Research* 211: 105005.

Kassem, I., Ablouh, E.-H., El Bouchtaoui, F.-Z. et al. (2024). Polymer coated slow/controlled release granular fertilizers: Fundamentals and research trends. *Progress in Materials Science* 144: 101269.

Kawa, N.C. (2021). A "win-win" for soil conservation? How indiana row-crop farmers perceive the benefits (and trade-offs) of no-till agriculture. *Culture, Agriculture, Food and Environment* 43 (1): 25–35.

Khan, M.I., Gwon, H.S., Alam, M.A. et al. (2020). Short term effects of different green manure amendments on the composition of main microbial groups and microbial activity of a submerged rice cropping system. *Applied Soil Ecology* 147: 103400.

KhokharVoytas, A., Shahbaz, M., Maqsood, M.F. et al. (2023). Genetic modification strategies for enhancing plant resilience to abiotic stresses in the context of climate change. *Functional & Integrative Genomics* 23 (3): 283.

Khoza, T., Masenya, A., Khanyile, N., and Thosago, S. (2025). Alleviating plant density and salinity stress in moringa oleifera using arbuscular mycorrhizal fungi: a review. *Journal of Fungi* 11 (4): 328.

Kimani, S.M., Bimantara, P.O., Kautsar, V. et al. (2021). Poultry litter biochar application in combination with chemical fertilizer and Azolla green manure improves rice grain yield and nitrogen use efficiency in paddy soil. *Biochar* 3 (4): 591–602.

Kocira, A., Staniak, M., Tomaszewska, M. et al. (2020). Legume cover crops as one of the elements of strategic weed management and soil quality improvement. A review. *Agriculture* 10 (9): 394.

Kong, Z. and Liu, H. (2022). Modification of rhizosphere microbial communities: a possible mechanism of plant growth promoting rhizobacteria enhancing plant growth and fitness. *Frontiers in Plant Science* 13: 920813.

van Koppen, B. (2024). Sustainable development goal 6: Ensuring the availability and sustainable management of water and sanitation for all. In: *Handbook on public policy and food security*, 165–173. Edward Elgar Publishing.

Kumar, K., Gambhir, G., Dass, A. et al. (2020a). Genetically modified crops: current status and future prospects. *Planta* 251 (4): 91.

Kumar, S., Meena, R.S., Datta, R. et al. (2020b). Legumes for carbon and nitrogen cycling: an organic approach. *Carbon and nitrogen cycling in soil* 337–375.

Lam, S.K., Wille, U., Hu, H.-W. et al. (2022). Next-generation enhanced-efficiency fertilizers for sustained food security. *Nature Food* 3 (8): 575–580.

Lei, B., Wang, J., and Yao, H. (2022). Ecological and environmental benefits of planting green manure in paddy fields. *Agriculture* 12 (2): 223.

Li, X. and Li, Z. (2024). Global trends and current advances in slow/controlled-release fertilizers: a bibliometric analysis from 1990 to 2023. *Agriculture* 14 (9): 1502.

Li, F., Ren, J., Wimmer, S. et al. (2020a). Incentive mechanism for promoting farmers to plant green manure in China. *Journal of Cleaner Production* 267: 122197.

Li, Z., Zhang, X., Xu, J. et al. (2020b). Green manure incorporation with reductions in chemical fertilizer inputs improves rice yield and soil organic matter accumulation. *Journal of Soils and Sediments* 20: 2784–2793.

Liang, K., Wang, X., Du, Y. et al. (2022). Effect of legume green manure on yield increases of three major crops in China: A meta-analysis. *Agronomy* 12 (8): 1753.

Liao, L., Wang, X., Wang, J. et al. (2021). Nitrogen fertilization increases fungal diversity and abundance of saprotrophs while reducing nitrogen fixation potential in a semiarid grassland. *Plant and Soil* 465 (1): 515–532.

Liu, S., He, F., Kuzyakov, Y. et al. (2022). Nutrients in the rhizosphere: A meta-analysis of content, availability, and influencing factors. *Science of the Total Environment* 826: 153908.

Lyu, H., Li, Y., Wang, Y. et al. (2024). Drive soil nitrogen transformation and improve crop nitrogen absorption and utilization-a review of green manure applications. *Frontiers in Plant Science* 14: 1305600.

Maja, M.M. and Ayano, S.F. (2021). The impact of population growth on natural resources and farmers' capacity to adapt to climate change in low-income countries. *Earth Systems and Environment* 5 (2): 271–283.

Malik, S.A. (2024). SDG 2: zero hunger, food and plant-related intellectual property, and access to plant genetic resources. In: *The elgar companion to intellectual property and the sustainable development goals* (ed. M. Rimmer, C.B. Ncube, and B. Amani), 50–73. Edward Elgar Publishing.

Mendonca Cidreira, A.C., Wei, L., Aldekhail, A., and Islam Rubel, R. (2025). Controlled-release nitrogen fertilizers: A review on bio-based and smart coating materials. *Journal of Applied Polymer Science* 142 (3): e56390.

Michael, P.S. (2021). Role of organic fertilizers in the management of nutrient deficiency, acidity, and toxicity in acid soils—a review. *Journal of Global Agriculture and Ecology* 12 (3): 19–30.

Moradi, S., Babapoor, A., Ghanbarlou, S. et al. (2024). Toward a new generation of fertilizers with the approach of controlled-release fertilizers: a review. *Journal of Coatings Technology and Research* 21 (1): 31–54.

Muindi, M.M. (2021). *Exploiting belowground biodiversity of rhizobia and arbuscular mycorrhizal fungi to upscale cowpea production in the semi-arid zones of eastern Kenya*. Kenyatta University.

Nawadkar, A., Chaskar, J., Tiwari, A., and Chaskar, A. (2024). Controlled release of nanofertilizers: a promising approach for efficient delivery of nutrients in plants. *Journal of Crop Improvement* 38 (3): 189–218.

Ning, Q., Hättenschwiler, S., Lü, X. et al. (2021). Carbon limitation overrides acidification in mediating soil microbial activity to nitrogen enrichment in a temperate grassland. *Global Change Biology* 27 (22): 5976–5988.

Nosheen, S., Ajmal, I., and Song, Y. (2021). Microbes as biofertilizers, a potential approach for sustainable crop production. *Sustainability* 13 (4): 1868.

Osumba, J.J., Recha, J.W., and Oroma, G.W. (2021). Transforming agricultural extension service delivery through innovative bottom–up climate-resilient agribusiness farmer field schools. *Sustainability* 13 (7): 3938.

Paillat, L., Cannavo, P., Barraud, F. et al. (2020). Growing medium type affects organic fertilizer mineralization and CNPS microbial enzyme activities. *Agronomy* 10 (12): 1955.

Pandey, N., de Coninck, H., and Sagar, A.D. (2022). Beyond technology transfer: Innovation cooperation to advance sustainable development in developing countries. *Wiley Interdisciplinary Reviews: Energy and Environment* 11 (2): e422.

Pearce-Higgins, J., Antão, L., Bates, R. et al. (2022). A framework for climate change adaptation indicators for the natural environment. *Ecological Indicators* 136: 108690.

Prajapati, D., Pal, A., Dimkpa, C. et al. (2022). Chitosan nanomaterials: A prelim of next-generation fertilizers; existing and future prospects. *Carbohydrate Polymers* 288: 119356.

Prajapati, S.K., Dayal, P., Kumar, V., and Gairola, A. (2023). Green manuring: a sustainable path to improve soil health and fertility. *International Journal* 1 (2): 24–33.

Prajapati, C.S., Priya, N.K., Bishnoi, S. et al. (2025). The role of participatory approaches in modern agricultural extension: bridging knowledge gaps for sustainable farming practices. *Journal of Experimental Agriculture International* 47 (2): 204–222.

Prasad, R. and Shivay, Y. (2021). A brief history of the fertilizer nitrogen. *Indian Journal of History of Science* 56 (1): 60–64.

Priya, E., Sarkar, S., and Maji, P.K. (2024). A review on slow-release fertilizer: Nutrient release mechanism and agricultural sustainability. *Journal of Environmental Chemical Engineering* 12: 113211.

Quintarelli, V., Radicetti, E., Allevato, E. et al. (2022). Cover crops for sustainable cropping systems: a review. *Agriculture* 12 (12): 2076.

Quiroga, S., Suárez, C., Santos-Arteaga, F.J., and Rodrigo, J.M. (2024). Do common agricultural policy subsidies matter for the market-environment trade off? An evaluation of R&D objectives and decisions across farmers. *Journal of Agriculture and Food Research* 15: 101047.

Rani, T.S., Ramulu, C., Kumar, T.S., and Reddy, R.U. (2024). Identification of superior breeding lines for soil chemical propertis in dhaincha (Sesbania aculeata L.) and sunhemp (Crotolaria Juncea L.). *Ecology, Environment & Conservation (0971765X)* 30: 7.

Raza, A., Zahra, N., Hafeez, M.B. et al. (2020). Nitrogen fixation of legumes: Biology and physiology. In: *The plant family Fabaceae: biology and physiological responses to environmental stresses* (ed. M. Hasanuzzaman, S. Araújo, and S.S. Gill), 43–74.

Rehman, A., Farooq, M., Lee, D.-J., and Siddique, K.H. (2022). Sustainable agricultural practices for food security and ecosystem services. *Environmental Science and Pollution Research* 29 (56): 84076–84095.

Sah, U., Singh, S., and Pal, J.K. (2021). Farmer-To-farmer extension (F2FE) approach for speedier dissemination of agricultural technologies: A review. *The Indian Journal of Agricultural Sciences* 91 (10): 1419–1425.

Salim, N. and Raza, A. (2020). Nutrient use efficiency (NUE) for sustainable wheat production: a review. *Journal of Plant Nutrition* 43 (2): 297–315.

Saquee, F.S., Norman, P.E., Saffa, M.D. et al. (2023). Impact of different types of green manure on pests and disease incidence and severity as well as growth and yield parameters of maize. *Heliyon* 9 (6): e17294.

Scavo, A., Fontanazza, S., Restuccia, A. et al. (2022). The role of cover crops in improving soil fertility and plant nutritional status in temperate climates. A review. *Agronomy for Sustainable Development* 42 (5): 93.

Selim, M.M. (2020). Introduction to the integrated nutrient management strategies and their contribution to yield and soil properties. *International Journal of Agronomy* 2020 (1): 2821678.

Shah, D. (2021). Central government agricultural subsidies in india: public sector expenditure, issues and policy implications.

Shahane, A.A. and Shivay, Y.S. (2021). Soil health and its improvement through novel agronomic and innovative approaches. *Frontiers in Agronomy* 3: 680456.

Shakoor, A., Shahbaz, M., Farooq, T.H. et al. (2021). A global meta-analysis of greenhouse gases emission and crop yield under no-tillage as compared to conventional tillage. *Science of the Total Environment* 750: 142299.

Sharma, G.K., Khan, S.A., Shrivastava, M. et al. (2022). Phycoremediated N-fertilization approaches on reducing environmental impacts of agricultural nitrate leaching. *Journal of Cleaner Production* 345: 131120.

Shrestha, J., Subedi, S., Timsina, K.P. et al. (2021). Sustainable intensification in agriculture: an approach for making agriculture greener and productive. *Journal of Nepal Agricultural Research Council* 7: 133–150.

Siedt, M., Schäffer, A., Smith, K.E. et al. (2021). Comparing straw, compost, and biochar regarding their suitability as agricultural soil amendments to affect soil structure, nutrient leaching, microbial communities, and the fate of pesticides. *Science of the Total Environment* 751: 141607.

Singh, R.K., Prasad, A., Muthamilarasan, M. et al. (2020). Breeding and biotechnological interventions for trait improvement: status and prospects. *Planta* 252: 1–18.

Singh, S., Singh, R., Singh, K. et al. (2024). Smart fertilizer technologies: An environmental impact assessment for sustainable agriculture. *Smart Agricultural Technology* 8: 100504.

Sultana, S.N., Zubayer, M., Islam, M.S. et al. (2024). Sustainable and ecofriendly approach to weed management. In: *Weed management-global strategies* (ed. M. Aamir Iqbal), 83.

Tao, Z., Jing, Z., Tao, M., and Chen, R. (2022). Recycled utilization of ryegrass litter in constructed wetland coupled microbial fuel cell for carbon-limited wastewater treatment. *Chemosphere* 302: 134882.

Thakur, N., Nigam, M., Mann, N.A. et al. (2023). Host-mediated gene engineering and microbiome-based technology optimization for sustainable agriculture and environment. *Functional & Integrative Genomics* 23 (1): 57.

Tian, J., Ge, F., Zhang, D. et al. (2021). Roles of phosphate solubilizing microorganisms from managing soil phosphorus deficiency to mediating biogeochemical P cycle. *Biology* 10 (2): 158.

Toma, Y., Takechi, Y., Inoue, A. et al. (2021). Early mid-season drainage can mitigate greenhouse gas emission from organic rice farming with green manure application. *Soil Science and Plant Nutrition* 67 (4): 482–492.

Ulian, T., Diazgranados, M., Pironon, S. et al. (2020). Unlocking plant resources to support food security and promote sustainable agriculture. *Plants, People, Planet* 2 (5): 421–445.

Vanlauwe, B., Amede, T., Bationo, A., Bindraban, P., Breman, H., Cardinael, R., Couedel, A., Chivenge, P., Corbeels, M., & Dobermann, A. (2023). Fertilizer and soil health in Africa: The role of fertilizer in building soil health to sustain farming and address climate change.

Villalobos-López, M.A., Arroyo-Becerra, A., Quintero-Jiménez, A., and Iturriaga, G. (2022). Biotechnological advances to improve abiotic stress tolerance in crops. *International Journal of Molecular Sciences* 23 (19): 12053.

Wu, H., MacDonald, G.K., Galloway, J.N. et al. (2021). The influence of crop and chemical fertilizer combinations on greenhouse gas emissions: A partial life-cycle assessment of fertilizer production and use in China. *Resources, Conservation and Recycling* 168: 105303.

Yadav, A.N., Kour, D., Kaur, T. et al. (2021). Biodiversity, and biotechnological contribution of beneficial soil microbiomes for nutrient cycling, plant growth improvement and nutrient uptake. *Biocatalysis and Agricultural Biotechnology* 33: 102009.

Yadav, R., Singh, R., & Jat, M. K. (2024). Organic matter dynamics and carbon sequestration. In Soil, 1.

Yang, W., Yao, L., Zhu, M. et al. (2022). Replacing urea-N with Chinese milk vetch (Astragalus sinicus L.) mitigates CH4 and N2O emissions in rice paddy. *Agriculture, Ecosystems & Environment* 336: 108033.

Yusuf, S.A. (2021). Refocusing agricultural subsidy for food security in Nigeria. *Economic and Financial Review* 59 (4): 159–196.

Zhang, J., He, W., Wei, Z. et al. (2025). Integrating green manure and fertilizer reduction strategies to enhance soil carbon sequestration and crop yield: evidence from a two-season pot experiment. *Frontiers in Sustainable Food Systems* 8: 1514409.

Zheng, H., Wang, X., Qiu, J., and Zhou, S. (2025). The application of gene editing technology in reducing the use of pesticides and chemical fertilizers and mitigating the environmental impact of agriculture. *Advances in Resources Research* 5 (2): 987–1003.

Index

o

p

t

u

v

w

x

z

Printed and bound by CPI Group (UK) Ltd, Croydon, CR0 4YY
17/09/2026
14964297-0003